Cap Gemini Ernst & Young Guide to Wireless Enterprise Application Architecture

Cap Gemini Ernst & Young Guide to Wireless Enterprise Application Architecture

Adam Kornak
John Distefano

Wiley Computer Publishing

John Wiley & Sons, Inc.

NEW YORK · CHICHESTER · WEINHEIM · BRISBANE · SINGAPORE · TORONTO

Publisher: Robert Ipsen
Editor: Carol A. Long
Developmental Editor: Adaobi Obi
Managing Editor: Micheline Frederick

Text Design & Composition: North Market Street Graphics

Designations used by companies to distinguish their products are often claimed as trademarks. In all instances where John Wiley & Sons, Inc., is aware of a claim, the product names appear in initial capital or ALL CAPITAL LETTERS. Readers, however, should contact the appropriate companies for more complete information regarding trademarks and registration.

This book is printed on acid-free paper. ⊗

This publication is designed to provide accurate and authoritative information in regard to the subject matter covered. It is sold with the understanding that the publisher is not engaged in professional services. If professional advice or other expert assistance is required, the services of a competent professional person should be sought.

Library of Congress Cataloging-in-Publication Data:

Kornak, Adam
    Cap Gemini Ernst & Young technology guide to wireless enterprise application
    architecture / Adam Kornak, John Distefano ; co-authors, Frank Jary . . . [et al.] ;
    contributing authors, Ken Norland . . . [et al.].
        p.   cm.
    Includes bibliographical references and index.
    ISBN 0-471-20951-1 (pbk. : alk. paper)
    1. Wireless communication systems.   2. Computer network architectures.   3. Wireless
Application Protocol (Computer network protocol)   4. Business enterprises—Computer
networks.   I. Title: Technology guide to wireless enterprise application architecture.   II.
Distefano, John, 1959–   III. Title.

TK5103.2.K69   2001
004.6—dc21   2001046613

Printed in the United States of America

10 9 8 7 6 5 4 3 2 1

This book is dedicated first and foremost to our families, who give us their never-ending support day in and day out. It's also dedicated to the members of the Cap Gemini Ernst & Young workforce and our clients, without whom the words of this book would not be possible. Our collaborative efforts in designing best-of-breed IT solutions were what made this book a reality.

Author Team of *Cap Gemini Ernst & Young Guide to Wireless Enterprise Application Architecture*

Adam Kornak

John Distefano

Frank Jary

Michael J. Ryan

Jorn Teutloff

Contributing authors: John R. Anderson, Phaseal Freckleton, Ken Norland, Stefan Olsson, and Heini Schulz

Contents

Acknowledgments

There are numerous people we would like to thank, without whom this book would not have been possible.

Executive Sponsors

Some time back when this book was just an idea, there were many people who believed in its vision. We wish to thank these individuals in the CGE&Y executive leadership for helping to make this book a reality:

Leslie Blair	Troy D. Smith
Mike Cantor	Steven Sparks
Colette Coad	Hans Torin

Subject Matter Expertise/Reviewers

Throughout the writing of this book, there were many individuals who graciously shared their knowledge and expertise. We would like to thank these individuals and the many unnamed others who contributed their experiences through successful projects over the years.

Jacob Andersson	Michael Botos
Behzad Andishmand	Peter Coffey
William Beshilas	Jeanine Dahling
Rich Bonugli	Edgar Fong

William Fournell

John Gillardi

Jason Gola

Kos Jha

Ulf Johanneson

John-Pierre Kamel

Ivo Kukavica

Mohamed Ladha

Frederik Lundborg

Vasantha K. Maheswarappa

Jayaram Padmanabhan

Michael Welin-Berger

Jim Wohlford

Other Contributors/Supporters

Last, but not least, we'd like to thank the many others who contributed their efforts and support in getting this book to our readers.

Julie Barnett-Nell

Jack Dugan

Geoffrey Kitson

Paul Li

Carol Long

Russell Riggen

Dan Russ

Brent Whitman

Roger Willis

Erica Young

About the Authors

Adam Kornak

Adam Kornak is a Senior Manager with Cap Gemini Ernst Young, U.S., LLC, practicing in the mobile commerce/critical technologies solutions area. Adam's professional background is comprised of over eleven years of IT solution design experience concentrating on eCommerce and mCommerce in enterprise, groupware technologies, and business process re-engineering. His industry expertise ranges from financial services and insurance, telecommunications, and legal. He has co-authored several books in the past on the topics of Lotus Notes/Domino and Microsoft technologies. Adam is the lead author and developer of "The CGE&Y Guide to Wireless Enterprise Application Architecture" publication.

Adam & his wife Julie reside in the windy city of Chicago. Adam can be reached via email at adam.kornak@us.cgeyc.com.

John Distefano

John Distefano is a Vice President at Cap Gemini Ernst & Young specializing on designing and delivering advanced technology solutions. He has more than 20 years of experience in the information technology field, and he leads the firm's mobile commerce and wireless solutions initiative. John has assisted companies in a variety of industry sectors, including healthcare, oil & gas, consumer products, pharmaceuticals, telecommunications and media/publishing, and in planning and deploying mobile technologies. He presents frequently on a broad range of technology topics, and has authored numerous articles and white papers regarding mobile enterprise solutions.

John holds a Bachelor of Business in Quantitative and Information Science and received a Master of Information Management from Washington University. John, his wife Kim and children Logan and Tom make their home in St. Louis. John can be reached via email at john.distefano@us.cgeyc.com.

Frank Jary

Frank Jary is a Manager with Cap Gemini Ernst & Young in Stockholm, Sweden. Frank has over 6 years of experience in the IT consulting field working as a Java developer and systems architect. In the last three years, he has focused on architecting multichannel systems for clients in various industries and conducting detailed investigations into various related technologies. Recently, Frank has been focusing on wireless business services and how existing systems can be adapted to use the required technologies effectively.

Frank and his wife Rebecca live in the Stockholm area where they enjoy traveling and the outdoors. Frank can be reached via email at frank.jary@capgemini.se.

Michael J. Ryan

Michael Ryan is a software solutions architect working for CGE&Y in the New York metropolitan area, specializing in data architecture and XML related technologies. He comes to CGE&Y after spending several years at IBM Research under the tutelage of Dr. Stephen Boies, noted researcher and human factors expert, during which Michael received his Masters degree in computer science.

He lives in the beautiful Hudson Valley region of New York State with his wife, Micheline, and daughters, Fiona and Ava. When he's not busy saving the world from sub-optimal software solutions, Mr. Ryan can be found on area golf courses, albeit too infrequently. Michael can be reached via email at Michael.J.Ryan@us.cgeyc.com.

Jorn Teutloff

Jorn K. Teutloff is a Manager in the Los Angeles office of Cap Gemini Ernst & Young's High Growth, Strategy and Transformation practice. For over six years, Mr. Teutloff has led project teams in the areas of eCommerce and wireless strategy development, application conceptualization, and business process reengineering at clients in various industries, including high technology, telecommunications, real estate, commercial print, consumer products, apparel, education, entertainment, and hospitality. In addition to project management, Mr. Teutloff has been instrumental in the development and deployment of CGE&Y's proprietary eC strategy and wireless consulting methodologies.

Mr. Teutloff holds an MBA from the University of Southern California and a Bachelor's of Science degree from the University of Florida where he graduated summa cum laude. Jorn can be reached via email at jorn.teutloff@us.cgeyc.com.

Foreword

Few things have changed the way we do business more than the Internet. New services, new business channels, and even entirely new business models have appeared along with the pervasive adoption of the Web. But here's the really good part: The boundaries of the Internet are now extending beyond the desktop and fixed locations. Through wireless technology and mobile computing platforms, companies have new, virtually limitless opportunities to continue to evolve the connected business model brought to life by the Internet. Applications and services based on mobile and wireless technology allow a business to radically change the way it works—how it interacts with its customers, how it connects with its workforce, how well it leverages partnering relationships.

The speed and scope of change driven by the Internet economy has placed unprecedented demands on key business processes, functional activities, and IT infrastructure. Enterprises today must operate at "Internet speed" to survive. This implies flexible, adaptable operations, and an agile IT capability, which we call Adaptive IT, providing the foundation to scale capacity up and down quickly to compress cycle times from weeks, to days, to hours, to minutes. Wireless technology and enterprise mobile solutions afford companies the opportunity to redefine Internet speed, thereby reinventing the Internet's value proposition to the marketplace and restructuring relationships among its key business constituencies. With a mobile solutions platform, anytime, anywhere access to corporate information, and true real-time operations, collaboration and execution are possible. Wireless technology will take a prominent position in promoting a continuous processing business model with core fundamentals such as real-time tracking and location systems, cross-enterprise collaborative planning and dynamic inventory management to improve customer service, quick response, and demand and supply matching. In today's supply chains, the interactions are simultaneous and many-to-many; operations are comanaged, products are codeveloped, and material must be visible and tracked in real time. The adoption of wireless technology will soon be an essential element in the business solutions devised to respond to these challenges that are rewriting the rules of the competitive marketplace.

Wireless technologies also present compelling opportunities for companies to establish new channels to their customers. The winners in the next-generation economy will be those who can best create and manage a mobile customer experience. Retailers will provide time-pressed shoppers with promotions, electronic coupons, comparison prices on favorite brands, and product information delivered in real time, wherever and whenever the consumer needs it. Mobility will allow new ways for any enterprise to weave itself more seamlessly into the daily life of their customers, whether they are riding home from work, rushing to a kid's ballgame, or sitting at the movie theater. By leveraging the wireless and interactive nature of mobile devices, leading companies will establish personalized channels to interact with each one of their valued customers.

Companies will also invent a new class of service that leverages consumer and product information and applications specifically designed for mobile use. One such market dynamic, which is being simultaneously driven and enabled by wireless technology, is mobile informatics, the exchange of real-time information between intelligent devices, consumers, enterprises, and service providers, often without human intervention. Mobile informatics will be widely implemented in vending machines, household appliances, surveillance equipment, cargo pallets, industrial doors, health monitoring equipment, commercial transportation fleets, and virtually all other classes of hard assets. Control and collection of data from these interface points are being made possible and practical by the advancements in intelligent embedded software and wireless communication. Correspondingly, these advancements will drive an enterprise's infrastructure requirements to new levels as this next-generation user base of intelligent contact points reaches the tens, if not hundreds, of millions.

This book is fundamentally about identifying strategies for defining and executing an Adaptive IT architecture which supports the enormous potential of a wireless enterprise. The wireless architecture you adopt and the vendors you partner with will directly influence your organization's ability to truly deliver on the promise of an untethered world. Key infrastructure decisions must be made regarding handheld devices, mobile data management, and enterprise architecture. It is very likely that not just one, but a portfolio of mobile devices will find their way into your enterprise wireless landscape—PDAs, smart phones, handheld PCs, embedded components. Your architecture will need to consider support for multiple-messaging formats, different levels of intelligence, and various screen sizes and graphics capabilities from multiple equipment classes. Intelligent wireless devices will need to support a data model that corresponds to the enterprise model of the corporation. Conversely, less-intelligent devices are forcing design constraints, demanding new techniques around data vocabulary, standard wireless tags, and abbreviated data formats. How your architecture anticipates and accommodates this disparity of data management capability will be central to effectively delivering information in a real-time world. Further, device constraints and wireless bandwidth limitations make content management in a pervasive computing world more complex, and less robust, at least for the next couple of years. Yet in an Adaptive IT world, where you architect to evolve, the potential is still very real for exciting applications that enable true mobility across your enterprise.

Today's marketplace is driving companies toward an Adaptive IT orientation and motivating an evolution from the wired to the wireless world. But where do you start? Should you begin by mobilizing your sales force or would you find greater benefit in enabling your order processing systems for wireless transactions? What about your supply chain—perhaps your competitive edge can best be improved by further streamlining your logistics and distribution processes through new wireless applications and technology. The answers regarding where to start and where to go are clearly unique to your business. This book aspires to assist you in considering an effective entry point into the wireless world, and to envision a sustainable road map for evolving your mobile business applications architecture. The authors of this book would be most gratified if you choose to bring this technology guide along with you as you unwire your business. We believe if you act quickly and plan fully to capitalize on this emerging space, your organization can move a step closer to your customers and a step ahead of the competition.

The Authors,
October 2001

The Wireless Landscape

The Wireless Internet Evolution

Introduction

If you're an enterprise architect or application developer, the terms *mCommerce* and *wireless Internet* have probably already hit your radar screen. Even more likely, you've already developed or started developing some wireless applications for your organization. Wireless technology is certainly not a completely new concept. In fact, many large organizations have used some form of wireless applications for more than 10 years and some even longer. Large shipping companies such as FedEx started to use wireless devices and expensive proprietary networks in the mid-1980s to track packages, route deliveries and pickups, store addresses, capture signatures, and, in turn, send data *wirelessly* to central systems within the company. However, the Internet has brought about a change in the wireless space that adds an entirely new value proposition to the term *eCommerce:* Electronic commerce gave us the ability to access and act on real-time information, as long as we were sitting in front of a wired PC with a live connection to the Web. The wireless Internet gives the user—including the company, its customers, suppliers, and partners—the ability to access and act on information virtually *anytime* and *anywhere.* Today, wireless technology and the Internet extend that ability and, in most cases, generate many additional opportunities to leverage eCommerce capabilities, several of which will be discussed in detail in the remainder of this book.

In this first chapter, you will learn about:

- The rapid evolution of wireless technology, from the origin of the wired Internet to today's wireless Web. In other words, how did we get from the ARPANET to WAP?

- The technologies surrounding devices, networks, and protocols required for the design of applications for the wireless Internet. The objective here is to provide a high-level overview focusing on the most basic elements of the wireless environment.

- The impact of the wireless Internet on traditional telecommunications service providers and how their operations will change with the introduction of new technologies and business models.

The Internet from Mosaic to WAP

Though you may already have learned about the history of the Web on your own or through other sources it is still important to provide a brief history of the Internet to show the progression from a *wired* environment to the *wireless Web* that we use today.

Perhaps the best way to illustrate the evolution of the Internet is to present you with a table (Table 1.1) that shows the key milestones in the development of the Internet, from the network's beginning through today.

One question you may be asking yourself is, "What does the history of the Internet have to do with wireless architecture?" However, the more appropriate question probably is, "What *doesn't* the Internet have to do with wireless architecture?" There are possibly better ways of phrasing that question, but you get the point. Keep reading and the answer will become clear.

As the timeline indicates, the Internet began in the days of Sputnik and ARPANET. The Advanced Research for Projects Agency developed

> **NOTE**
>
> *WAP* stands for Wireless Application Protocol and is just one avenue for wireless devices to access real-time data. WAP's foundation is the Wireless Markup Language (WML) inherited from Hypertext Markup Language (HTML). This book focuses on architecture, as opposed to specific development languages. Thus, we will not go into great detail about the development tools, but will concentrate on how the application is actually designed.

Table 1.1 The Timeline of the Internet

DATE	EVENT
1957	USSR launches Sputnik.
1968	First packet-switching network.
1969	ARPANET starts.
1972	First public demo of ARPANET; Internet mail invented.
1973	Kahn and Cerf present paper on Internet; first international Internet connection. Together they are considered the coinventors of TCP/IP.
1975	Microsoft founded.
1976	Apple founded.
1979	UseNet starts.
1983	ARPANET changes over to TCP/IP; ARPANET splits into ARPANET and MILNET; Microsoft introduces Windows.
1984	Internet exceeds 1,000 hosts; William Gibson writes *Neuromancer*; Domain Name Server introduced.
1986	NSFNET created.
1987	Internet exceeds 10,000 hosts.
1988	Worm attacks 6,000 of Internet's 60,000 hosts.
1989	Internet exceeds 100,000 hosts.
1990	ARPANET dismantled; Archie starts.
1991	WAIS started; Gopher started; NSF lifts commercial ban.
1992	Internet exceeds 1 million hosts; Web Invented by Tim Berners-Lee; Veronica introduced.
1993	MOSAIC developed by Marc Andreesen; InterNIC founded by NSF.
1994	ATM (Asynchronous Transmission Mode, 145 Mbps) backbone is added on NSFNET.
1995	The National Science Foundation announced that as of April 30, 1995, it would no longer allow direct access to the NSF backbone. The NSF contracted with four companies to be providers of access to the NSF backbone, which in turn would sell connections to groups, organizations, and companies.
1996–present	Most Internet traffic is carried by backbones of independent ISPs, including MCI, AT&T, Spring, UUNet, BBN planet, and others.

"The Internet's Brief History," by Jerry Honeycutt, from *Special Edition Using the Internet, Fourth Edition*, 1997. Que Publishing.

The History of the Internet, by Dave Kristula, March 1997 from www.davesite.com/webstation/net-history.shtml.

ARPANET, a communication network of computers to allow researchers to collaborate and exchange information and data more effectively. ARPANET was also intended to provide the military a way to communicate during a nuclear attack by utilizing a redundant infrastructure, so that the destruction of any one computer or *node* in the network would not prevent the other nodes from continuing operations. In addition, ARPANET introduced what is known as *packet switching*.

> **NOTE**
>
> By definition, *packet switching* refers to protocols in which messages are divided into packets before they are sent. Each packet is then transmitted individually and the packets can even follow different routes to their destination. Once all the packets forming a message arrive at the destination, they are recompiled into the original message. Since the Internet is simply a network of computers, packets or messages are sent that contain information about their path so that any one computer on the network knows where to forward (switch) the message.

The original ARPANET grew into the Internet as we know it today, connecting multiple independent networks of random design. As a comprehensive communication system, the Internet follows the theory of open architecture networking, which was first developed by Dr. Robert Kahn shortly after he arrived at ARPA's Information Processing Techniques Office (IPTO) in 1972. In Kahn's approach, the choice of any individual network technology was not dictated by specific network architecture but rather could be selected freely by a provider and made to integrate with the other networks through a metalevel *internetworking architecture*. In an open architecture network, the individual networks may be designed and developed separately. Each may have its own unique interface that it may offer to users and/or other providers, including other Internet providers. Each network can be designed in accordance with the specific environment and user requirements of that network.

As a result of employing an open architecture, we have witnessed the dramatic growth of the Internet over the last several decades. This expansion has been incredible to say the least. Figure 1.1 illustrates just how quickly the number of Internet hosts has grown over time.

As we move into the more recent decades, another key development in the Internet growth period was the design of the World Wide Web in 1989 by Tim Berners-Lee while at CERN, the European Organization for Nuclear Research and the world's largest particle physics lab. At the time, the Web was used as a research tool that involved rich text documents embedded in text body links to related material. These *enhanced* documents essentially were format-

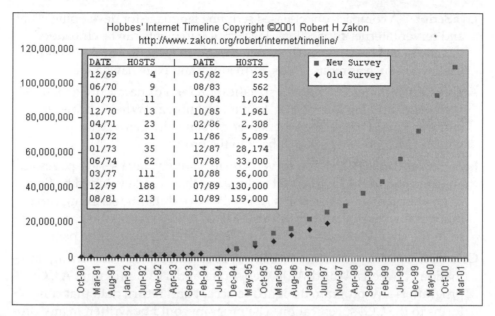

Figure 1.1   Growth of the Internet.

ted text that could include embedded objects, such as other text or pictures. Sound familiar? Very quickly, the invention of the World Wide Web spawned several protocols and development languages such as SGML, HTML, Java, ASP, and JSP, as briefly described in the following:

Hypertext Transfer Protocol (HTTP).   A protocol that defines the communication between a Web server and a client.

Standard Generalized Markup Language (SGML).   A system for organizing and tagging elements of a document. SGML itself does not specify any particular formatting; rather, it specifies the rules for tagging elements. These tags can then be interpreted to format elements in different ways.

Hypertext Markup Language (HTML).   The authoring language used to create documents on the World Wide Web. HTML defines the structure and layout of a Web document by using a variety of tags and attributes.

Java.   Developed by Sun Microsystems, the programming language Java was originally called *OAK*, designed for handheld devices and set-top boxes. Oak was unsuccessful, so in 1995 Sun changed the name to Java and modified the language to take advantage of the fast-growing Web. Java is a simple, robust, dynamic, multithreaded, general-purpose, object-oriented, platform-independent programming environment. Developers can write custom mini-applications called *applets*, which will provide Internet sites with a huge range of new functionality—animation, live updating, two-way interaction, and more.

JavaScript.   A compact, object-based scripting language for developing client and server Internet applications. JavaScript statements can be embedded directly in an HTML page. These statements can recognize and respond to user events such as mouse clicks, form input, and page navigation.

Active Server Pages (ASPs).   A specification for a dynamically generated Web page utilizing Microsoft's ActiveX technology. When a browser requests an ASP page, the Web server dynamically generates a page with HTML code and sends it back to the browser.

Java Server Pages (JSPs).   A server-side technology, Java server pages are an extension to the Java servlet technology that was developed by Sun as an alternative to Microsoft's ASP. JSPs have dynamic scripting capability that works in tandem with HTML code, separating the page logic from the static elements—the actual design and display of the page.

Common Gateway Interface (CGI).   A specification for transferring information between a World Wide Web server and a CGI program. A CGI program is any program designed to accept and return data that conforms to the CGI specification. The program could be written in any programming language, including C, Perl, Java, or Visual Basic.

Extensible Markup Language (XML).   A human-readable, machine-understandable, general syntax for describing hierarchical data, applicable to a wide range of applications (databases, eCommerce, Java, Web development, searching, etc.). Custom tags enable the definition, transmission, validation, and interpretation of data between applications and between organizations.

Practical Extraction and Report Language (Perl).   A programming language especially designed for processing text. Because of its strong text-processing abilities, Perl has become one of the most popular languages for writing CGI scripts.

Personal Homepage (PHP) tool.   Server-side scripting language used to provide dynamic content. Its syntax is reminiscent of the C programming language.

In 1993 Marc Andreesen, then a student and part-time assistant, teamed with programmers at the National Center for Supercomputing Applications (NCSA) at the University of Illinois at Urbana-Champaign and created a revolutionary method for displaying enhanced documents in a graphical environment. This tool, the first Web browser, was called *Mosaic*, and at the time was distributed for free on the Internet, available to anyone who wanted it. In 1994, Andreesen partnered with Jim Clark, cofounder of Silicon Graphics, and started Netscape. The rest is history.

Soon, several companies were developing some form of a browser-based tool to access the World Wide Web. In the summer of 1995, Microsoft published its first, barely working version of a browser known as Internet

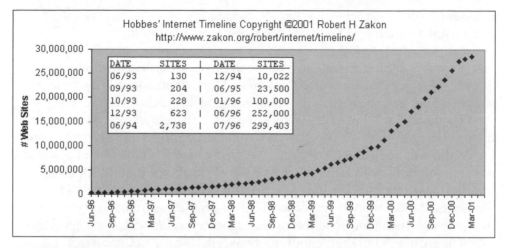

Figure 1.2   Web site growth from 1996 to 2001.

Explorer. Microsoft packaged the application as an accessory pack shipped with the Windows 95 operating system released in August of that year. As a direct result of Internet browsers being available for free, a growing number of Web-based software applications being developed, and the rapid adoption by both consumers and businesses, the number of Web sites has grown exponentially over time. Figure 1.2 shows the unparallel explosion of Web sites over the last several years.

Without the innovative work of network technology pioneers and visionaries, including Kahn, Cerf, Berners-Lee, Andreesen, and Clark, the Internet would not have evolved as rapidly as it did, and network protocols and Internet programming languages would have little meaning. As the Internet grew by leaps and bounds, there were some exciting developments down the road for the Web-browsing community. Enter the *wireless Web*.

The Wireless Web

The next wave to hit the U.S. market in late 1997 was the mobile Internet, or the wireless Web. One way of looking at the transition from the traditional Internet or WWW to the wireless Web is to determine how its users access the Internet. In the wireless environment, you obviously can browse the Net via your WAP phone or PDA while on the go. The true innovation here is the ability to access the Internet anywhere or anytime you desire—on the train, walking through the park, at home, or anywhere that a wired computer is not accessible. Although wireless applications had been developed and were available to professionals within select industries for several years, these solutions were running on expensive proprietary networks, using custom devices, effectively shutting out the mass market. Now, with the Internet in full swing, end users were increasingly asking for access to the Web while out and about. And the cost for such access had to be reasonable.

Anticipating an exploding demand for wireless services, several online content providers started the race to make their pages accessible to wireless devices. However, it is important to note that the wireless Web is more than simply a front end to an existing Web site. The reality is that most of today's wireless devices still have very small screens and therefore don't have the ability to show an entire Web page, as you would expect to see on a 17-inch monitor. As diligent application architects, we are charged with the task of designing wireless solutions with factors such as device limitations in mind. Other factors, besides screen size, include security and privacy issues, as well as the requirement for content to be accessible from a multitude of independent devices, running on incompatible operating systems. We will touch upon these topics in greater detail throughout this book, specifically in Chapters 3 and 4.

The Growth of the Wireless Internet

As demonstrated earlier, the growth of the Internet has been staggering. Even more amazing is the immense growth expected of wireless Internet services, targeting a rapidly growing wireless user base. As Figure 1.3 illustrates, the compound annual growth rate (CAGR) of mobile Internet users is expected to reach 207 percent, for a total of almost half a billion users by 2005. This rate

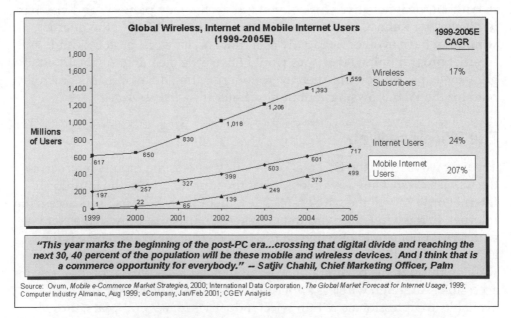

Figure 1.3 Dramatic growth in number of mobile Internet users over the next few years.

is eclipsing the expected growth in the number of all Internet users by almost nine times!

In addition, several industry reports published by leading research firms indicate that the mobile Internet is here to stay. For those of you who relish big numbers, have a look at Figure 1.4 along with the following forecasts and quotes:

- The number of wireless subscribers in the United States in 2000 totals 115 million. By 2005, this number is estimated to reach 208 million (IDC, 2000).

- In Europe, the number of wireless subscribers in 2000 totals 133 million. By 2003, the forecast targets 270 million (IDC, 2000).

- On a worldwide basis, wireless subscribers in 2000 numbered 650 million. By 2005, the subscriber base will hit around 1.5 billion (Ovum, 2000).

- By 2003, 50 percent of all business users will be mobile, and each will typically use three or four different information appliances, seamlessly connected via personal area networks (PANs) to mission-critical applications and the Internet (Meta Group, 2000).

- Handheld devices accessing the Internet will outnumber PCs by 2003 (Forrester, 2000).

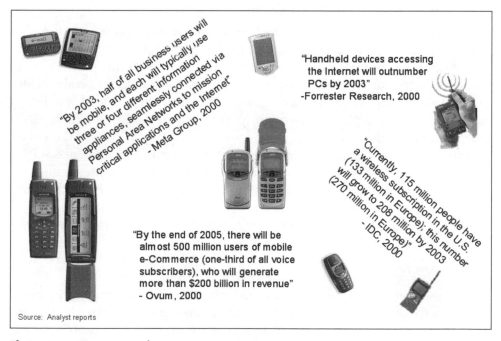

Figure 1.4   Forecasts and quotes.

- mCommerce revenues worldwide in 2000 amounted to about $4 billion. By the end of 2005, there will be almost 500 million users of mobile eCommerce who will generate more than $200 billion in revenue (Ovum, 2000).

As you can see, several studies indicate that the number of wireless Internet users is on a steep trajectory and expected to soon outnumber those that access the Internet through a wired device. To keep up with the incredible demand for wireless services, new applications will need to be developed and deployed in ever shortening time frames. This book will provide you with a detailed road map to allow you to successfully participate in the wireless realities to come by showcasing some very specific applications that leverage wireless technologies in real-life business environments.

Wireless Technologies

Now that we have given you an overview of the history of the Internet and introduced the rapid growth of wireless services and subscriber bases, let's briefly look at some of the technologies that make wireless work. At the highest level, the wireless environment entails three basic elements:

1. The mobile device.
2. Accessing specially formatted content.
3. Delivering via a wireless network.

Let's first look briefly at the *handheld hardware* currently on the market.

Users all around the world are accessing the Internet and/or their company's applications via hardware that continues to dramatically shrink in size while becoming increasingly more powerful. A mobile device can come in all kinds of shapes and flavors, be it a personal digital assistant (PDA), a handheld computer, a cell phone, a two-way pager, or other device (see Table 1.2). However, due to limitations regarding content delivery and presentation, the selection of a device usually depends on the user's communication needs. The two dimensions that need to be taken into consideration are time criticality and application complexity. Whereas simple yet time-critical information such as a stock quote or an inventory alert can effectively be delivered to a small-screen device such as a cell phone, more complex data sets that require further analysis are better suited for PDAs or handheld computers that pack more memory and number-crunching power. Table 1.2 provides a short overview of today's most common hardware available in the United States.

With the appropriate mobile device in hand, the next element within the wireless environment is content. Due to limitations in screen size, resolution, and other factors, the experience of accessing the Internet via a mobile device is far from what we are used to when sitting in front of our large-screen mon-

Table 1.2   Popular Handheld Devices

Web-enabled cellular phones	Nokia Ericsson Qualcomm Motorola Kyocera Nextel	Voice messaging, Short Message System (SMS), address book, phone list Access WAP-enabled Web sites
Two-way pagers	RIM Blackberry Motorola Timeport	Email, SMS, address book, phone list Access WAP-enabled Web sites
PDAs	Palm Handspring Visor	Address book, phone list, scheduler, notebook, calendar, calculator, games Synchronized with a desktop computer Access to a select set of Web sites Expandable via expansion slot
Handheld computers	HP Jornada Casio Cassiopeia Compaq iPAQ	Address book, phone list, scheduler, notebook, calendar, calculator, games, enhanced email, media player, Microsoft Word and Excel email attachments Synchronized with a desktop computer Access to all Web sites Expandable via expansion slot

itors. Yet as the technology is evolving, so is our ability to replicate the wire line experience to the untethered world. Whereas PDAs and handheld PCs are marginally effective at allowing us to browse the Web, cell phones are the most handicapped devices at this time.

Thus, Ericsson, Nokia, Motorola, Openwave and other companies are looking for a standard that would allow Internet pages to be accessible via cell phones without the need to customize each Web page for display on every mobile phone on the market today and in the future. AT&T Wireless, Motorola, Qualcomm, and others funded Unwired Planet. The telecom group (AT&T Wireless, Motorola, and Qualcomm) founded the WAP Forum, an industry association that grew to more than 500 members over three years and includes everyone from handset manufacturers to network carriers, infrastructure providers, software developers, and other organizations.

WAP is a relatively secure specification that allows a user to access Internet content via a phone that runs a *microbrowser*—an Internet browser with a very small footprint that fits into the tiny memory of a cell phone or other small handheld device. The WAP stack is a set of protocols that covers the whole process of wireless content delivery, from the creation and layout of actual content and the specification of security measures, to the actual transport of content. The protocol supports most wireless networks, and is being supported by leading operating systems, including PalmOS, EPOC, Windows CE, FLEXOS, OS/9, and JavaOS.

Table 1.3 illustrates the definitions, protocols, and wireless technologies surrounding WAP.

Table 1.3   WAP Technologies

WML	Wireless Markup Language—An application of XML used as the default markup language for WAP content. WML is the equivalent of HTML, the language used to develop Web pages. Pages written in WML can be read by a WAP phone or similar device. Although WML is similar to HTML, there are fewer tags.
WMLScript	Wireless Markup Language Script—A client-side script language for use with WML. Written in a separate file and called from a WML card.
Deck	In WML, a single page is called a deck. A deck is the container for one or more cards. The deck provides necessary information about the document type, markup language, and navigational templates for the cards incorporated in it.
Card	In WML, a card is the subpage that is displayed to the user. A card must be contained in a WML deck.
HDML	Handheld Device Markup Language—A subset of HTML supported by some microbrowsers, it can be compared to WML, but is not widely used now.
SMS	Short Messaging Service—A technology for sending text to and from mobile phones, using 160 characters or less. The message is sent through an SMS Service Center that pushes it to the recipient(s). If the phone is turned off for a shorter time (normally 1 to 24 hours), the message can be received after it is turned on again.
WBMP	Wireless BitMaP—WBMP is the default picture format for WAP.
WSP	Wireless Session Protocol—A layer in the WAP protocol stack, similar to the HTTP protocol. It controls the connection between a client and the server.
WTA	Wireless Telephony Application—Provides interfaces to the telephony part of a mobile phone.
WTAI	Wireless Telephony Application Interface—Makes it possible to develop telephony applications within WAP.
WTLS	Wireless Transport Layer Security—A layer in the WAP protocol stack that handles security and extends SSL (Secure Sockets Layer) to the mobile device.

For more information, see www.wap.com.

Now that we have reviewed mobile devices and quickly looked at the technologies involved with rendering Internet content via WAP, let's look at the third basic element, the networks that connect the two. Table 1.4 provides a high-level overview of wireless network technologies currently deployed and on the horizon. In the 1980s, first-generation cellular networks, or 1G networks, provided for analog voice transmissions only, while in the 1990s, second-generation networks (2G) moved toward digital delivery, adding data and messaging capabilities at improved security levels while using multiple standards including GSM, CDMA, and TDMA (see Table 1.4). Introducing higher data transmission rates via GPRS and EDGE, 2.5G networks of the late 1990s bridged the gap between the second- and third-generation (3G) net-

Table 1.4   Wireless Telephony

GSM	Global Standard for Mobile Communications (originally *Groupe Spécial Mobile*) is a second-generation digital mobile telephony system used mainly in Europe, South Africa, Australia, and many countries in the Middle East and Asia, and in commercial service since 1991. Requires a GSM phone, and runs at speeds of 9.6 kilobits per second (Kbps).
TDMA	Time Division Multiple Access is a 2G digital wireless access standard mainly used in the United States and Canada. Frequencies are normally 800 and 1900 MHz. TDMA works by dividing a radio frequency into time slots and then allocating slots to multiple calls. In this way, a single frequency can support multiple, simultaneous data channels.
CDMA	Code Division Multiple Access is a 2G digital wireless access standard that uses spread-spectrum techniques, developed by Qualcomm. Unlike competing systems such as GSM—which uses time division multiplexing—CDMA does not assign a frequency to each user. Instead, many conversations are carried over one frequency, while each individual conversation is encoded with a pseudo-random digital sequence. CDMA is the most common technology used within the United States, Canada, and Korea. The current transmission rate for CDMA is 14.4 Kbps.
GPRS	General Packet Radio Service is a 2.5G digital standard that runs at speeds up to 150 Kbps, and provides always-on access.
EDGE	Enhanced Data GSM Environment, is a 2.5G technology for high-speed data transfer over existing GSM networks without the need to install new equipment. A network whose software has been upgraded to EDGE allows speeds of up to 384 Kbps.
UMTS	Universal Mobile Telecommunications System is a 3G technology that will deliver broadband information including voice, data, audio, and video at speeds up to 2 Mbps.
WCDMA	Wideband Code Division Multiple Access (also known as IMT-2000 direct spread) is a 3G cellular technology that can reach speeds from 384 Kbps to 2Mbps, enabling broadband information delivery.

works. Using 2.5G technologies, users will be able to access the Internet at speeds comparable to high-speed wireline access. True broadband applications will be available once third-generation networks are in place. Yet to be deployed in the United States, 3G technologies include UMTS and WCDMA, which promise to eliminate incompatible standards in favor of a consistent, worldwide approach to delivering data transmission rates of up to 2 megabits per second (Mbps), thus enabling multimedia services such as video and audio transmissions.

Before we continue to explore wireless technologies by looking at some very specific wireless applications in Chapter 2, let's take a quick moment to reflect upon what the recent developments mean for the traditional providers of telecommunication services as they are most deeply affected by the new technologies.

The Future of the Telecom Service Provider

(Excerpts from "The Future in Your Hands—The Mobile Internet Drives the Future of Wireless," Cap Gemini Ernst & Young Telecom Media & Networks).

As demonstrated in this chapter, the Internet and mobile telecommunications were the two most significant breakthroughs of the 1990s and beyond. Until recently, however, they were two separate developments. Managing the convergence of these two stunning technological success stories into a ubiquitous, integrated service is the challenge telecom companies must meet today, for soon this combination will be treated as a necessity by millions of people around the world. As future-generation network technologies (GPRS, EDGE, UMTS, etc.) and software applications (location-based services, customer intelligence, etc.) move forward, end users will accelerate their demands on service providers for products they can use any time, anywhere. Mobile Internet devices will become essential tools for business and residential consumers alike.

Understandably, everyone in the industry wants a piece of this pie. The projected ability to reach more than a billion people via mobile devices by 2005, combined with the $200 billion in expected revenues generated from mobile commerce in that year, is a compelling prize that successful mobile operators will attempt to win by developing and implementing well-thought-out mCommerce strategies. The contest between existing mobile and fixed operators, equipment manufacturers, content providers, and retailers— all scrambling for their share of this potentially lucrative market—has already begun. In addition, mobile start-ups are also beginning to challenge traditional operators. The current market hype surrounds the new 3G licenses that are being auctioned off at astronomical rates, illustrated by the recent U.K. auction raising £22.5 billion (approximately $31 billion U.S.). Yet 3G is only one of the various future-generation technologies, and its full-scale deployment is at least two to three years away. New business models, not new technology, will drive mobile growth—and to succeed, tomorrow's operators will need to reinvent their businesses.

Seizing the Opportunity

Owning the means of transmission will no longer be enough to guarantee success for tomorrow's operator. Simple carriage of telecom traffic is quickly becoming a commodity business as the market reaches mass proportions. While mobile commerce holds the strongest potential for generating revenues, most operators are flying blind, not quite sure about the best course to set. Should they expand into content? Can they be a wireless portal? With which companies should they partner? Where will established Internet companies play? And how can they make money? For mobile operators the Internet represents a significant business opportunity as well as a new way to deliver value in the network economy. But the rules are changing on a daily basis. Carriers need to respond by changing their business model to fit the Internet world—multiple partnerships of various types, multiple sources of revenue, multiple distribution channels, and multiple types of business models, all changing at Internet speed. And because the Internet lets customers interact directly with content creators, aggregators, and others in the value chain, carriers are at risk of not controlling the customer relationships the service involves. Success will not be easy. Defining a clear market strategy and a long-term service portfolio road map is the first step toward identifying which services can be launched and, just as important, which services will be exciting and user-friendly—and generate customer demand. Operators will need to help customers overcome resistance to new technology. They will also need to expand their own competencies by negotiating alliances with content, retail, Internet, and software players.

One of the toughest challenges will be the quick introduction and migration of services to take advantage of the rapid technology migration path (from WAP to GPRS to EDGE to UMTS). Operators need to develop portfolio road maps that maximize network capabilities while ensuring fast rollout of services focused on customer needs. This rapid launch may be the most difficult task. Before launch, furthermore, operators will need to change their processes (such as billing) across the customer experience, and convert their organization structures from a functional to an integrated model. Without first addressing these issues, operators run the risk of suboptimizing investments and confusing customers.

Looking Ahead

The market will continue to evolve as providers enhance their data capabilities and move from pure voice to mobile information to mCommerce to broadband services. And as new competitors enter from left field, the battle to dominate the mCommerce market will be intense. Competitive attacks will come not only from traditional mobile operators themselves, but also from established portals such as Yahoo! and AOL, together with mobile service start-ups, which are positioning themselves to win their share of the converging fixed and mobile communications market. Speed is essential. Many organizations are already moving quickly, investing heavily at the front end. Multiaccess portals that users can access from several points, ranging from a mobile phone to a PC to a PDA, will be essential, along with new partnerships and distribution channels. Open access will win in the long term, with the winners focusing on products that drive usage, such as location-based services. As the new converged offerings reach the marketplace, they will soon begin to cannibalize operators' profitable offerings while introducing little incremental revenue. Operators will be forced to seek new sources of revenue from partners rather than customers. The new business models will be based on a web of alliances. Only those who move quickly and change their traditional views of control will win.

Looking forward, the next chapter will answer questions such as why an enterprise should go wireless, what types of business models mark today's Internet landscape, and how mobile enterprises relate to them.

CHAPTER 2

Mobility Applications

What Is a Mobility Application?

Mobility applications are applications that handle the transmission of data and user interaction between wireless devices and a central repository. The transmission can occur in real time, by means of synchronization, or via a hybrid of the two. Let's analyze what each approach entails.

Synchronization Applications

Synchronization-based applications are applications that handle data transfers in situations where the user is not connected to the central database on a real-time basis. Instead of an instant data transfer, the user occasionally synchronizes the data residing on his or her mobile device with either a PC application or, potentially, a server. In this scenario, the user must schedule when the synchronization occurs and must be connected to a network of some type at that time. Synchronization applications work with a *thick* client that contains most or all of the data. During the synchronization process, the device receives new or updated data, or the device sends its data to the PC or server. There are several vendors on the market who deliver synchronization solutions with varying degrees of complexity.

Common examples of synchronization applications are those that are running on Palm Pilots that have no wireless connection to a network. These applications connect with the back end only during the time they are placed in a synchronization cradle or when the user applies other synchronization mechanisms, such as a direct cable or infrared connection. Synchronization applications are clearly very effective because they allow the user to access and carry critical information while on the go. Synchronization in many cases should be the preferred means of data transmission, given our current wireless networks' limited bandwidth and connection reliability, especially for those business applications that are not time critical or that do require large amounts of data to be transferred back and forth from the device.

Real-Time Applications

Other types of mobile applications include those that connect in real time only. Examples of these include WAP connections over cell phones, PDAs with wireless connectivity, or applications that are available if and only if there is a connection at the exact moment that the information is needed. The distinction here is that limited data is kept on the device, and data that the user wants to see is accessible only at the time that a reliable network connection can be established.

This type of application is the one we commonly think of when using the term *mobile commerce,* as the term implies that the activities are happening in real time exclusively. Many applications that contain time-critical elements and low data volumes, such as financial trading applications, bidding in auctions, and status tracking, require such connectivity to function reliably and meaningfully.

Hybrid Applications

The third type of application is one that is only sporadically connected. We're using the term *sporadic* to distinguish it from the preceding synchronization discussion. These applications typically have a thick client, one that will process interactions in real time if the real-time connection can be established, but queue the transaction and do something else (in the context of the business process) when the connection is not available. When the connection becomes accessible again, the queued interactions are processed at that time.

Many of the most valuable hybrid applications that we've seen automatically detect the availability of a connection and then choose whichever means of connectivity is most applicable. The important part here is that the user of theses applications does not have to know whether the applications are connected or not. The software performing the applications will take care of the connectivity and render it mostly invisible to the user. Obviously, if you need real-time access to satisfy a query but you're not real-time connected, you will receive an error message. However, most information can be fetched ahead, it can be synchronized, and it can be available with some degree of currency even though the wireless connection may not be available at the exact time of processing.

Each application being designed today falls into one of the three categories that we've listed in the preceding discussion. It depends on the situation and the real business needs that drive the application itself. If a user can satisfy his or her needs with an occasionally connected application, developing such an application is usually much less expensive and complex than building a real-time solution. Having the wireless, real-time connectivity certainly has some benefits in terms of the data being current, but one must weigh the disadvantages of such an approach, mostly stemming from the fact that connectivity is not always available in key locations. All of us have experienced dropped calls in certain indoor environments or in locations where there was no wireless coverage. In this book, a key aspect of architecting a meaningful solution is to use the simplest and least expensive option that still provides all required functionality, is effective from a usability perspective, and allows for future growth.

What Devices Do We Consider for Mobile Applications?

Handheld devices come in all shapes and sizes—with many more currently being developed in manufacturers' research labs around the globe. A simple categorization separates the Core Devices, such as cell phones and PDAs, from peripherals, including bar code scanners, GPS transceivers, and mobile printers.

Core Devices

We divided the core devices for mobility applications into four categories. The boundaries between these categories are flexible, as new devices are emerging continually and redefine the art of the possible. Some of today's devices blur the distinctions we're putting forth, and we expect that reality to accelerate as the industry continues to develop new gadgets. The categories are as follows:

1. Unintelligent gadgets.
2. Cellular phones.
3. Smart phones.
4. Devices with operating systems.

Let's describe each of these briefly.

Unintelligent Gadgets

This category includes devices such as sensors and radio frequency identification (RFID) tags. Unintelligent gadgets typically contain little or no processing power. They can, however, participate in mobile applications in the sense that they use connectivity to provide a real service for someone. A common example of an RFID device used on many highway systems is the small plastic decal that you place inside your windshield that allows you to pass a tollbooth without having to stop. The booth reads the decal and adds the amount to your monthly bill. There is very little intelligence in that device, but the application is a very effective one, as it provides the key to speeding up traffic in many locales.

Cellular Phones

The second category is cell phones. These phones show increasing signs of intelligence. And even though early cell phones were not sufficiently capable of providing applications because they mostly did not have data connectivity and were relegated to voice only, the emerging phones are embedding data capability of varying degrees. Data capability is required to exploit the 2.5G networks currently being deployed as well as the 3G capabilities that are on the horizon. Cell phone applications are typified by the iMode offerings provided by NTT DoCoMo in Japan and the WAP capabilities in Europe and the United States. Although WAP phones are somewhat primitive, as they offer only limited functionality due to their early stages in the technology life cycle, it is anticipated that over the next few months and years these capabilities will increase dramatically, paving the path for increased application complexity. The increase in capabilities we are talking about include not only functionality, bandwidth, and constant connectivity, but also the type and amount of processing that the devices will be capable of.

Smart Phones

As cell phones mature in terms of functionality, connectivity, and processing power, they start to become what we call *smart phones*. The category of smart phones includes devices such as the Kyocera phone that includes the Palm operating system along with voice and data capabilities. Another device in this category that combines PDA and email capabilities with cell phone functionality is the Stinger phone, developed by Microsoft in collaboration with various hardware providers. This marriage of phone features and data access technology is the basis for the convergence of devices that we see going forward. The key notion here is that these devices typically use the cellular networks that are made available by telecom carriers as opposed to other forms of wireless connectivity. Many of these devices also support the disconnected or occasionally connected modes of operation described previously. They are capable of supporting thick as well as thin clients, which allows them to operate with a wide range of applications.

Devices with Operating Systems

The fourth category consists of devices with operating systems. While it is technically true that smart phones have operating systems and paging devices may have operating systems, the distinction we're drawing here is to separate devices that are primarily of the PDA and personal computer ancestry rather than that of the phone. Some of these operating systems include the Palm OS, EPOC, the Pocket PC, Linux, and the various versions of Windows and Macintosh. For many of these devices, the native mode of access would be assumed to be over a wireless LAN or proprietary network rather than the public cellular data networks, although either or both is certainly a viable choice. This is a new point of distinction, since the smart phones, as seen today, are not capable of this type of connection.

In addition to the four categories of wireless gadgets described here, combination devices are now hitting the market. These combination devices may have several characteristics that are similar to the ones we've described, yet some of these combinations are rather arbitrary. For instance, we now can purchase cell phones that contain integrated voice recorders and MP3 players. In addition, there are phones on the market that will replay slow-motion video and display pictures. There are PDAs that have a phone or digital camera capability that's been added on to them. We have PCs and Pocket PCs that are starting to grow all sorts of appendages to allow for MP3 playback, for bar code readers, for GPS location finders, scanners, printers, and many more gizmos that are rapidly making their way from the R&D departments to the production floor.

The challenge in this evolution of devices is to predict which innovative widget is next so that we can prepare to successfully maneuver the increasingly complex landscape. The variety and the opportunity this evolution represents will keep application designers on their toes to come up with the best device (or at least a workable device) that contains the right combination of capabilities to satisfy the end user.

Peripherals

In addition to the core devices we've described, there are quite a number of peripherals that are starting to find their way to the marketplace. Some of these were mentioned before. As an example, the bar code reader is a peripheral that's been available now for quite some time on mobile devices. These scanners, often used in warehouses and other locations or situations that track inventory, are one of the most requested devices used in mobile applications. A number of companies manufacture these scanners, and they have been integrated onto a number of platforms. Going forward, we expect more and more bar code scanners to be integrated with wireless gadgets, including PDAs and cell phones, or to be available as a simple add-on, thus decreasing the device's overall cost. These scanners are continuously evolving. For example, the functionality is now being extended to phones that include bar code readers so that you could scan the UPC code at your local electronics retailer (or other location) and then use the connectivity to the Internet to check the price at neighboring or Internet-only stores. Other bar code scanners provide the capability to read text and store it for some later recognition and analysis.

Another peripheral that is being used more and more is the GPS Location Finder. While these appliances have been sold for some time as stand-alone add-ons to hikers, campers, skippers, and other adventurers who are out and about, they are now becoming integrated with wireless devices and are far more ubiquitous. The E911 service in the United States is driving some of this development, even though GPS isn't the only solution that satisfies the E911 requirements. Beyond the need for these GPS peripherals to identify your location in the event of an emergency, there is an increasing need or desire for standard mobile users to know where they are and be able to get services based on their whereabouts. The general need is to be located at a finer level of detail than the zip code, without the user having to know or supply his or her latitude or longitude.

Yet another group of powerful peripherals are mobile printers—useful in many applications. Examples range from the rental car receipt that you get when you return your car at the rental car drop-off location, to the boarding pass you receive from the mobile kiosk at the airport that you use to check in for your flight. For many years, the challenge has been the printers' power consumption. As the battery life is getting longer, we expect even more portable printers to be incorporated into mobile devices or added on to them.

This list of peripherals is certainly not an exhaustive one, but it should provide a good introduction to an increasing array of add-ons that will soon come to a mobile device near you. We expect this list to grow, with devices expanding both in capability and specialization as many of their current limitations, such as excessive power consumption, are overcome.

Network Access

For the wireless portion of mobility applications, access to some networks is almost imperative. A variety of these networks have evolved over time, categorized as follows:

1. Custom radio frequency (RF) networks.
2. Specialized RF networks.
3. Carrier-provided networks.
4. Optical networks.

Custom RF Networks

Custom RF networks include those that are typically restricted to limited areas. Several companies have deployed these networks, including transmitter towers and other necessary equipment, in an effort to get coverage within some particular location. These networks typically have been expensive, proprietary, and subject to many local operational issues. In addition, expensive maintenance required for these custom networks has been difficult for individual companies to sustain over the long run. Going forward, we would expect the importance of these networks to decline as ubiquitous access to high-speed digital networks becomes more common.

Specialized RF Networks

By specialized RF networks we mean those that were originally built for a particular purpose, although they may have been generalized over time to add other capabilities. Specialized RF networks were built usually not by the standard telecom operators, but by companies that focused on delivering digital data. Examples of such networks include paging networks such as Micro Burst. In recent months some of these paging networks have been leveraged to provide digital data in a nearly synchronous mode, delivered by wireless products such as the Blackberry. While the bandwidth is low and latency may be high, the benefits of these networks to their users have been significant, as their availability over a large area was not possible with the custom RF networks mentioned previously.

Carrier-Provided Networks

This category includes the extension to existing cellular networks that allow the transmission of digital services. An example of a carrier-provided network would be CDPD. As we go forward, these cellular networks are being enhanced to provide digital capability, leveraging GPRS, 2.5G, 3G, and other technologies. Important characteristics of these networks include that they are rapidly changing, requiring us to keep close watch. In addition, we expect that over the next few years the majority of digital communications will move to these networks, provided they are, in fact, built out as planned.

Optical Networks

Today, companies are often using optical networks as backbone supplements to existing ones. Terabit speeds are available between line-of-sight locations from a number of providers. This network option is interesting, because it doesn't entail tariffs, and organizations can buy the equipment and set it up relatively inexpensively in comparison to other solutions on a cost per megabit per second basis. These networks are typically used as replacements for or supplements to microwave networks and other point-to-point networking capabilities that have been needed for very high-speed data transmission. While it is not expected that these networks will have direct interaction with consumer devices, the low-speed brother of this technology, the InfraRed networks that have evolved in places like factories, are frequently being used by consumers to synchronize their PDAs with their PCs.

Enterprise Mobile Applications

A fundamental trend within enterprises today is extending access to the existing corporate information to mobile users. This access must be provided in a controlled fashion, as access will not be limited to the company premises. We'll discuss several mobile application types in a number of separate categories. Several of these may apply to any one individual in a company.

The first group of enterprise mobile applications is in some ways also the simplest: extending simple office productivity applications to mobile workers. Extending the reach of office functionality starts with making enterprise email remotely available to the organization's employees. Inherent in this solution is the ability to filter email and make selections as to what should be displayed on the mobile devices. Access to email is sufficiently important to corporations that several software providers have sprung up to enable this specific functionality. In addition, some of the more popular enterprise applications today support devices that are focused on enterprise email access.

In order to allow access to the corporate email server, different types of networks have been used. These include networks provided by the cellular companies, the paging networks, and custom networks, as we've described. In addition, these applications allow for instant messaging, and generally enable various types of communication applications to function from mobile devices. We should note that this capability is not restricted solely to the enterprise user. It is also available in a number of the common mobile portables that are consumer driven. Messaging is one of the major applications of the Internet, and the feature has also been one of the most important for mobile workers. In addition to real-time messaging, PDA users without the wireless connectivity can use a synchronization version that periodically connects with the network to retrieve and respond to email.

The next group of enterprise mobile applications includes those that are used heavily within a corporation and now are being specifically designed to reach the mobile workforce. The longest-standing example is field service automation. Field service automation has been used for a number of years, provided by a number of specialized networks and carriers. The recent changes in networking technology have allowed field service automation to be much more accessible for many corporations. Many of these applications function quite efficiently, with the majority of the interaction being handled via synchronization, rather than being network connected. Another advantage of these applications is that the return on investment can be calculated relatively easily. They frequently provide an extremely quick payback period, measured by additional calls serviced per day or a decrease in time spent idle between customer visits. Another benefit of this application is the ability to dynamically reroute a field person's schedule as emergencies arise, provided that the field service rep has a wireless/messaging connection.

The next group that we frequently encounter enhances the capability of an organization's sales force personnel to view data in an organized fashion. This group of applications, largely known as *sales force automation* tools, is rapidly growing. The reason for this expansion is the increasing need to gain access to real-time data, such as inventory levels, customer records, billing records, price levels, and other key data. As business becomes more dynamic, the need to access such information has increased accordingly. That trend, coupled with lower prices for devices that enable wireless data access, has led to a major market for these applications. Typical functionality includes access to customer records, access to the history of the customer with the company, current payment records, current listing of outstanding customer service issues, breaking new headlines related to a customer, and the ability to place orders in real-time. Wirelessly enabled sales force automation tools can extend the traditional capabilities of the Internet to anytime, anywhere functionality.

Going forward, there are a number of other areas where the ability to have mobile access to back-end data is critical. Figure 2.1 illustrates some of the key mobility solutions and industries where they apply.

An additional example would include the business-to-consumer and business-to-business trading exchanges whose services can be significantly enhanced by providing wireless access. As the end of an auction comes closer, it is important that the participants can view the latest information and be notified when their bids are no longer valid. Similarly, the entire supply chain is ripe for wireless and mobile automation. As business becomes more and more real-time data dependent, the ability to access this information across the entire supply chain becomes a paramount business imperative. Supply chain applications will require greater back-end integration within the enterprises themselves, a normal outgrowth of the need to drive business to a much more dynamic basis. Another consequence of the increas-

Mobility Solutions - Applications

Industry	Wireless Portal / Commerce	Location Based Services	Telematics / Fleet Mgmt	Sales Force Automation	Remote Field Service	Mobile Networking
Financial Services	✓			✓		✓
Manufacturing			✓	✓	✓	✓
High-Tech Manufacturing	✓		✓	✓	✓	✓
Telco, Media & Networks	✓	✓	✓	✓	✓	✓
Retail/ Consumer Products	✓	✓	✓	✓		✓
Heathcare		✓		✓		✓
Life Sciences & Chemicals				✓		✓
Energy & Utilities					✓	✓

Source: CG E&Y Analysis

Definitions:
Wireless Portal / Commerce – involves the facilitation of product/service transactions in exchange for money
Location Based Services – involves the two-way transmission of location-sensitive information (e.g., promotions, telemedicine)
Telematics – involves the remote monitoring and management of mobile assets (e.g., fleet management, machine-to-machine)
Sales Force Automation – involves the remote access/manipulation of information stored on a common network (e.g., e-detail, CRM)
Remote Field Services – involves remote access and input of information (e.g., manuals, part re-order)
Mobile Networking – remote access to corporate networks to access email, Internet and other systems

Figure 2.1   Mobility solutions and industry sectors.

ing integration of the wireless aspect into the supply chain is the extension of access to data outside the enterprise. A basic capability for B2B exchanges, it will become even more prevalent in many supply chain implementations. As enterprises continue to virtualize themselves, additional companies will have to be integrated into a single view, from the information technology perspective. That view will then have to be available to the mobile user as well as to normal management structures within each of the involved organizations.

Consumer Applications

While this book concentrates to a large extent on enterprise applications for the mobile user/employee, there are a large number of opportunities to provide consumers with wireless capabilities. As Figure 2.2 illustrates, these opportunities will vary based upon the application.

Some of these applications allow users to simply view data that resides in a company's database that pertains to them. Others involve the added functionality of being able to conduct transactions and dynamically modify their information kept in those corporate repositories. Key examples here include the ability to trade in securities and execute other transactions against one's financial accounts. These are, and can be, done wirelessly, with good security and reasonable degrees of nonrepudiation and authentication. There are many folktales of the horrible things that could happen as wireless data net-

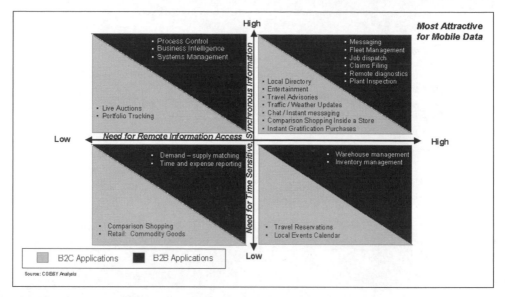

Figure 2.2   B2C vs. B2B applications.

works spread across the landscape. However, the purpose of this book is not to speculate, but to observe with a keen eye that we are at the beginning of a technology wave whose implications nobody can predict with certainty at this time. What we do know, however, is that the applications we know of today already provide real value to their users. Many more are undoubtedly to follow.

Delivering Mobile-Sensitive Applications

Introduction

The purpose of this chapter is to map out the terrain of the wireless application environment. The chapter looks at what we must consider in order to frame out, at a high level, the delivery of mobile applications. First and foremost, we need to understand the unique characteristics of mobility, as wireless applications entail much more than just fronting the Internet onto wireless devices. We will elaborate on this concept in the section titled *The Death of the Killer App*. It is the mobile process itself and its unique qualities that provide the momentum.

The aspects of mobility that are primary in an application will influence the technical infrastructure systems that must support it. Therefore, while it may seem peculiar, the first part of this book defines the unique aspects of the mobile environment rather than the technical components. This is analogous to the way a building architect will first consider the environment in which the structure will be placed before designing a composition to work harmoniously within the surroundings.

The architectural design specification of any technology must also consider individual and organizational needs, both of which drive its application. It might be easy to imagine the benefits of a new technology. But, it is not as easy to understand the environment of needs in which the new technology will play. Again, to help frame this understanding we have to consider the unique attributes of mobile applications. After we have gained insights about the environment, we will:

- Take a high-level view of the wireless technical architecture and infrastructure in its entirety from end to end.

- Look at the mobile components that will need to be added into an existing enterprise system. In this part of the chapter the complexity of mobile systems and the need for precise systems integration become apparent.

- Review the evolution of mobile services and technologies and examine global considerations.

- Catch a glimpse of where technologies are likely headed and the issues of scale and flexibility that must be addressed in the initial solution design.

Attributes of Mobility

The unique characteristics of mobility are what differentiates this emerging concept from its predecessors. This section should give you a good idea of what these unique attributes to mobility are and how they are applied to the enterprise.

The Death of the Killer App

The myth of the killer app often floats during the introduction of a new technology. A *killer app* is typically defined as the key application that drives the explosive use of a technology. For example, from certain industry points of view, Web browsers were once viewed as the killer app driving use of the Internet. However, upon closer examination, the browser was only a means, or a tool, for navigating across connected systems. The users of these systems developed the real applications, such as search engines, auctions, joint purchase complementariness, and so on, based on the unique attributes afforded by the Internet, such as open connected systems. The unique attributes of the Internet are what drive its use and application to specific needs.

A similar illusion permeates the space of mobile computing. Here, the mobile fronting of Web applications is often viewed as the killer app that will drive wireless computing. It is not. The killer application is the mobile process itself and the ways in which its unique characteristics, injected into a public or private enterprise, enable the accomplishment of individual and organizational needs. Let's consider some unique characteristics of mobility and then look at the systems supporting the concept's application.

Figure 3.1 shows the various aspects of mobility that can add value to any enterprise's endeavors. Many of these features are unique to mobility and add value to organizational needs in ways that cannot be achieved without them. Let's examine these features in detail.

Ubiquity

Ubiquity is perhaps the most promising aspect of mobile computing. *Ubiquity* can be defined as the ability, or the appearance thereof, to access or apply applications anywhere, at any time. This is a key attribute of mobility. Implicit in this concept is the embedding of network access into the environment in a pervasive way. The importance of ubiquity cannot be understated. It allows for new gains in effectiveness and efficiency that cannot be achieved in a tethered system. Ubiquity is therefore a chief architectural consideration.

Reachability

The idea of ubiquity offers little without the ability to reach or connect with the desired contact or source of information in a coherent, relevant fashion. This is what we mean by reachability. The seamless connectivity we expect when using a telephone is a good model for such reach. Internet connectivity runs a close second when useful information is present in a consistent and

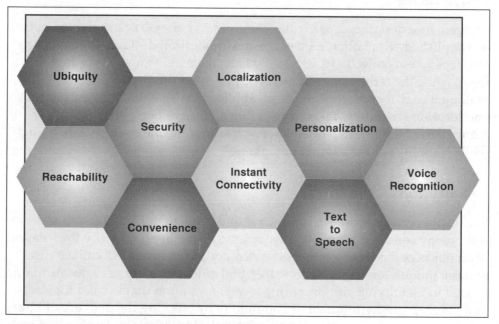

Figure 3.1   Attributes of a mobile-sensitive application.

coherent manner. When talking about reachability we consider ubiquity, connectivity, and coherence in terms of reaching or acquiring information from a stationary or mobile source while we're on the move.

Security

Mobile communication, including voice and data, must be secured to be of any use to most individuals or organizations. There are two aspects of security we would like to point out at this time. First, the very nature of telecommunication protocols used in digital wireless systems provides a certain level of security. Many of these techniques were first developed by the military to ensure safe transmissions. This is not to say that wireless system protocols are failsafe. They are, however, tough to break and inherently more secure than other methods. Telecommunication protocols add another layer to the security equation, which will be addressed in more detail later.

A second aspect of security is that mobile devices remain relatively within the sphere of individual control; that is, they are often worn or carried by the individual. Because of this, mobile devices can be adapted to reflect security information unique to the person owning the device. Adapting a user's devices can even be applied in securing tethered systems where the proximity of the person permits access.

Convenience

Convenience describes a particular comfort level or ease of use that must be accomplished with both the interface and application designs. Convenience also facilitates connectivity. Especially here, the magic must be applied up front, while the technology should be kept behind the curtain. In addition, making it easy for someone to use an application can entail elements of time and location sensitivity. Taken from a human factors perspective, convenience drives usage. How often have you seen someone use a mobile phone while standing next to a wired pay phone?

Localization

Localization offers some very attractive and challenging architecture and technology considerations. Here we refer to the ability to identify the location of an object or a user of a wireless device, and/or to provide localized information, unique services, conveniences, and efficiencies. As such, localization is used in identifying and providing local information that is valid for a certain geographic environment or community by tailoring the delivery of information on a personalized and proximity basis. Furthermore, localization is often applied in the areas of Global Positioning Systems (GPSs) and Geographic Information Systems (GISs) to navigate a physical space. Thus, localization applications can also be leveraged in supply chain, warehousing, and delivery systems.

Viewed as a subset of ubiquity, localization carries importance and weight as a unique aspect of mobility. This is so much so that the attribute requires the implementation of new and unique information systems in the enterprise.

Instant Connectivity

From the perspective of mobile computing, instant connectivity means *always on*. This suggests that while people's mobile devices are switched on, they are connected to the network. In order to achieve this, wireless network providers must implement a packet-based connectionless network. This is different from what exists today in the United States, where the circuit-switched model prevails, requiring one to dial into a network connection. This is not the case in certain parts of Europe and Asia. NTT DoCoMo, for example, implemented a connectionless network called iMode in February 1999 and grew its service to more than 10 million subscribers in one year. In addition, fixed wireless solutions offer instant connectivity, especially in the small office/home office (SOHO) environment. These transmission technologies include two-way satellite communication, IEEE802.11b wireless Ethernet, IEEE802.15 Personal Area Ethernet, as well as the Bluetooth protocol. As these technologies proliferate, it will be interesting to see what pressures they exhort on wireless network providers in the United States.

Personalization

Each of us has people we care about, things we like to do, and places we like to go. We all have our own personal worlds and unique styles. Businesses are recognizing this and driving fast into the area of customer relationship management (CRM) and the personalization of services based upon an individual's unique requirements. They recognize that smart agents and integrated personalized services are two important ways to increase the stickiness of their services to improve customer loyalty.

In a truly wireless world, people can take their personal spheres with them wherever they go. Developers need to think about how users will use a wireless device as the *universal remote* while going about their daily activities. For example, imagine a person throughout the day relying on wireless technology to do his job, make travel plans, and even arrange for flowers to be delivered to the home of his significant other, arriving at her place at the same time as the user of the wireless device. This could even be done through voice-activated conversations with a personal wireless butler. Services must be capable of being personalized. A wireless service must be everything to everyone. It must be location aware. Personalization means the delivery of tailored services and information in real time over a network that is always on.

Voice Recognition

Voice recognition is a very important human factor in the mobile space. Several companies have developed suites of voice recognition servers that now span more than 22 languages. Here we are not discussing speech transcription services, although those, too, can be a useful mobile application. Instead, speech recognition can offer significant cost savings in call center applications. More complex functions include commands for dialing or navigating portals, handling email, voice mail, calendar, and other personal applications over a handheld device. Voice recognition can also be used in the area of security and voice verification. The objective of voice recognition is convenience and ease of use where typing, jotting, or pointing in an effort to enter information into a system becomes a nuisance and prohibits use.

Text-to-Speech Conversion

Text-to-speech (TTS) conversion is the other side of the voice recognition coin. Again, the same goals of convenience and ease of use drive usage in the area of handheld devices. When screens are very small, or when users are interacting with information systems through a voice interface, text-to-speech conversion can become critical. In addition, TTS is essential in some applications for the handicapped, and increasingly finds applications in situations where the user is occupied with other activities, such as operating an automobile or other machinery, which requires the use of the hands and eyes.

Technology Systems for Mobile-Sensitive Applications

Now that we understand the unique qualities of mobility we will look at technology components that deliver *mobile-sensitive* applications. By *sensitive* we are referring to the qualities of responsiveness and fit. What we need are systems that are receptive, aware, and insightful, and have the ability to respond in an exact, delicate, and finely tuned way within the framework of mobility.

Figure 3.2 depicts the components of a connected wireless system enterprise. What we see are mobile devices connecting to mobile-sensitive applica-

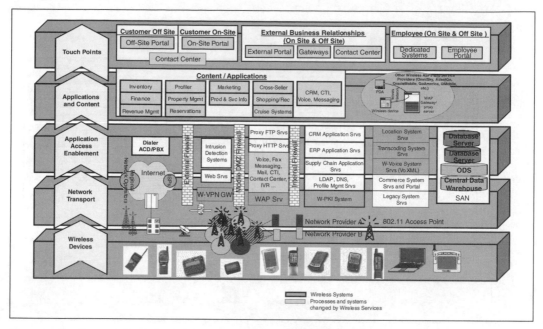

Figure 3.2 The mComputing connected enterprise architecture.

tions via a wired and wireless network infrastructure reaching into an enterprise's integrated information systems environment. This process is called *application access enabling*. It is important to understand the depth of systems integration that takes place in each of these five areas.

Wireless Devices

In the untethered world, there are a great many wireless devices that we need to understand. This section looks at some of those devices and their components.

Device Providers

As consumers, most of us think of wireless in terms of our handheld devices. Handheld devices are at the forefront of the wireless revolution. Vendors of handheld devices, including the ever shrinking notebook computer, are critical in the value chain.

Generally, customers do not shop for a particular service provider or network operator, but rather for the device brand and its features (see Figure 3.3). The emergence of the mobile phone not only as a consumer electronic device, but also as a personal item such as a pen or a watch, has created lots of value for the handheld brands. Manufacturers will continue to develop a wider variety of products, and future applications will require different com-

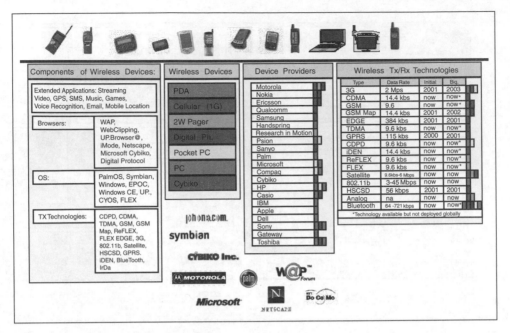

Figure 3.3   Wireless devices.

binations of features. Handheld devices, optimized for extended applications such as music download and listening, video streaming and watching, computing, game playing, or just managing one's life are becoming more widely available. At the same time, mobile handset manufacturers are coming closer to the traditional PDA manufacturers, while those companies encroach upon the territory of cell phone providers. The lines between both industries are blurring, as both are increasingly offering devices with combined phone and organizer functionality.

Manufacturers

A well-known, but not inclusive, list of handheld device manufacturers includes Motorola, Nokia, Ericsson, Qualcomm, Samsung, Kyocera, Siemens, Handspring, Psion, Symbol, Sanyo, Palm, Microsoft, Compaq, Cybiko, Hewlett-Packard, Casio, IBM, Apple, Sony, Dell, Gateway, and Toshiba. These manufacturers produce a variety of devices, including personal digital assistants (PDAs) such as the Palm Computer or the Pocket PC, cellular and digital phones, two-way pagers, and custom devices such as rugged, industrial-strength wireless PDAs with built-in bar code scanners. All of these devices work over a plethora of network transmission protocols. Perhaps not so well known is the abundance of different operating systems and various microbrowsers implemented in these intelligent devices.

Components of Wireless Devices

When we implement a wireless application we must consider the components of the various devices from a systems integration perspective. Following is a list of many of the receive and transmit network protocols, data rates, operating systems, browsers, microbrowsers, and extended applications found on a variety of different mobile device types.

Cellular Digital Packet Data (CDPD).   Digital portion of traditional Cellular Analog 1G systems. Rates are 9 to 14.4 Kbps.

Code Division Multiple Access (CDMA).   Each conversation is digitized and then tagged with a code. The mobile phone is then instructed to decipher only a particular code to pluck the right conversation off the air. Data rates are at about 14.4 Kbps.

Time Division Multiple Access (TDMA).   A method of digital wireless communication transmission allowing a large number of users to access (in sequence) a single radio frequency channel without interference by allocating unique time slots to each user within each channel.

GSM Map.   Global System for Mobile Communication (GSM) operates in the 900-MHz and the 1,800-MHz (1900 MHz in the United States) frequency band and is the prevailing mobile standard in Europe and most of the Asia-Pacific region. GPRS is derived from GSM Map.

ReFlex Enhanced Data Rates for GSM Evolution.   A technology that will enable even higher HSCSD and GPRS speeds. EDGE is another short-term improvement that is being developed for TDMA/GSM systems. EDGE will provide multimedia and other broadband applications over a bandwidth of 384 Kbps by 2002.

Flex.   One-way protocol developed by Motorola for its text pagers.

Enhanced Data Rates for GSM Evolution (EDGE).   A technology that will enable even higher HSCSD and GPRS speeds. EDGE is another short-term improvement that is being developed for TDMA/GSM systems. EDGE will provide multimedia and other broadband applications over a bandwidth of 384 Kbps by 2002.

3G.   The third generation of wireless industry technology represents the next major upgrade in functionality. Examples of 3G technologies include Universal Mobile Telecommunication System (UMTS) and wideband-CDMA (WCDMA).

UTMS.   Universal Telecommunications System/International Mobile Telephony 2000, the third generation of mobile telecommunications. Enables wideband connections using WCDMA (wideband code-division multiple access). Data rates will range from 384 Kbps to 2 Mbps.

High-Speed Circuit-Switched Data (HSCSD).   HSCSD = multi-slot. Enables higher speeds using several time slots, typically 2 to 4 (giving speeds up to 28.8 Kbps to 57.6 Kbps). Usually asynchronous, for example, 43.2-Kbps downlink and 14.4 Kbps for the uplink. Many operators have chosen not to launch HSCSD but instead wait for GPRS.

General Packet Radio Service (GPRS).   A new service that provides actual packet radio access for mobile GSM and TDMA users, essentially always on. It's a non-circuit-switch solution that reduces cost to carriers. Rates can range from 9.6 to 150 Kbps.

Wideband CDMA (WCDMA).   The third generation standard for CDMA.

Unstructured Supplementary Services Data (USSD).   A means of transmitting information via a GSM network. It is to some extent similar to SMS, but in contrast to SMS, which is basically a store-and-forward service, USSD offers a real-time connection during a session.

802.11b.   IEEE protocol for wireless Ethernet technology compatible with 802.3 protocols.

Global Positioning System (GPS).   A system that consists of 24 satellites circling the earth in a particular constellation in relation to each other so that several satellites fall within the line of sight for any GPS receiver on Earth. Because the satellites are continuously broadcasting their own position and direction, the GPS receiver can calculate its position very exactly.

Cell Broadcast (CB).   A technology that is designed for simultaneous delivery of short messages to multiple mobile users within a specified region or nationwide.

PalmOS.   The PalmOS has a particular wide acceptance in the United States, where the Palm VII, with its wireless connectivity and Web clipping technology, has hit the market already.

Symbian.   A consortium of leading mobile handset manufacturers Nokia, Motorola, Ericsson, Matsushita, and U.K. PDA manufacturer Psion, established in June 1998. The operating system, which is based on Psion's earlier software, is called EPOC.

Windows.   Microsoft's operating system for the IBM personal computer. It appears in ever smaller laptop PCs. This includes all the sub-notebook-sized equipment, such as the Sony Vaio.

Linux.   Unix operating system developed for the Intel x86 architecture.

Unix.   Operating system, originally developed at AT&T's Bell Laboratories, that runs on a variety of platforms such as Sun, HP, IBM, and Apple.

Windows CE or Pocket PC.   Microsoft has developed a lighter version of its Windows operating system, called Windows CE (aka Pocket PC), that has been created especially for palm-size, handheld PCs and other consumer electronics devices. A large number of handheld computer/PDA manufacturers, mostly coming from the PC industry, such as HP, Casio, Philips, and Compaq, have developed their devices around CE.

Unwired Planet (UP).   (Now Openwave.) OS for phones.

Wireless Application Protocol (WAP).   An open, global standard for mobile solutions, including connecting mobile terminals to the Internet. WAP-based technology permits the design of interactive, real-time mobile services for smart phones or communicators.

Web Clipping.   A Palm proprietary format for delivery of Web-based information to Palm devices via synchronization or wireless communication to the Palm VII.

UPBrowser.   The microbrowser made by Phone.com that has dominated the microbrowser market. The company was recently renamed Openwave.

iMode.   DoCoMo's iMode service, although it provides a bandwidth of only 9.6 Kbps, is the first always-on wireless Internet service, because of its packet-based technologies. As a result, users are always connected and do not need to establish a dial-up connection to access data services.

Short Message Service (SMS).   Since 1992 SMS has provided the ability to send and receive text messages to and from mobile phones. Each message can contain up to 160 alphanumeric characters. After historically finding it tough going in the GSM markets, during the year 1998 SMS suddenly started to explode.

SIM Application Toolkit (SAT).   This technology allows network operators to send applications over the air as SMS or as Cell Broadcast messages in order to update SIM (Subscriber Identification Module) cards with changed or new services. SIM Toolkit applications are built in Java for a client/server environment.

Mobile Station Application Execution Environment (MExE).   This is, essentially, the incorporation of a Java virtual machine into the mobile phone. The purpose of MExE is to provide a framework on mobile phones for executing operator- or service provider-specific applications. It allows full application programming. The protocol is integrating location services, sophisticated intelligent customer menus, and a variety of interfaces, such as voice recognition. MExE will incorporate WAP, but also provides additional services exceeding the WAP functionality. We believe that MExE might be built into future UMTS phones, which will have the processing power to run the Java programs.

Bluetooth.   Bluetooth is a low-power radio technology that is being developed to replace the cables and infrared links for distances up to 10 meters. Devices such as PCs, printers, mobile phones, and PDAs can be linked to communicate and exchange data via a wireless transceiver that fits on a single chip.

Does building applications that touch across a variety of network protocols, operating systems, and microbrowser technology while supporting extended applications and the evolution of mobile service technology seem like a daunting task? Well, it sure can be. Fortunately, dominant players are now starting to emerge. Still, there already exist more than a handful of devices and networks that the application systems must bridge. To start diminishing the overwhelming task at hand is where transcoding systems come into play, while the carriers and device manufacturers address the network protocols. Besides the efforts undertaken at that level, we should stay cognizant of the data throughput and synchronization as well as the connection or connection-less orientation of the networks. For example, certain applications such as SMS and video packet data will not work over a 2.5 CDMA network. So let's look at the real work at hand: transcoding content and developing applications to work across several device types and browser technology.

Wireless Transcoding

The key to solving complexity is to seek its lowest common denominator. XML (eXtensible Markup Language) and XSL (eXtensible Style sheet Language) are such keys. The ability to translate standard HTML/XML content into the various flavors of wireless content is based on the way that the XML/XSL standards operate. Due to the definition of these standards, XML is able to function as a meta language, flexible enough to recreate most other markup languages and use them as subsets. Technologies such as Java Server Pages (JSP) and Active Server Pages (ASP) are also being extended to support multimode clients. Figure 3.4 illustrates some examples of transcoding systems. Additionally, the following are three types of transcoding mechanisms that are in use:

Clipping.   The clipping mechanism is useful when there is HTML content to be made available to wireless users. Items in an existing Web page are selected, or *clipped*, for salient information and data and then formatted to fit the handset display.

Fragmentation.   The fragmentation mechanism breaks an existing HTML page into a series of smaller, linked pieces, like supercards. When a user accesses a URL, the WAP server can send the entire set of cards in one transmission and the user then navigates through them.

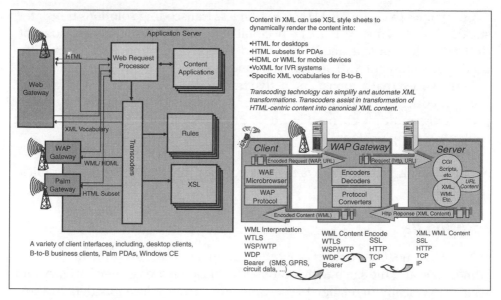

Figure 3.4   Transcoding system services.

Style sheets.   The style sheet mechanism is the most versatile. It requires XML source content. Different style sheets can be defined for any client device, and business rules support dynamic and custom content transformations.

A good transcoding system should supervise the channel, looking at the specific types of devices and correlate information to be sent out. Applications, built on dynamic content and/or resulting from multiple sources, require a highly developed management and control system of the source data and documents. This will require the upfront identification of the logic, engines, and data structure. The advantage of the extra work is that it reduces deployment time by leveraging the existing business logic, application logic, transaction engines, trading engines, commerce engines, and data structures that are already built. Additionally, any new functionality of the enterprise system can be mobilized for device access, further ensuring consistency across all types of devices. The key here is to provide consistency across multiple clients.

For example, imagine a situation where a customer plans to buy a certain IPO stock on the first day it's offered. She checked the stock price through her electronic financial service provider before leaving the office but decided to wait for the price to go down. About an hour later this person decides to check the IPO stock price from her mobile device using the same financial service. Unfortunately, the wireless application on the smart phone did not have the IPO stock information.

Companies that provide HTML-to-wireless transcoding solutions translate standard content into XML that reflects all of the information flow and logic of the transcoded content, and by browser detection determines for each mobile user which subset of XML to translate content into.

Thus we see that a robust and active transcoding system goes a long way toward reducing complexity and applying consistency throughout the enterprise across multiple clients. Figure 3.5 provides another view of transcoding content and applications.

However, this covers only the up-front component of delivering mobile applications. Let us now move to other components that bring these applications to life.

Location ID

When we come to the subject of location ID, many scenarios come to mind. Almost one and the same with location identification are issues of privacy and security. Typically, in relation to consumer applications, location identification constitutes a voluntary trade-off of information in return for benefits. In business applications, it becomes less voluntary. Finally, location services in some governmental areas are typically mandatory. Governmental applications include:

- E911
- Firefighter location/navigation
- Ambulance location/navigation

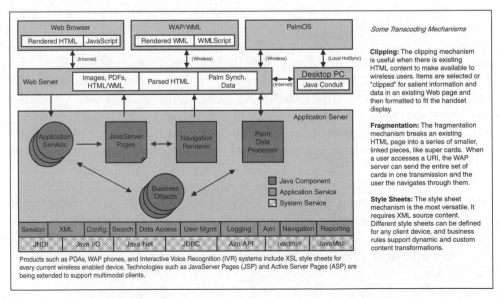

Figure 3.5 Transcoding system services, another view.

- Police location/navigation
- Patient tracking

People in general do not mind location identification in the areas concerning their safety, protection, and health. However, emergency services providers have been unable to respond quickly when no specific location information is available. To this end, the FCC has mandated implementation of location identification by the carriers in two phases.

1. Phase I—already in effect.
   a. Callback number.
   b. Approximate location.
2. Phase II—ready to implement October 1, 2001.
   a. Callback number.
   b. Location within 50–300 meters 67–95% of time.
   c. Carriers were to choose technology by October 2000 and be ready to implement October 2001.

It is interesting that some location ID technology will be available as a result of governmental regulations rather than customer or business demands. However, governmental regulations will provide capabilities that can then be extended to nongovernmental needs.

Within the commercial realm, location ID is viewed as a necessary course of doing business and a nonissue when it comes to tracking products inventories and warehouse and supply chain management. Here, radio frequency identification (RFID), Bluetooth, and the IEEE 802.15 protocols will likely come into play. Sales force and field service automation will be aligned more with carrier solutions. Business applications include:

- Asset tracking
- Traffic information
- Fleet management and dispatch
- Field and sales force automation
- Emergency services
- Roadside assistance
- Navigation
- Location-sensitive billing
- Automatic tollways

The successful implementation of consumer location identification services will be achieved when the customer views the technology as beneficial. In this sense, consumer applications may be better positioned when they are based on pull activities—features a user initiates—rather than pushed functionality. Examples of consumer location identification include:

- Child tracking
- Pet tracking
- Parent locator
- Emergency services
- Roadside assistance
- Navigation
- Concierge service
- Mobile yellow pages
- Location- and time-sensitive purchases
- Traffic information
- Dynamic location- and time-sensitive event information
- Automatic tollways
- Home automation
- Messaging services

Location ID Technologies

The Global Positioning System (GPS) consists of 24 satellites circling the earth in a particular constellation in relation to each other so that several satellites fall within the line of sight for any GPS receiver on Earth. Because the satellites are continuously broadcasting their own position and direction, the GPS receiver can calculate its position very exactly. Anybody can use the GPS system for free with an appropriate receiver. Figure 3.6 illustrates GPS and location technologies in action. GPS has been developed in the United States for military use, but since the early 1990's it has been made available (with lower resolution) for civilian purposes.

The following is a brief discussion of hardware and network-based GPS technologies.

GPS requires additional equipment or some modification in the mobile device, so that it can become a GPS receiver. The technology is also used in car navigation systems. SnapTrack and SiRF are two companies currently developing GPS technology for mobile phones. In Europe, GPS is already used in Benefon dual-mode GSM/GPS handsets.

Microwave RFID systems have been in extensive use for more than 10 years in transportation applications. RFID systems, operating in the UHF and microwave frequency range, are separated into *active power* and *passive power* tag categories. Operational range and functionality can be extended with

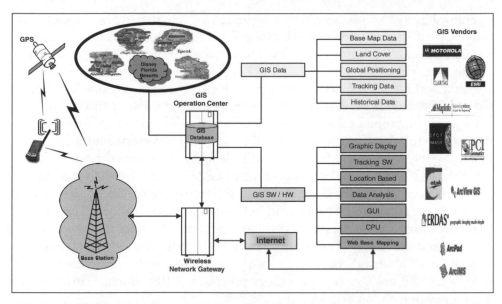

Figure 3.6   Location-based services and solutions.

active power tags that carry an embedded battery. Passive power, on the other hand, provides longer life and lower costs. RFID devices (tags and readers) may eventually operate in the microwave frequency range of 2400 to 2483.5 MHz since today's supply chains are global. Although UHF frequencies provide a greater range for RFID, they are not available on a global basis. Thus, microwave is the only practical, or, rather, realistic, place where adequate bandwidth is available to accommodate multiple, unlicensed, and unsynchronized systems.

The Enhanced Observed Time Difference (E-OTD) system works by using the existing GSM infrastructure to determine the mobile phone's location. When a user calls a selected service provider, E-OTD simultaneously sends data indicating the phone's position. The technology works by comparing the relative times of arrival of signals transmitted by the underlying mobile network base stations, at the handset and at a nearby fixed receiver. The E-OTD system overlays the existing mobile network. Suppliers for E-OTD solutions include CPS, Ericsson, and BT Cellnet.

Time Of Arrival (TOA) technology requires larger network modifications and is therefore not very cost effective. Rolling out TOA for an entire network is estimated to cost as much as 10 times the price of an E-OTD system.

Cell Of Origin (COO) can be used as a location-fixing scheme for existing customers of network operators, but it is not as exact as the three other methods. COO requires no modification to the mobile terminal, but the network operator has to do some significant upgrade work. In urban areas, COO might be sufficient to determine location fairly accurately, because the cell size is very small. In more rural areas, where the cell radius is larger, it might not be exact enough.

SignalSoft has developed its own proprietary wireless location solution that can use any or several of these methods to determine position. The software is installed on the operator's network and is able to combine the position with relevant content. SignalSoft has also developed a tool for provisioning the service and fixing the latitude against a defined zone.

CellPoint (formerly Technor), a recently Nasdaq-listed Swedish start-up, has developed yet another approach. The technology is handset based, and it works on the standard GSM network without any modifications. The solution needs only a proprietary server, and works on triangulation between the handset and the nearest base stations from both ends. Thus, the system is quickly installable and cost efficient when compared with competing technologies. CellPoint's system provides a precision of 100 meters in urban and 200 meters in rural areas.

We believe that GSM is a key technology that supports many of the technologies discussed. However, it is too early to determine which technologies will lead the pack and dominate the market. This is even more complicated in the United States, where multiple carriers provide different wireless network protocols and location solutions. It is also important to note strategically that due to cost issues carriers will at first implement only minimum 911 functionality, while looking for applications that will pay for the further development of the infrastructure "Car 54, where are you?".

Telematics will fundamentally change the nature of current automotive products and is expected to fulfill latent customer needs. In the realm of telematics, two sides have to be integrated: the electronic world inside the car and its connection to the outside world. Figure 3.7 illustrates classifications of telematics solutions in the automotive industry.

While location identification offers significant potential benefits, the extent of their exploitation depends on the competitive position that a company can achieve. As companies form their location identification strategy, they must focus on some vital, critical success factors. Here the trend is clearly aiming at an integration of all electronic components into a comprehensive architecture that addresses user needs while paying careful attention to issues of privacy and security.

Personalization Systems

Just as localization systems provide content dependent on time and geography-sensitive information, personalization enables services to be tailored to the particular needs of the customer as determined by their stated

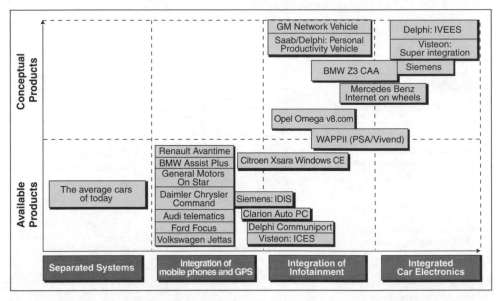

Figure 3.7   Automotive industry classification of ongoing telematic projects.

preferences or observed behaviors. Customer ownership is key to personalization systems, and convenience optimizes its value. Mobile computing allows firms to service customers from a distance. Customer interaction shifts from administrative tasks to relationship building (see Figure 3.8).

Using personalization systems, customer contact is focused on improving the customer relationship, ensuring that the customer is consistently presented with a high-quality experience. Personalization systems capture and utilize information about the customer. The goal is to enhance the experiences of all clients at critical points of contact in such a way that they are motivated to return again and again.

Here are some key attributes of mobile personalization systems:

- They demonstrate an understanding of customer value, and serve as the basis for real-time decisions made in marketing, sales, and customer service.

- They represent awareness of a customer's history and relationship with the entire enterprise.

- They involve the customer in the specification, design, and/or delivery of personalized services.

- Multiple channels are aligned with the customer's needs and values.

- They solve customer needs the first time.

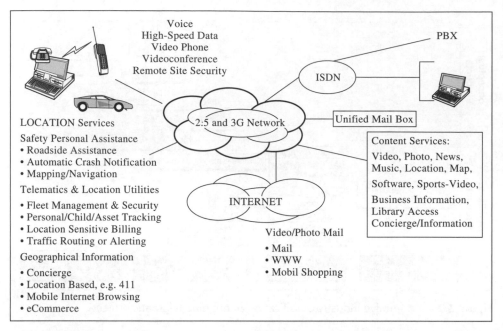

Figure 3.8 Data, networks, and location convergence increase data speeds for more robust applications.

- The right information gets in the right hands at the right time.
- Integrated front- and back-office information closes the loop and updates all systems.
- They listen and apply customer input, integrating it as a part of the way of doing business and sharing it with customers.
- They deliver a consistent, quality, value-added experience across all mobile and tethered touchpoints.

Figure 3.9 depicts the types of profiles, server engines, and event and knowledge managers that are desirable in a comprehensive personalized service.

First come the profiles. We need to understand many things about the customer in terms of preferences and demographics. For an obvious example, consistent with our view of privacy concerns, when does the customer want to be contacted? What do we know that is unique to the individual that can help us better match solutions to his or her needs? What type of devices are they using; what are the capabilities of the device, the screen format, and the available bandwidth? We also need to know many of the parameters about the time and geocentric events to match to their preferences. Here we are talking about such things as calendars, unified messaging, interest, and historic activities.

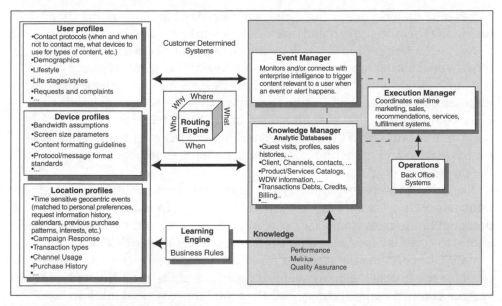

Figure 3.9   Personalization services.

These systems are customer determined. Two primary engines are required to facilitate appropriate customer relationships. These are the routing and learning engines. The routing engines are constantly determining the what, why, how, and where of delivering personalized content to the user. These engines also make the correct system linkages with the back-office systems. The knowledge engine is constantly learning new information about the customer and seeing that it gets recorded.

Finally, in the back-office systems we find event, knowledge, and execution managers. The event manager will work with the knowledge and execution managers to link enterprise intelligence with content relevant to the user based upon defined events and alerts. The knowledge manager analyzes the knowledge databases, and the execution manager coordinates real-time services that link into back-office systems such as sales, marketing, and customer support.

Voice Recognition Systems and Text-to-Speech

Simply put, voice recognition systems take the user's voice as an input device. Early systems were applied to dictating text to a computer or pulling down menus and saving work. They would typically require that each spoken word be separated by a distinct pause. Later systems recognized speech in a more fluent manner. Natural language speech systems enabled users to access data they needed using their own everyday language rather than restricting them to keypad entries or single-word answers.

As speech recognition technology improved in terms of accuracy, language types, vocabularies, and their ability to understand natural language, so, too, did our concept of talking to the electronic devices we encounter every day. For example, Clarion has already developed an AutoPC that responds to voice commands. So the idea illustrated in the 1980s television series *Knight Rider* and its famous KITT is now real and we can use it in a variety of ways for navigation and gathering information, perhaps even about the person in the car next to us.

A number of large corporations and aggressive start-ups have invested immense budgets in speech recognition technology. In order to avoid the potential hindrances from a lack of standards, several of them combined their efforts to establish the Voice eXtensible Markup Language (VXML).

VXML has become the standard for which voice applications are developed for the Internet, gaining wide acceptance and the promise of ubiquity in various enterprise applications. VXML is combining several markup languages that were already based on XML. These include:

- Java Speech Markup Language (JSML) by Sun Microsystems
- Speech Markup Language (SML) by IBM
- VoxML by Motorola

The specification was defined for making it easy to add scripts to a Web site that can be retrieved by phone or another device. A site with a VXML document is accessed through a voice portal, enabled with VXML and speech recognition software. The aim of VXML is to bring the rewards of Web development to voice applications. However, it should be noted that VXML systems typically incorporate dedicated voice recognition servers to employ finer levels of control.

The VXML architecture includes a document server, a VXML preprocessing context (recognition client) server, and an implementation (recognition and text-to-speech) system. The document server typically has documents written in VXML and HTML. It defines the site's communication roles and information access. Figure 3.10 illustrates the VXML systems services.

The preprocessing context server contains a recognition client for the document and interfaces to the speech recognition and text-to-speech system, as well as call control services. The recognition client can run procedures separate from the document, such as detecting and answering an incoming call and getting a VXML document from a site. It reads the VXML page and renders the markup as dialog.

The current specification does not choose an explicit grammar for VXML documents, although it puts forward, as an example, the use of the Java

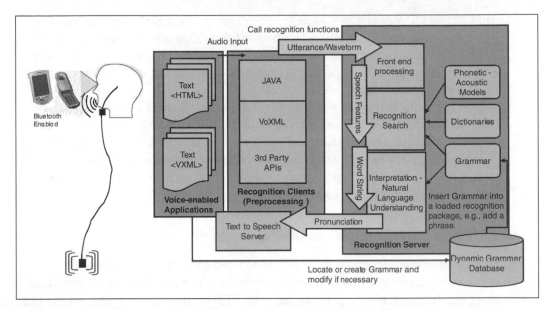

Figure 3.10   W-Voice system services.

Speech Grammar Format. Also, it does not specify the recognition and text-to-speech system.

> What is *grammar* to begin with? Grammar defines a set of rules regarding how words are used and combined when interacting with the application. These expressions can be defined within a single grammar, but often are categorized into separate grammars with limited scopes. For example, an application may provide instructions for navigating menus, getting more information, or getting support. The commands for each type of action are defined in separate grammars. This makes it easier to maintain and control the scope of each set of grammars.

Grammar scope refers to the component in which it is defined or referenced. A grammar is in force only while the context is within the logic of the grammar's component. The following are examples of such use.

A grammar defined or referenced within an application root document has application or universal scope. The grammar is active in any VXML file that references it through that document's VXML application feature. This can be used to ensure that users have a consistent experience when navigating between various enterprise systems and the application, thus reducing the number of commands they must learn.

A grammar defined within the VXML or *link* component has document scope. The grammar is active at any point in the document. A grammar defined within the *form* component has form scope. The grammar is active only in the form in which it is defined. While the user is within the form, all grammars with application and document scope are also active. A grammar defined within the *field* element has field scope. The grammar is active only in the field in which it is defined. While the user is within the field, all grammars with higher scopes are also active.

By limiting a grammar's scope, the number of allowed verbal communications against which the speech recognition server must compare are reduced. This increases performance by reducing the time before the system acts on a recognized match or nonmatch. This also provides greater control over the user's response by limiting scope, diverting the user from a selection that does not fit the particular context.

Wireless Portals (ISPs and ASPs)

One of the many decisions organizations need to make is how and what mechanism they will use to access the wireless Internet. Moreover, how will their applications leverage these services. This section reviews the concept of portals and how they play into the wireless architecture.

Internet Service Providers (ISPs)

Wireless Internet service providers are growing rapidly. We often think of the network carriers as the ISPs for wireless services and certainly these carriers would like us to view them as such. However, for now, they are providers of the infrastructure with a strong lock only on telephone devices. This walled-garden approach has limits that are already being traversed. ISPs such as GoAmerica, PalmNet, Omnisky, and Aether Systems are buying time from the carriers and offering connectivity for other devices such as notebooks and PDAs. In addition to this, there are broad areas in *fixed* wireless ISP activity that have taken off in a big way in certain local markets using 802.11b. One also needs to look at two-way Internet access via satellite and the providers of those services, which include Hughes, Starband, BreezeCOM, Cisco, and Wireless Internet Services.

Mobile wireless ISPs (carriers) have been quick to capitalize on the unique attributes of mobile computing for the consumer. Here content is king, with a focus on providing local and personalized value services with informational benefits. Without this approach, many users would not be drawn to wireless services because of the bandwidth issue. The ISP not only provides access, it also acts as a portal to such services as MapQuest, Travelocity, the Weather Channel, Yahoo!, Fidelity.com, and ESPN.com, as well as other entertaining activities such as Variety.com and Ticketron.

Less known providers are offering fixed wireless Internet services of 11 Mbps and higher, using technologies such as IEEE 802.11b on various frequency bands. Some of these sites cover large portions of states—for example, Wireless Internet Services providing coverage in Florida and another ISP providing national coverage in several key metropolitan areas. Wireless ISP frequency band spectrums of interest include:

ISM bands (900 MHz, 2.4 GHz, 5.8 MHz). The Industrial, Scientific, and Medical radio bands are the industrial equivalent of the citizens band. No license is required as long as only type-approved equipment is deployed. The main limitations are that there is only 1 watt of output power and that only spread-spectrum modulations are allowed. The amount of spectrum is limited, and each band eventually fills up, forcing new users to higher bands.

UNII band (5.2 GHz). The Unlicensed National Information Infrastructure band overlaps with ISM-5.8. It does not require spread spectrum but does have limitations on spectral power density limits that effectively correspond to ISM flavor. The band effectively expects high data rates to perform the spreading.

MMDS (2.496 GHz–2.644 GHz). The Multipoint Multichannel Distribution Service band was intended for wireless cable systems, and the spectrum allocations were sold by auction in 1995 and 1996. With the advent of Direct Broadcast Satellite systems, wireless cable vanished and the spectrum is being retargeted to data applications. The original rules specify a one-way system, but the FCC is now allowing two-way service. The spectrum is part of a slice originally designated for Instructional Television Fixed Service and is divided into 6-MHz channels. Each service area has two operators with four channels each, and channels alternate between the two operators, thus enforcing channelization. Many of the MMDS ISPs use a hybrid service, where the wireless link is used downstream, while upstream data goes on a V.34 modem link.

LMDS (27.5 GHz–29.5 GHz and 31.0 GHz–31.3 GHz). The Local Multipoint Distribution Service bands were allocated by auction in early 1998. Each coverage area has two licenses, one for 150 MHz and the other for 1150 MHz. The rules are specifically set up for two-way data service. Generally, the subscriber equipment will have 10 mW feeding a 35 dBi antenna, and vendors claim distances up to 5 miles at 150 Mbps or 10 miles at 10 Mbps.

While the wireless cable bands are not directly available to the smaller users that tend to use ISM band communications, they are of interest nevertheless, for several reasons:

- They are used for similar services (multipoint data applications).
- Where the MMDS operators are active, they are competitors of the independent wireless ISPs.
- MMDS equipment is often collocated in the same antenna sites as ISM band equipment.

- Where the MMDS license holder is not actively using the band, there may be interest in leasing channels from the license holders.

MDS, ITFS, and MMDS are really one set of licensed spectrum, sharing a common history and frequencies close enough to have similar characteristics. These channels were originally allocated for wireless cable television-encrypted local rebroadcast of commercial video networks to paying subscribers (MMDS) and closed-circuit feeds of university lectures to educational classrooms in the community (ITFS). The rapid rise of direct-broadcast satellites soon after these bands were licensed but before significant build-out had occurred made the original purpose of the bands not viable commercially. As a result, the FCC amended the rules to allow two-way communication.

The two old MDS channels at 2150 to 2162 MHz have been reallocated as MMDS (except that MDS channel 2 has been truncated to 4MHz—the 2156 to 2160 band is now identified as channel 2A) and the 2160 to 2162 band is reserved for emerging new services. The 20 channels of ITFS and 11 channels of MMDS are really one service, where the ITFS channels are reserved for nonprofit and educational use, while the MMDS channels were auctioned off in 1996. Because the channels are interleaved, the 6-MHz channel width is stringently enforced. Each user gets up to four 6-MHz channels.

The detailed rules of what can be transmitted in ITFS and MMDS bands are outlined in NPRM 97-360 (101-page PDF file), which can be found at *www.fcc.gov/mmb/vsd/itfs/itfshot.html.*

A challenge to the digital cable and DSL industry is two-way satellite services. One provider, Starband, is currently available and another will soon be available from Hughes. More players are entering the field, including Microsoft and Gilat Satellite Networks, as well as Globalstar.

Starband Services offers one-antenna, two-way, high-speed Internet and Satellite TV. It uses an always-on Internet access network model. StarBand uses a single satellite dish antenna for receiving and sending information. The antenna can accommodate both the Internet and EchoStar's DISH Network satellite TV programming, bringing the Internet and hundreds of channels of television into your home, through a single dish.

StarBand users can expect download speeds of up to 500 Kbps and upload speeds at about 150 Kbps. The speed can burst higher depending on the time of online and current traffic. The company's goal is to provide 150-Kbps download speeds and upload speeds of 50 Kbps during the busiest hours on the Net. The service is available virtually everywhere in the continental United States.

Hughes is also offering a two-way satellite broadband service. The company plans to offer two-way broadband satellite services targeted at remote workers and offices and small businesses. This adds yet another option for companies considering high-speed phone lines, digital subscriber lines (DSLs), or cable modems to serve their scattered workforces and offices.

As of the print date for this book, Hughes had not yet set pricing for what it calls the Satellite Return service of its DirectPC Broadband Everywhere package. For consumers, the upstream speed for the new service is anticipated to be between 128 Kbps and 256 Kbps, and the downstream will support bursts up to 400 Kbps for each user. The new two way terminal will support small office/home office (SOHO) and enterprise applications such as IP multicasting and content delivery. The two-way DirecPC will also be offered with a DirecDuo antenna system, allowing consumers to receive both DirecPC and DIRECTV on the same antenna.

Application Service Providers (ASPs)

As we saw earlier, enabling your mobile workforce to stay connected from anywhere, at anytime, can seem like a daunting task, especially if it's not part of the company's core competencies. Wireless data hosting requires an integrated knowledge of communications, networking, and hosting, in addition to expertise in wireless applications. While traditional Web hosting facilities are capable of delivering content to wireless devices, the expertise developed in a wired environment does not automatically transfer to the wireless domain. Bandwidth, latency, security, user application, and experience differ greatly between the wireless and wired worlds. In addition, economies of scale can often be realized and passed back to the enterprise by leveraging the experience and environment of a qualified ASP.

Many enterprise systems and application solutions take advantage of an ASP offering. By doing so, they gain in several key areas, including the following:

Quicker implementation.   With deep, qualified experience to rely upon, many wireless data centers are able to quickly configure and implement communications solutions that fit unique business needs.

Reduced IT resource requirements.   Many offer 24×7 support from their operations center, utilizing their own equipment to provide redundancy and maintenance services.

Minimized capital expenditures.   Require no up-front hardware costs and no recurring outlays for software upgrades.

Quicker return on investment.   Network solutions are often up and running more quickly and offer the advantage of immediate productivity savings.

Greater flexibility.   Solutions can be modified as a business changes and they can scale as the business grows.

Reduced technology risk.   Applications can be delivered using the latest software upgrades and the most advanced hardware, reducing the risk of technology obsolescence.

It's sensible to look to ASPs for wireless applications. They already have the technology running and can retool and reconfigure applications quickly. However, selecting or becoming a good wireless ASP is not a small challenge. There are big differences between a traditional wired ASP and a wireless ASP, including the applications and attendant infrastructure involved in the integration between the Web and wireless networks.

As a result, wireless ASPs need to deal with a wide range of hardware devices, mobile operating systems, and wireless networking protocols, many of which are just starting to work their way into mainstream business. As with the ASP model in general, the right partnerships in the wireless space can enhance a company's skill set, increase its customer count, and add credibility to its business model.

Building a wireless solution requires lots of pieces. Wireless ASP partnerships extend to the Web, security, voice recognition, middleware, devices, locations technologies, ERP, and CRM applications, as well as the wireless infrastructure providers.

The wireless ASP business is still in its formative stage, but companies from an assortment of backgrounds are already entering the arena. Exodus and IBM, for example, are moving toward the wireless ASP space coming from their legacy in Web hosting. Other companies, such as Everypath and Aether Technologies, are pure play wireless ASP start-ups. As another example, MobileLogic is marrying partnership value with wireless network and data center infrastructure providers, including AT&T, BellSouth, and Verizon Communications. MobileLogic looks to Ericsson and IBM for its core software technology. Partnerships are the only way at present for ASPs to effectively offer a total wireless solution to customers.

Finally, for the worker who is never in the office, wireless access to business applications may sound like an ultimate answer to vexing problems regarding access to information. At the same time, trying to run every aspect of an application through current wireless devices strains belief. The real promise of wireless applications lies with utility and the right sizing of an application.

Companies will invest in applications and solutions that have a demonstrated impact on the revenue of their businesses. The first question that must be asked is: Which applications are likely candidates? We see applications that are designed around time-sensitive processes as leading the way.

Within the next year, many applications should be adaptable to reasonable wireless utility without the use of heavy graphics. They should address location- and time-sensitive needs and be portable across a variety of networks and devices. They should be solutions that create connections to established enterprise resource planning, customer relationship management, sales, field service, and other systems for wireless phones, PDAs, laptops, and other devices.

Services too focused on the wireless aspect and not geared to business processes underlying the application might prove unsuccessful. Wireless services should be applied to the enterprise where they make good business sense. It's a delicate balance. Often, whether in relation to inventory, customer care, or finances, the utility of applications lies with simply sharing and updating information in a timely fashion.

Fundamentally, application developers must heed the limits of the technology and human factor issues. Applications should be very clean and not require a lot of information to be entered. Content-based applications for the consumer space will likely not prove successful in the enterprise. As we'll see later in this chapter, these constraints will dissolve as the technology services evolve.

Security Systems

In the first iterations of wireless Internet services, security was not a critical concern, as most services were low risks—news, weather, directions, and so on—and users paid for connection-oriented circuit-switch data access. The existing levels of network security were sufficient to mitigate the low-level risks that operators faced.

As more sophisticated services and applications roll out, security will become a pressing concern. The network operators may not have an incentive to increase their level of security, but the content providers and enterprises, which are experiencing greater risks, will certainly desire enhanced security measures. Security is absolutely critical to the success of *any* wireless computing environment. This section will therefore consider in more detail some of the issues and the prospective solutions to these problems.

To provide secure transactions in any medium, there are four components that must exist: confidentiality, authentication, integrity, and nonrepudiation. These components are illustrated in Figure 3.11. Let's examine these terms in order to understand the criteria for secure systems.

   Confidentiality.   This requires that only the parties of a transaction are aware of the private details of that transaction. In mobile computing this suggests that the server and the terminal are the only points that can examine the real contents of a transmission. The primary tool for confidentiality is cryptography. Plaintext is encrypted at the origination point and decrypted on receipt to provide privacy.

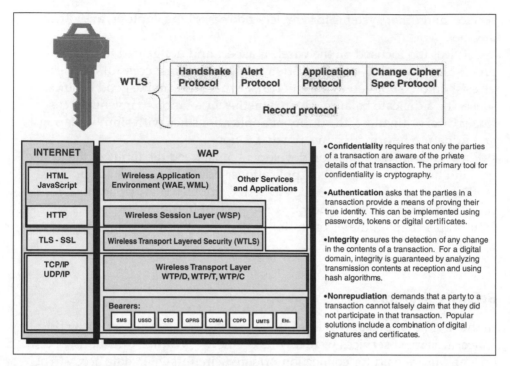

Figure 3.11   Mobile security system services.

Authentication.   This asks that the parties in a transaction provide a means of proving their identity. In the brick-and-mortar world this is done through forms of trusted identification or by asking for pre-arranged secret information. In wireless data realms this is also provided by a means of trusted identification. Server authentication provides a way for users to verify that they are really communicating with the entity with whom they believe they are connected. Client authentication verifies that the user is who they claim to be. This can be implemented using passwords, tokens, or digital certificates.

Integrity.   This ensures the detection of any change in the contents of a transaction. In commerce this has been accomplished by sealing documents and, in extreme cases, by providing a chain of custody. For a digital domain, analyzing transmission contents at reception and using algorithms that determine if the contents have been altered guarantee integrity. In addition, a digital signature may be used to provide a stronger test for integrity.

Nonrepudiation.   This demands that a party to a transaction cannot falsely claim that they did not participate in that transaction. In common transactions, this is accomplished via signatures, seals, and notaries. This is more difficult to realize in a mobile computing environment. Popular solutions include a combination of digital signatures and certificates.

Figure 3.12 lists many of the layers that organizations build into their security architecture, and most of these are found in the wired portion of the enterprise network. It should be noted that no single layer would, in and of itself, prevent intrusion from unfriendly sources. However, a multilayered approach can go a very long way in thwarting unauthorized access. In addition, many of the greatest threats to security come from within an organization due to organizational and procedural issues.

Referring back to Figure 3.11, we show how in the initial mobile environment security was viewed as adequate through the carrier's infrastructure. However, once these systems become more open, security becomes a major concern—and for several reasons. In the world of mobile computing there are some unique challenges that must be satisfied in order to have effective levels of security.

Some of the key challenges are the existence of multiple communication media, a range of device types, and different internetworking protocols. To

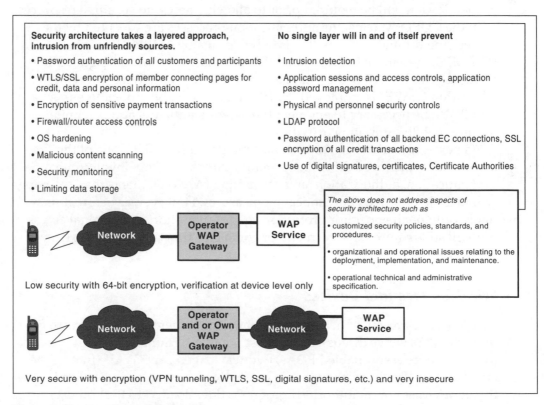

Figure 3.12   Security architecture layers.

properly examine these issues, each layer (network, device, and application) needs to be considered.

There are presently a number of different types of networks in operation that support the mobile environment. These boil down to a handful of different technologies: ARDIS from Motorola, MOBITEX, Circuit Switched Data (CSD) for digital cellular systems, Cellular Digital Packet Data (CDPD), and General Packet Radio Service (GPRS).

Although the network operator has a whole host of security concerns ranging from unauthorized network access to the use of services without a subscription, for mobile computing the real concern is authentication. This is a starting point for a transaction and should be successfully executed prior to any access to wireless applications or services.

In a CSD model, the authentication is largely the same as that which exists for voice communications. The strength and efficacy of this authentication varies across technologies, from 56- to 64-bit encryption, but is generally not as strong as the standard for Internet transactions of 128-bit encryption. Also, the authentication performed by the network is only verifying the rights of the device. It does not ensure the identity of the actual user. This suggests that for enterprise data or mobile computing transactions a second layer of authentication will be required prior to allowing access to sensitive resources.

In the CDPD systems, the authentication processes are executed by the underlying cellular network that is either an analog (AMPS) or digital (D-AMPS) system. In the case of AMPS, there is a very low strength authentication. With D-AMPS, the Cellular Authentication and Voice Encryption Algorithm (CAVE) is employed, which involves a 64-bit cipher. In either case, it will also be necessary to augment the network authentication techniques for any mobile computing activities.

As next-generation cellular systems emerge, GPRS will be one of the first architectures deployed. These new architectures will likely involve enhanced authentication, authorization, and accounting (AAA) elements that, we hope, will include improved authentication measures. However, as before, this will likely provide the operator protection against unauthorized terminal access but will still require reinforcement for commerce transactions and private data access.

Wireless Terminal Devices

As there are a variety of networking technologies, there is also a diverse field of devices. As we've discussed, the most popular terminal devices in use today are wirelessly enabled PDAs, two-way pagers, and WAP-(Internet-) enabled digital cellular phones. The greatest concern at the device level is the threat of viruses or unauthorized programs downloaded to the end user's equipment.

Recently, PDAs running the Palm OS and Windows CE (Pocket PC) plat-forms were compromised. It is expected that even greater threats will arise as mobile handsets begin to incorporate the J2ME Java virtual machines. To mount an effective counterstrategy, mobile operators will have to coordinate their efforts with the terminal equipment manufacturers (TEMs), mCommerce merchants, and enterprises with connected private networks.

Internetworking Protocols

The final piece in the wireless Internet puzzle is the technology used to connect to the IP network (whether public Internet or private intranet). There are also a variety of interfaces supported by wireless devices today, including HTML, CHTML, HDML, WML, and VXML. These interfaces and layers are illustrated in Figure 3.13. As a case study, WAP, which is WML based, will be considered in depth.

WAP Security Model

WAP is a layered protocol stack that contains an application protocol, a session protocol, a transaction protocol, a security protocol, and a datagram protocol. This stack isolates the application from the bearer when used as a transport service. WAP enables a flexible security infrastructure that focuses on providing connection security between a WAP client and server. WAP can provide end-to-end security between WAP protocol endpoints. Actually, the endpoints for the WAP security layer are the mobile terminal and the WAP

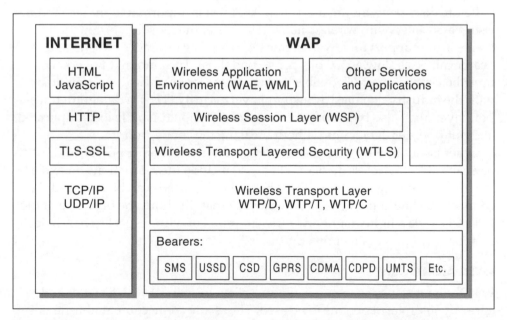

Figure 3.13   Interfaces and internetworking protocols in the wireless internet.

gateway. When the WAP gateway makes the request to the origin server, it will use the SSL below HTTP to secure the request. This means that the data is decrypted and again encrypted at the WAP gateway. If a browser and origin server desire end-to-end security, they must communicate directly using the WAP protocols. The complete secure connection between the client and the service can be achieved in two different ways.

The safest way for the service provider is to place a WAP gateway in his or her own network. Then the whole connection between the client and server can be trusted because the decryption will not take place until the transmission has reached the service provider's network rather than the mobile operator's network. When placing the WAP gateway outside the mobile operator's network, a remote access server is needed. The origin server could be created to include the functionality of the WAP gateway. This gives the highest security solution available.

The service and content providers can also trust the mobile operators' gateway and use virtual private networks to connect their servers to the WAP gateway. But then they do not have the possibility of managing and controlling the parameters used by the WTLS at the WAP gateway. These scenarios can be further summarized on the basis of security, control, and cost. The negotiating parties are able to decide the security features that they want to utilize during the connection. According to the security requirements, the applications enable and disable the WTLS features. For instance, privacy may be left out if the network already provides this service at a lower layer. The connection between two terminals can also be secured by the WTLS.

Before discussing the architecture of WTLS, it is important to understand the issues associated with wireless networks. One of the most important requirements is the support for low data transfer rates. For instance, the SMS as a bearer can be as slow as 100 bits per second (bps). The amount of overhead must be kept as small as possible because of the low bandwidth. Compared with the industry-standard Transport Layer Security (TLS), a datagram transport layer must also be supported. The protocol should handle lost, duplicated, and out-of-order datagrams without breaking the connection state.

Other issues are the slow interactions, limited processing power, and memory capacity of wireless devices. They also include the restrictions on exporting and employing cryptography. The round-trip times can be long and the connection should not be closed because of that. In short, the objective of the WTLS is to be a lightweight and efficient protocol with respect to bandwidth, memory, and processing power.

WTLS Architecture

WTLS connection management allows a client to connect with a server and to agree upon protocol options to be used. The secure connection establishment consists of several steps, and either client or server can interrupt the negotiation at will—for example, if the parameters proposed by the peer are not acceptable. The negotiation may include the security parameters such as cryptographic algorithms and key lengths, key exchange, and authentication. Either the server or the client service user can also terminate the connection at any time.

The complete connection and data transmission process is defined using a finite state machine model. The WTLS record protocol takes care of message transmission, data compression, encryption, and result notification. The WTLS Record Protocol is a layered protocol, which accepts raw data from the upper layers to be transmitted and applies the selected compression and encryption algorithms to the data. A WTLS connection state is the operating environment of the WTLS Record Protocol. The Record Protocol is divided into four protocol clients as shown subsequently.

All the security-related parameters are agreed upon during the handshake. These parameters include attributes such as used protocol versions, used cryptographic algorithms, information on the use of authentication, and public-key techniques to generate a shared secret. The handshake starts with a "Hello" message. The client sends a "Client Hello" message to the server. The server must respond to the message with a "Server Hello" message. In the two hello messages, communicating parties agree on the session capabilities.

For example, the client announces the supported encryption algorithms and the trusted certificates known by the client. The server responds by determining the session properties to be used during the session. If the client does not provide the required parameters, a property server must determine one.

After the client has sent the "Client Hello" message, it starts receiving messages until the "Server Hello Done" message is received. The server sends a "Server Certificate" message if authentication is required on behalf of the server. Moreover, the server may require the client to authenticate itself. The "Server Key Exchange" is used to provide the client with the public key, which can be used to conduct or exchange the premaster secret value.

After receiving the "Server Hello Done," the client continues its part of the handshake. At request, the client sends a "Client Certificate" message, where it authenticates itself. Then the client sends a "Client Key Exchange" message containing either a premaster secret encrypted with the server's public key or the information that both parties can use to complete the key exchange. Finally, the client sends a "Finished" message that contains verification of all the previous data, including the calculated security-related information. The server must respond with the "Finished" message, where it also verifies the exchanged and calculated information. In addition to these messages, a "Change Cipher Spec" message is sent to make sure that the parties decide to start the use of the negotiated session parameters.

The WTLS also defines an abbreviated handshake, where only the "Hello" and the "Finished" messages are sent. In this case, both parties must have the shared secret, which is used as a premaster secret. If the client and the server decide to resume a previously negotiated session, sending a "Client Hello" message where the "Session Identifier" is initialized with the identifier of the previous session may start the handshake. If both parties share a common session identifier, they may continue the secure session. The parties may start to use the connection after they have confirmed the session and informed the other party with the "Change Cipher Spec" message.

The Record Protocol also provides a content type of alert messages. There are three types of alert messages: warning, critical, and fatal. Error handling in the WTLS is based on the alert messages. The connection is closed using the alert messages. Either party may initiate the exchange of the closing messages. If a closing message is received, then any data after this message is ignored.

Authentication

Authentication in the WTLS is carried out with certificates. Authentication can occur between the client and the server, or the client only authenticates the server. The latter procedure can happen only if the server allows it to occur. The server can require the client to authenticate itself to the server. However, the WTLS specification defines that authentication is an optional procedure. The WTLS certificate is optimized for size. The authentication procedure immediately follows after the client and server hello messages.

When the authentication is used, the server sends a "Server Certificate" message to the client. The server may also send a "Certificate Request" message to the client in order to authenticate it. At request, the client sends a "Client Certificate" message back to the server. The client end certificates follow the same structure as the server certificates. Both certificates have certificate version, signature algorithm, certification authority's number, valid date period, owner of the public key, public-key algorithm type, parameters for public key, and the public key.

In order to ensure a secure communication channel, encryption keys or initial values to calculate keys have to be exchanged in a secure manner. If the client has listed the cryptographic key exchange methods, which it supports, the server may choose whether it is going to use one based on the client's suggestions or define another method. If the client has not proposed any method, the server has to indicate it.

Privacy

Privacy in the WTLS is implemented by means of encrypting the communication channel. The used encryption methods and all the necessary values for calculating the shared secret are exchanged during the handshake. Currently, the most common bulk encryption algorithms are supported, such as RC5 with 40-, 56-, and 128-bit keys, DES with 40- and 56-bit keys, 3DES, and IDEA with 40-, 56-, and 128-bit keys. All the algorithms are block cipher algorithms; no streams ciphers, except NULLs, are supported.

Integrity

Data integrity is ensured using the message authentication codes (MACs). The MAC algorithm to be used is decided at the same time as the encryption algorithm. The client sends a list of supported MAC algorithms, where the preferred algorithm is the first in the list. The server returns the selected algorithm in the "Server Hello" message. The WTLS supports common MAC algorithms, such as SHA and MD5. Figure 3.14 illustrates the components of security and the steps to authentication.

Private Key Infrastructures

Most of the vendors in the wireless security market are the established Internet security companies, such as VeriSign (www.verisign.com), Entrust (www.entrust.com), Baltimore (www.baltimore.com), Certicom (www.certicom.com), Diversinet (www.dvnet.com), and iD2 (www.id2tech.com). Surprisingly, Verisign and Entrust, who are leaders in Internet security, came a little late into the wireless market. Instead, companies like Baltimore and Certicom were the first to implement WTLS 1.1 with their WAP security toolkits. However, VeriSign and Entrust joined the WAP bandwagon from the implementation of WTLS 1.2 onward. Vendors like Diversinet provide solu-

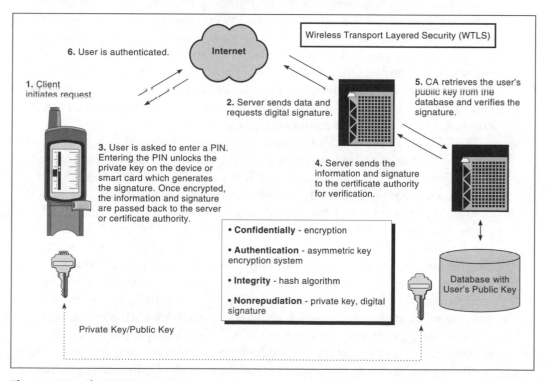

Figure 3.14 The WTLS process.

tions primarily for wireless and mobile commerce security only. Diversinet's solution does not follow the conventional approach of deploying certification authority (CA) and some other features of PKI infrastructure. It looks very interesting because it uniquely provides digital permits, which can carry application-specific information such as rights and privileges to authorize users.

All vendors are concentrating at the deployment of various levels of PKI. For PKI implementation, WTLS uses two types of certificates: WTLS server certificates and WTLS client certificates.

WTLS server certificates, defined as part of WAP 1.1, are used to authenticate a WTLS server or WAP gateway to a WTLS client (handset) and to provide a basis for establishing a key to encrypt a client/server session. They are like SSL server certificates, except that two different certificate formats are defined: X.509 certificates (as in SSL) and WTLS *mini-certificates*, which are functionally similar to X.509 but are smaller and simpler than X.509 to facilitate their processing in resource-constrained handsets. The mini-certificate is mandatory to implement and the X.509 certificate is optional to implement.

WTLS client certificates, defined as part of WAP 1.2, are used to authenticate a WTLS client (handset) to a WTLS server. They also can be formatted as either X.509 certificates or mini-certificates.

WAP 1.2 also defines an interesting PKI-based function that is not part of WTLS. This function, which allows a WAP client to digitally sign a transaction, is known as the WML Script SignText function, and it is intended for applications that require nonrepudiated signatures from clients.

The WAP model does not support certificate revocation as a result of some decisions to reduce the size of the certificate. WAP compensates by keeping the certificate validity period short, which requires that the certificate be renewed or refreshed frequently. Short-lived certificate service is an excellent solution for certificate revocation. In the event that a server is compromised or decommissioned, users cannot unwittingly continue to execute what appear to be valid, secured transactions with a rogue server. WAP server/gateway certificates are relied upon and processed by the mobile client devices, which do not have the local resources or the communication bandwidth to implement revocation methods used in the wired world such as certificate revocation lists (CRLs) or the Online Certificate

Status Protocol (OCSP). With the short-lived certification approach, a server or gateway is authenticated once in a *long-term credentials* period—typically one year—with the expectation that the one-server/gateway key pair will be used throughout that period. However, instead of issuing a one-year-validity certificate, the certification authority issues a new short-lived certificate for the public key with a lifetime of about 25 hours, every day throughout that year. The server or gateway picks up its short-lived certificate daily and uses that certificate for client sessions established that day. If the certification authority wishes to revoke the server or gateway, it simply ceases issuing further short-lived certificates. Clients will no longer be presented with a currently valid certificate, and so will cease to consider the server authenticated.

Network Transport (Wireless-to-Wired Network Infrastructure)

The majority of our discussions have focused on the unique attributes of mobility to consider when delivering mobile applications. We looked at the key technology systems for mobile-sensitive applications. We examined the components of wireless devices, wireless transcoding systems, location ID technologies, and personalization systems. We've seen how voice recognition and text-to-speech can be implemented to add new effectiveness and efficiencies to the mobile process. We've taken a look at wireless Internet service providers, application service providers, and security systems. It is important, as wireless system architects, to consider all of these components.

We will now take a brief look at the carrier network infrastructures that come into play in delivering mobile applications. We will look at the composition of the cellular wireless-to-wired infrastructure. The intent is to provide a general conceptual understanding of the infrastructure. Our coverage will include the transport systems, the protocols, the evolving wireless protocols and services, the global perspectives, and the implications for delivering mobile applications.

Cellular Infrastructure Components

The development of the cellular concept enabled the advancement of mobile communications and was first articulated in the *Bell System Technical Journal* by D.H. Ring (AT&T Bell Laboratories) in 1947. The term *cellular* comes from the honeycomb shape of an area into which a covered region is divided. Cells are base stations transmitting over small geographic areas represented as hexagons. A frequency spectrum is divided into blocks of frequencies and those blocks are assigned to the cells. The reason we tend to call everything *cellular* is that most wireless services, including fixed wireless solutions at times, work from this concept. As a mobile unit passes from one cell site to the next, the call is handed off from one transceiver to another. Each cell size varies, depending on the landscape. Because of constraints imposed by natural terrain and man-made structures, the true shape of cells is not a perfect hexagon.

A *cluster* is a group of cells. Each cell within a cluster is allocated a distinct set or block of frequency channels, and the cells labeled with a given channel reuse that same channel set block. Channels are divided into seven blocks of frequencies and are assigned to seven cells that are adjacent in hexagonal shape as shown in Figure 3.15. Interference problems caused by mobile units using the same channel in adjacent areas proved that channels could not be used in an adjacent cell. Areas had to be skipped before the same channel could be reused. Even though this affected the efficiency of the original concept, frequency reuse was still a viable solution to the problems of mobile telephony systems.

Engineers discovered that the interference effects were not due to the distance between areas but to the ratio of the distance between areas to the transmitter power radius. By reducing the radius of an area 50 percent, service providers could increase the number of potential customers in that area. For example, sites based on areas with a 1-kilometer radius could have a hundred times more channels than sites with an area of 10 kilometers. This led to the idea that by reducing the radius of areas to a few hundred meters, millions of calls could be served.

The cellular concept uses a set of variable low-power levels. This allows the cellular network to be sized according to the subscriber density and demand of a given site. As the population grows, cells can be added to accommodate that growth through cell splitting. Economic considerations

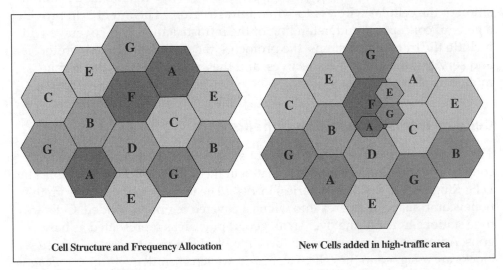

Figure 3.15   Cell structure.

make the concept of creating full systems with many small areas impractical. Therefore, cells start big and split as the system grows. As a service area becomes full of users, the approach is to split a single area into smaller ones. This way, urban centers can be split into as many areas as necessary to provide acceptable service levels in heavy-traffic regions. Larger, less expensive cells are then deployed to cover the remote rural areas. A cell site is also referred to as a *base station* (BS).

A BS houses several components of telecommunication equipment at the transmit/receive antenna's base, the center of a cell. The antenna is a free-standing tower structure with metal transmitter and receiver rods used to transmit and receive signals from the mobile unit. Antenna engineering is very important in wireless communications because the power and strength of the signal is based on the antenna design. An antenna sending and receiving signals in a spherical 360-degree field is a nominal omnidirectional antenna. In between hills or in a city with tall buildings, signals to and from the omnidirectional antennas are blocked. In such places, directional antennas are used toward specific directions known as *sectors*. These antennas are referred to as *directional antennas*. If an antenna covers one-third of its possible 360-degree range, it is known as a 120-degree directional antenna. If an antenna covers one-sixth of its possible 360-degree range, it is known as a 60-degree directional antenna. The significant advantage of directional antennas is that the signals can be concentrated on the sector. Also, the transceiver uses very low power that allows the cellular network to be sized according to the subscriber density and demand of a given site. The sector of the antenna is chosen depending on the location.

In the small coverage areas such as microcells and picocells, a remote antenna driver (RAD) and remote antenna signal processing (RASP) are used (see Figure 3.17). RADs are small antennas placed on top of street lamps or telephone poles. The RAD receives the signals and communicates with the RASP through the television cable. The RASP then translates the channel signals and communicates it to the base station of the cellular network.

The base station in Figure 3.16 consists of two major electronic units known as the *base station transceiver* (BST) and the *base station controller* (BSC). The transmitter and receiver rods of the antenna are connected to the signal transmitters and receivers of the BSTs. The BSTs are full-duplex transmitter and receiver units (transceivers) that signal to the mobile units and receive signals from them. The BST serves one cell in the cellular network and contains one or more transceivers. The BSC controls the transmission and reception of signals of the BST and switches calls between the BST and mobile units. The BS is also responsible for administration, call processing, maintenance, handoff, and audio signal compression and decompression as directed by the *mobile switching center* (MSC). Each BS is controlled by the MSC. The BS is connected to the MSC using voice trunks and data links.

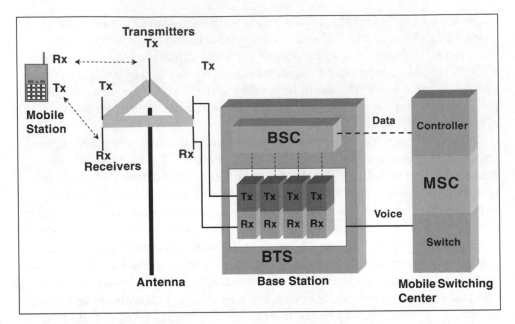

Figure 3.16   Cell site components.

The MSC serves as a central coordinator and controller for the cellular system and is also an interface between the wireless and wired infrastructure. The MSC is connected to the Public Switched Telephone Network (PSTN) using voice trunks and data links. The MSC can be considered the same as a landline-switching center in the central office (CO) with the enhanced capability to track each mobile unit. The MSC communicates with the PSTN using standard telephone signaling protocols. The MSC administers the radio frequency channels allocated to the base station controllers and base stations, coordinates the paging and handoff between the antennas, and maintains the integrity of the wireless system. Every MSC has a five-digit *system identification number* (SID). In the United States when you buy a cell phone, the nearest MSC to your home address is assigned to the cell phone number. This is called the *home system identification number* (home SID). This home SID is used to identify whether a person is roaming or in the home area. If a person makes a call within the home area, then the MSC serving the call is considered the *home MSC*. If the person is roaming outside the home area, then the MSC serving the call is known as the *serving MSC*. In that case, the SID of the MSC is known as *serving SID*.

The overall wireless-to-wired-to-wireless infrastructure is illustrated in Figure 3.18. It represents a traditional telephony voice view of the infrastruc-

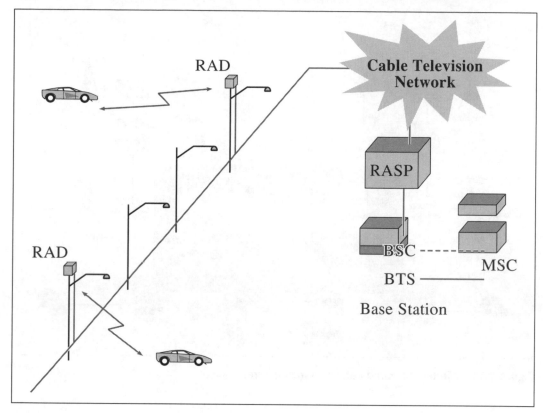

Figure 3.17   RAD/RASP for micro- and picocells.

ture. We will see later that this infrastructure is in a state of rapid change. However, for illustration of the traditional view, in addition to the previous components, we find the following.

Home Location Register (HLR)

This is the database owned and maintained by the service providers to store information on all subscribers registered in the home area. The HLR is available within the service provider's *metropolitan and rural statistical area* (MSA/RSA). The HLR contains information such as the subscriber's *mobile identification number* (MIN), the cell phone's *electronic serial number* (ESN), the home SID, the subscriber's profile and assigned features, and account status information.

Visiting Location Register (VLR)

This is the location database, also owned and maintained by the service providers, that holds subscribers' roaming location information. The VLR is a very dynamic database and keeps the very latest cell site information indicat-

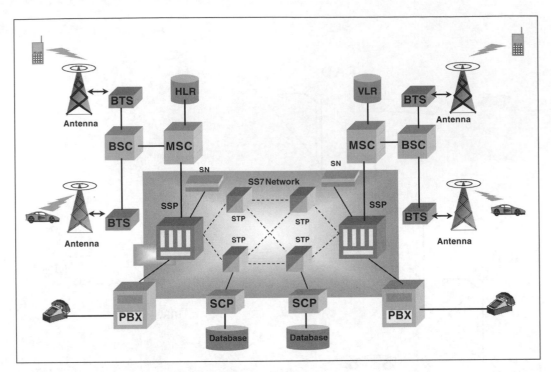

Figure 3.18   Wireless-to-wired network infrastructure.

ing where the subscriber was located last. Periodically, the cell phone sends active signals to the cell site along with the MIN number. If the subscriber is not in the home area (home SID), then the MIN and the subscriber's account information are stored in the VLR, and information is sent to the subscriber home MSC about the serving MSC (serving SID). The Home MSC stores the serving MSC (serving SID) information in the subscriber record. When a call arrives for the subscriber who is roaming, the call first goes to the subscriber's home MSC. The home MSC looks for the subscriber in the HLR. If it cannot find the subscriber in HLR then it looks at the VLR. The VLR provides the information on the cell site where the subscriber was last found. The MSC uses the cell site information and connects the call to the subscriber. The VLR has a time limit for storing subscriber information. If the VLR does not get an active signal from the subscriber's cell phone within a certain period of time, the VLR will automatically delete the subscriber's record.

Signaling System 7 Network (SS7)

When you dial a long-distance phone call using a destination phone number, your phone call is routed through very high-speed circuit switches with the help of the destination phone numbers. The signaling system actually refers to a set of communication protocols for the signaling and control of the call information for various network services. There are many standards used for the signaling system. The SS7 standard is used in the United States and CCITT#7 is used internationally. In the United States, AT&T uses its own standard, known as CCS7 (Common Channel Signaling 7).

The SS7 network is composed of a series of interconnected network elements, including switches, nodes, and databases. Each of the network elements is interconnected with data links. The components in the SS7 network architecture can be grouped into three signaling points.

1. Service Switching Point (SSP).
2. Service Transfer Point (STP).
3. Service Control Point (SCP).

Service Switching Point (SSP)

This is the functional point that contains the control logic to originate and receive an SS7 message. When the call arrives at the SSP, the SSP connects to the database using the destination phone number and obtains the routing information with the help of the SS7 node. The SSP retrieves the routing information from the database and determines the destination SSP. Using this routing information, the SSP generates an SS7 message that is passed to the transfer functional point. The SSP generates an SS7 message even if it cannot determine the destination node.

Signal Transfer Points (STPs)

This is an intermediate functional point for simply passing the SS7 message's traffic to another network element without any delay. All SS7 messages are destined for an SSP. In the SS7 message there is a *signal connection control part* (SCCP) that contains the destination phone number information. The STP connects to the database, known as the *subsystem*, through the SCP and obtains the routing information. The STP makes the routing decision by using only the destination area code and office switch code in the SCCP. This unique feature is called the *global title transition* (GTT). The advantage of GTT is that SS7 does not have to know or advertise the address of all databases in the network. The routing information can be determined using the destination phone numbers. The STP has the functionality to allow only certain SS7 message traffic that helps in network security for filtering unsecured message origination. This secured feature is especially useful when many different networks share SS7.

STP Pair

Two STPs are interconnected using cross data links to ensure redundancy in the network in the event of failure. If one STP fails to operate, then the SS7 traffic can be routed through its redundant STP cross data links.

Service Control Point (SCP)

The SCP is responsible for managing access to the database for SS7 messages. The SCP manages more than one subsystem. All SS7 network service elements must send queries to the SCP for retrieving information from the database. The process of identifying the correct database is done using the subsystem number. The subsystem number identifies the application served by the database. The SCP processes the queries and determines the exact database subsystem from which the information has to be retrieved.

Service Node (SN)

The service node has all the capabilities of the SCP along with voice interaction and control of voice resources. The SN is used mainly for low-volume SS7 message traffic and during special circumstances of call processing involving voice interaction.

Evolving Telecommunication Infrastructure

As we mentioned earlier, the telecommunications infrastructure is in a state of rapid change. It is a complicated landscape due to not only the traditional telephone local exchange carriers (LECs) but also the long-distance carriers, competitive local exchange carriers (CLECs), and new technologies such as SONET; asynchronous transfer mode (ATM); SS7; Internet Protocol (IP); Voice over IP; Voice over ATM; IP over ATM; cable, satellite and wireless IP networks; corporate data enterprises; and many new switching and aggregate devices supporting the throughput of tremendous data volumes and types of data. We will cover these to a certain extent only in order to provide a framework of understanding and tracking wireless voice and data across the new and evolving network infrastructures. First let's look at the traditional telephone network and concentrate on it at a high level.

The public switched telephone network (PSTN) is based on the circuit-switched network technologies. Figure 3.19 provides a high-level view of the PSTN. In a simplified description, a switch establishes a circuit connection known as a *line interface* between the source and destination telephone lines. A switch can also connect other switches that are known as *trunk interfaces*. A fabric connects the line interface to the trunk interface. The class 5 switch that is used in the telephone *central offices* (CO) is more intelligent and performs more functionality than a dumb switch. In terms of the PSTN, a *local switch* refers to the switch used in the local telephone exchange office in the city or area. This local switch allows the connection of telephones within certain distances in a city or area. A *digital access and cross-connect system* (DACS) is electronic equipment that connects all trunk interfaces and line interfaces. The main advantage of the DACS is to connect trunk interfaces through local switches with the help of electronics and software instead of physically

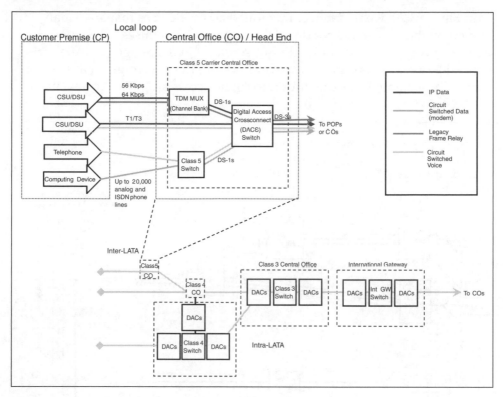

Figure 3.19   Traditional telephone central office, inter/intra LATA network.

wiring all line interfaces together. We refer to the carriers that control and maintain the local switches as *local exchange carriers* (LECs).

The office that connects all local home telephones to the local switches is called a *central office*. In the United States, carriers who provide local services are Ameritech, Pacific Bell, South Western Bell, and so on. Some of the manufacturers of local switches are Lucent Technologies, Alcatel, Siemens, and Nortel. A toll or (class 4) switch allows the connection of long-distance telephone calls. A toll switch connects local switches between LECs. The carriers that control and maintain toll (intra-LATA) switches are known as *long-distance carriers*. In the United States, some of the popular long-distance carriers are AT&T, Sprint, and MCI WorldCom.

Now let's look at the more advanced (point of presence) data networks evolving from new technologies such as the Internet, cable, satellite, and cellular systems and the introduction of competition in the local markets since the U.S. Telecommunications Act of 1996. The newer access technologies of enterprise corporate networks, cable, and wireless systems link to the new carrier networks. Here we see that the traditional telephone networks also link their access through the new networks. However, it is quite rare that the emerging technologies are linked through the traditional telephone network.

In Figure 3.20, we find newer integrated access devices that are usually owned by the carrier and reside on the customer's premises. These newer devices incorporate routers, quality-of-service (QoS) managers, virtual private networks (VPNs), and Voice over IP (VoIP) that converge what was traditionally separate infrastructures into single access systems. We also find

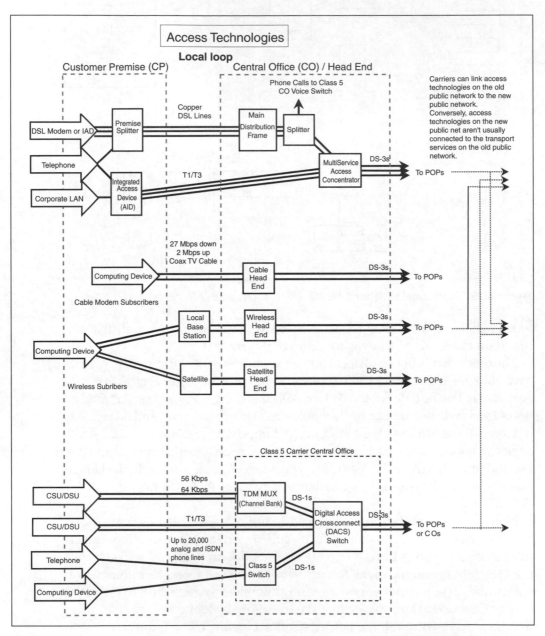

Figure 3.20   Integrated access devices and transport services of enterprise networks.

distribution frame splitters that split thousands of DSLs into circuit-switched voice and data circuits and concentrators that terminate DSL leased line services. There are terabit IP edge routers accepting packets from access networks and passing them to ATM switches and terabit IP core routers. We also find such things as monster remote access servers (RASs) that terminate

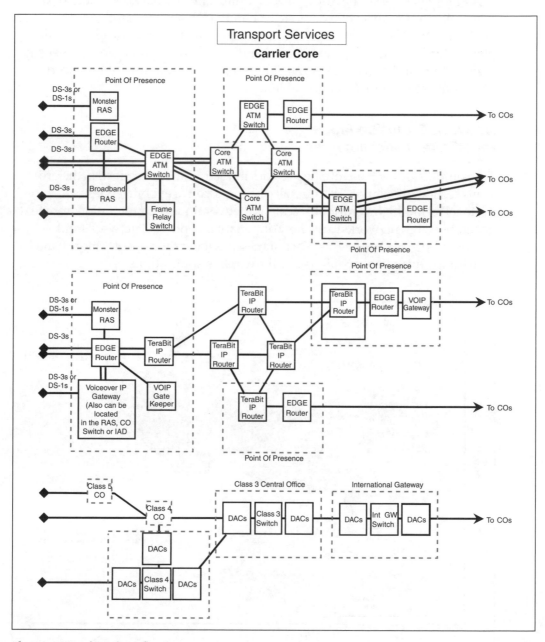

Figure 3.20 *(Continued).*

thousands of calls from analog modems and utilize SS7 protocols. There are similar devices for DSL lines that need IP addresses on the fly. Finally, for traditional voice applications we have VoIP gateways that terminate thousands of calls and convert them into IP, using SS7 to translate between phone numbers and IP addresses.

As these networks converge, they exit into singular backbone networks, where we find more exotic devices for bandwidth speeds ranging from DS-3s (45 Mbps) to OC-192 (10 Gbps). Two examples are SONET add-drop multiplexers for grafting lower-speed fiber and copper onto and off of high-speed logical rings and dense wave-division multiplexers that mux high-speed fiber connections onto a single cable.

Network Evolution Model for Wireless Technologies

The complexities of the evolving network infrastructures described in the previous section in isolation. We must also be cognizant of evolving wireless protocol technologies on a global scale as they continue their migration from first-generation (1G) networks to the now dominant 2G and 2.5G networks and onward to 3G, see Figure 3.21. Each successive generation enables more bandwidth and always-on capabilities for the wireless applications.

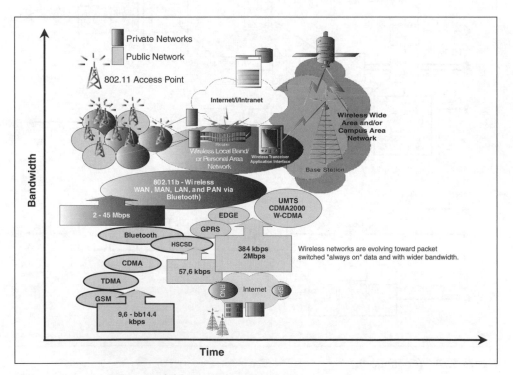

Figure 3.21   Evolution model for mobile technologies.

Today's 2G technologies are optimized for voice services and offer limited data capabilities. To address these restrictions, network operators are developing data-centric enhancements to the existing systems. These improvements are considered to be 2.5G, because they are improvements on the current systems.

Currently, the 2G CDMA protocol is being evolved to a 2.5G technology called CDMA2000. Phase 1 of CDMA2000 is being developed and was planned to be available for release in late 2001. This upgrade will allow network operators to provide always-on access to the wireless Internet at transmission speeds of over 100 Kbps, sufficiently fast to retrieve data and complete simple commercial transactions quickly. CDMA2000 is compatible with existing CDMA technology and, thus, does not require a complete overhaul to the existing wireless infrastructure. Networks can be upgraded satisfactorily with minimal hardware and software upgrades. Subsequent phases of CDMA2000 are expected to provide bandwidth of up to 2 Mbps.

Likewise, the 2G TDMA and GSM protocols are being evolved to a 2.5G technology called General Packet Radio Service (GPRS). GPRS will allow TDMA and GSM network operators to provide always-on access to the wireless Internet at transmission speeds of over 100 Kbps in 2001. GPRS is compatible with existing TDMA/GSM technology and, thus, does not require a complete overhaul to the existing wireless infrastructure. Minimal hardware and software upgrades are sufficient for upgrading the network.

Enhanced Data GSM Environment (EDGE) is another short-term improvement that is being developed for TDMA/GSM systems. EDGE will provide

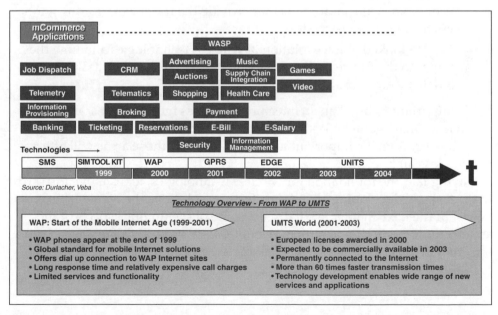

Figure 3.22   mCommerce applications linked to technology development.

multimedia and other broadband applications over a bandwidth of 384 Kbps by 2002.

These technologies represent real developments in wireless protocols that have a sizable impact on the types of applications that can be offered. This viewpoint is illustrated in Figure 3.22.

Summary

We have covered a lot of ground in our attempt to provide an adequate end-to-end view of wireless application technology. We showed that, first and foremost, is the need to understand the unique characteristics of mobility before we can begin to deliver mobile applications. Wireless applications entail more than simply rendering the Internet onto wireless devices. The architectural design specification of any technology must also consider the needs, both individual and organizational, that drive its application.

Some of the key lessons learned from this chapter can be summarized as follows:

- Mobility is much more than using your cell phone to make a phone call to your colleague or checking the latest sports scores on your PDA. Terms like *ubiquity, reachability, convenience,* and *localization* truly define what mobility and wireless technology mean.

- The components of a wireless infrastructure can be very complex and very confusing to someone unfamiliar with it. We reviewed some of the wireless devices, wireless network protocols such as CDMA, and other technologies like location ID that provide the framework for the wireless technical infrastructure.

- We also looked at the evolution of wireless technologies and how they play into the future of the wireless networks. Terms such as *2G, 2.5G,* and *3G,* were considered.

It's important to build an architecture that takes into account scalability and release management. With wireless technologies constantly changing and transforming, it's important to understand how those areas will affect the wireless architecture.

Having laid the foundation for wireless technologies, let us move forward to the tasks of developing sound architectural solutions the wireless business needs. In the next chapter, we'll discuss some of the unique aspects of the architectural methods and structure that CGE&Y typically leverages on projects. In that context, you'll see how all the aspects of architecture fit together to build effective and efficient wireless applications.

An Overview of Wireless Application Life Cycle

One of the biggest concerns with implementing any solution, whether it is a software application or an infrastructure in preparation of a software deployment, is its long-term viability. Five factors are known industrywide as the ones that determine our current situation. These factors can be found on any chief information officer's leading list of concerns: *costs, speed, size, complexity,* and *rapid change.* While the business ecosystem had centuries to develop its processes and models to deal with these factors, information technology (IT) organizations compared to this time frame are in their infancy and in general are still struggling with just developing software that meets user requirements.

These factors are especially true in the wireless environment. Advances in multiple areas are changing the capabilities of wireless clients. Legacy applications are linked through an enterprise application integration (EAI) construction to Web-enabled clients; network infrastructures globally and domestically deployed show uncountable variety. Wireless technologies reflected in today's 2G wireless Internet phones have a bandwidth of about 9,600 baud, limiting communication to text messages. Meanwhile, new 3G networks of over 1 megabit are already being deployed around the globe. The wireless devices themselves must evolve from pure oral communication devices to more intelligent ones because they need to store some logic locally to handle larger volumes of data and diverse media formats.

The only sword a CIO has in the battle with speed to market, up/down sizing, increasing complexities due to higher sophistication in business processes, and accelerating change is rigorously disciplined architecture and design.

In our opinion, a well-balanced architectural approach that seamlessly integrates system design is the foundation for an IT organization—or, to use a Zen image, the solid anchorage of any initiative. Utilizing a consistent methodology and applying industry standards wherever possible constitute the strengths and the agility of the IT organization. In the Zen picture, these are practice and knowledge of, as well as focus on, your abilities. Based on this foundation and agility, an IT organization will become a driving force within or for a business unit. Back in our Zen picture, the warrior is prepared to recognize the proper time and select the appropriate stroke to beat his or her competition. Granted, very few IT organizations can realistically claim to have reached this stage or be even close. On the other hand, all journeys start with the first step.

While not a silver bullet or a form of insurance to eliminate any and all potential problems, these techniques can help in managing the four factors that impact the status quo. Everyone is aware that eliminating these factors is not an option. The only realistic option is how well an organization prepares itself to manage these four factors. This positioning determines the value of the IT organization to the overall business purpose. Just as business models and ideas become obsolete if not evolved and enhanced, this threat is also valid for IT solutions. The objective is to create solutions that fit into business as well as IT strategies and that, as frameworks and principles, have a higher rate of survival and longer duration.

This chapter describes a typical project life cycle and its different phases, pointing out how the project life cycle may change due to the wireless focus. However, this chapter will not cover the areas of systems development and testing. For a more complete discussion on these areas, please refer to Appendix C.

The Project Life Cycle: From Idea to Operations

In the late 1970s and early 1980s, system development evolved from pure programming to software engineering. It was then that the magnitude of the challenge dictated a more structured approach to system development. It was also then that more sophisticated collaboration strategies were developed.

Growing with the magnitude of the challenge in system development, a new thinking and mind-set started to develop. It was then that we started to use terms such as *frameworks* and *reuse of components*. Still, the approach was project oriented and project driven. Return on investment (ROI) was measured in cost/benefit analysis and limited to individual projects. But how do you measure savings on an infrastructure project? Well, the typical answer was the declaration of these projects as strategic ones—or, in other words, beyond the need to measure success. This was the era of the first methodologies to guide the software development effort. However, this did not scale to an enterprise level. In the 1990s, two complex and dynamic elements—the Internet and wireless clients—added speed to market necessities. Suddenly corporations needed to rethink and rearrange their assets in order to meet the time lines requested by clients. Pure software development methods such as the famous *waterfall method* and the current dominance of *iterative development cycles* could no longer capture the broader spectrum of requirements of infrastructure, delivery mechanisms, version control across multiple platforms, diverse access channels, and so on. This marked the advent of the *architecture* approach that needed to align business needs, IT initiatives, and landscapes. Architecture translates business strategies and principles into IT strategies and principles. Subsequently, they drive a corporation's system landscape and identify initiatives required to follow within the scope of a transformation map, the path to the desired state. Actual initiatives such as development or infrastructure projects are the result of decisions now being made within the scope of an enterprise view. The term *enterprise architecture* was born.

Cap Gemini Ernst & Young has created a set of methodologies that are the extract of the best-of-breed approaches, and that are pragmatic and oriented only toward a single objective: value to its customers. These methodologies have evolved over time and have proven themselves effective and flexible. They have been successfully used worldwide on hundreds of large and small projects. As a result, they have been adopted by industry-standard groups such as the Institute of Electrical and Electronics Engineers (IEEE).

These projects have a multistep life cycle (see Figure 4.1) starting with analyzing the idea for a business opportunity and ending with an implemented system that is now becoming operational.

Any system implementation has to be based on a business strategy that defines a business case for how the system will be used, including the value of this initiative to the business. Though a project may be successfully implemented without a business strategy, the developed solution may not be applied by the project's users or may not be cost effective.

A good business case serves as the basis to begin addressing the technical issues of implementing the idea. The first step in a successful implementation, after alignment with the business purpose, is to develop a good architecture. The architecture should define how the business concept will be implemented and all the services and products that are used to do so. The architecture will have to be aggregated on several levels to manage the real-life complexities and to visualize collaborations and dependencies of compo-

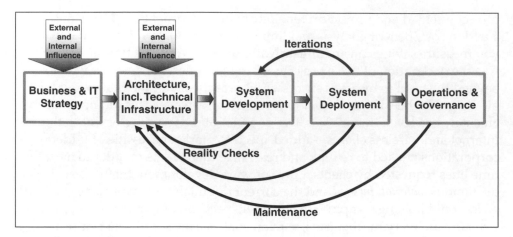

Figure 4.1 The project life cycle.

nents. The architecture phase may include some prototyping to ensure the feasibility of the solution. Architecture is not *completed* after the initial definitions are finalized. Architecture is a mind-set and a set of rules that define and structure approaches starting at the enterprise level and progressing down to project levels, where architecture defines reusable components, data entities and their access methods, and even the development environment itself. Architectures incorporate and are home to all nonfunctional requirements—security, governance, maintainability, deployability, and so on. Architecture is also home to the technical infrastructure that is transparent to system users but is so critical to success. Architecture lives and evolves over time and needs to be reviewed periodically to be really effective rather than just something to which lip service is paid.

Once the foundation is in place, the solution must be implemented. The architecture work results in a logical and physical view of the solution, while the development phase drives into the details and builds actual products that can be implemented. Another mind-set and toolset are necessary to successfully finish this phase. Typically, the most critical steps in this phase are the initiation and the closeout of projects. Once projects are well defined and properly initiated, they don't show symptoms of overtime and budget overruns. A smooth transition from architecture to system development fulfills exactly this requirement. Projects that do overlap between the architecture study and the elaboration or initiation phase usually will finish successfully. Granted, there are other factors to successful system development, which will be discussed in subsequent chapters of this book.

The second critical factor—the definition of when a project is finished—is addressed by the development methodology itself. CGE&Y utilizes the Rational Unified Process (RUP), which is based on visual modeling and iterative development and deployment. The end of each iteration—per definition—

can be seen as such a project closeout. All criteria such as user requirements, time, and budget have to be met by each of these iterations.

The deployment phase also includes the test phase of the solution. This point of view is highly open to discussion. The test phase can be included in the development cycle, although it may also stand as an individual phase. We prefer to incorporate testing into the deployment phase for the following reasons:

- Developers should not be testers.

- Users need to perform testing in accordance with specifications.

- Deployers and support personnel need to be trained and become familiar with the applications.

An operations organization takes over the solution once the implementation is deployed. This organization ensures that the solution is performing to the service-level agreements that include criteria for system availability, system performance, and system response times. The operations organization is responsible for the creation and deployment of maintenance packages.

In conjunction with all these activities, the engagement must be managed from start to finish to ensure the success of the project. Engagement management ensures that the transition between the different project phases is smooth and that each phase is completed successfully. It also ensures that quality assurance objectives are met and therefore periodically reviews projects and follows up on identified corrective actions.

The following sections describe each of the project phases in detail.

Business and IT Strategy

To be very clear up front: The objective of this phase is not to discuss in any detail the actual business strategy. Nor do we try to develop any strategy or change the strategy, which is the responsibility of the business or process owners.

The goal of this strategy phase is analyze how an existing business strategy can be transformed into an IT strategy with detailed IT principles. The focus of this book is wireless technology, so we must assume the existence of an IT strategy that reflects IT principles. At the beginning of this phase, a client may wish to add only this IT principle: *Use wireless technologies.* CGE&Y uses a four-step approach to help clients synthesize these ideas into detailed business and IT initiatives.

Business Assessment

During this first step, CGE&Y records existing business strategies as they apply to the area of business that is most likely affected. Following are some examples of business strategies:

- 5% annual autonomous growth in revenue
- 6% growth in market share annually
- Administrative force down to 7% in 3 years
- In-stock products down by 5% annually
- Ordered products in stock up from 80% to 95%
- 95% of orders delivered in 24 hours
- Easy acquisition and integration of business

Business principles are derived directly from business strategies. Business principles, and later, IT principles, can and will be used to evaluate potential alternative solutions. These principles will be the decision factors against which all solutions will be measured. Following are some examples of business principles:

- Branches must be able to operate autonomously.
- The corporate office monitors and guides product assortment.
- We must be able to exploit new distribution channels.
- Management of IT should be centralized as much as possible.
- External access to systems must be secure but may not hamper efficiency.
- Acquired companies must be operational within the corporation context within one month.

IT strategies typically include statements about the six aspect areas of architecture: information, business support, information systems, governance, security, and technical infrastructure. The strategies are then transformed into IT principles. Examples of principles are shown in Figure 4.2.

Based on this collected information, an impact analysis illustrates which of the business and IT strategies and principles will be impacted by the new requirement, which states: *Use wireless technology.* Typically, this effort also includes identifying applications that need to be modified to enable and support this change in business policies.

The subsequent analysis based on the impact findings focuses on four goals:

1. Determine desired usage of wireless solutions for this client.

2. Define initial options through connectivity points (suppliers, customers, partners, and employees) and strategic objectives (improvement and innovation).

3. Define a project work plan for the strategy effort.

4. Validate ongoing initiatives.

Applications Portfolio Principles

Develop buy vs. build strategies based on competitive requirements identified in the Business Strategy

    Buy applications for competitive parity

    Integrate and build applications for competitive advantage

Maintain global consistency of applications where possible (use identical code-base for identical requirements)

    Regional exceptions based on business requirements and appropriate value proposition

Maintain a vendor-neutral application architecture that is based on industry & market-leading standards

Data Management Principles

Maintain operational data (data used by transactional systems) separately from informational data (data used for historical analysis)

Use real-time transactional processing where possible, including with external partners; drive for 'Zero Latency' across extended enterprise

Avoid independent data duplication across stand-alone applications by using a centralized 'information router' to keep all systems linked

    In cases where stand-alone data duplication is unavoidable, establish a master source for the common data

Centralize system metrics and management information (meta-data)

Infrastructure Technology Principles

Align with EAA Architecture in migrating to a plug-and-play architecture that uses infrastructure services for application integration rather than point-to-point connections

    Develop an infrastructure that enables new applications to be integrated with existing applications by integrating just once with a central integration service rather than individually to each existing application via point-to-point connections

Leverage Covisint and Virtual Private Networks (VPNs) on the Internet to establish connectivity with business partners for application and data integration, and to provide users access to externally hosted applications (i.e., apps-on-tap services)

Limit number of points for external integration to ensure tight control over information integrity & security

Figure 4.2 Example IT principles.

At this point, the decision will be made about which opportunities to pursue further. The next three phases focus on breaking down the findings of the assessment and creating business cases.

Environmental Analysis

The strategy effort now focuses on gathering more information from and about the client, the industry, the external requirements such as regulatory restrictions, and all critical factors necessary to drive the idea forward. This step will also identify hurdles that need to be surmounted along the path to realization. One of the critical parts of this step is to gather information about the competition. Overall, the environmental analysis phase focuses on seven goals:

1. Assess defined connectivity points, needs, and expectations.
2. Assess client capabilities (business models, technology, and organization).

3. Assess external requirements (external forces that have to be taken into account).

4. Review/research leading practices and market trends.

5. Perform competitive analysis.

6. Identify initial eCommerce opportunities.

7. Define the investment needed to pursue initial eCommerce opportunities.

At the end of this phase, much of the research needed to determine how to move forward is complete. The next phase focuses on the analysis of the research and the development of value propositions based on it.

Value Proposition Definition and Strategy Synthesis

This phase synthesizes all the information collected so far and creates value propositions for implementing wireless solutions. More specifically, the business value and a solution's feasibility are assessed. This process is highly client dependent, as very specific and individual measurement criteria are used for each client. The value proposition definition and strategy synthesis phase focuses on the following goals:

- Determine prioritization criteria for potential conflict situations or alternatives.

- Review the impact of wireless initiatives on the business and IT organization.

- Prioritize wireless initiatives.

- Develop value propositions for each alternative and initiative.

At the end of this phase, the decision to move forward with specific value propositions is made. This could include one or multiple initiatives and/or projects. The next phase focuses on writing business cases for implementing the chosen initiatives.

Business Plan Development and Transformation Map

Now that the opportunities have been decided, it's time to write business plans for them. The business plan development phase focuses on the following goals:

- Identify financial measures and potential risk factors to develop business cases for key opportunities.

- Develop metrics to track.

- Develop a high-level initiative implementation schedule.

- Determine the investment needed to pursue initial eCommerce opportunities.

This process step can be highly complex and overwhelming. Our collective experience shows that, as a first artifact, a transformation map that is accepted throughout the organization is essential and helps those involved to visualize the corporation's action plan in simple terms. Figure 4.3 is an example of such a transformation map. Each step within one business domain represents a key milestone with measurable success criteria.

The end of the business plan development phase completes the strategy work. Now we move into the implementation of the business case, which will start with an architecture definition.

Architecture Definition

This section demonstrates the process CGE&Y uses to develop architecture definitions. In 1996, the IEEE, an international standards body, selected the Cap Gemini Group approach to developing infrastructure architecture as the international standard for developing open systems architectures (IEEE 1003.23,

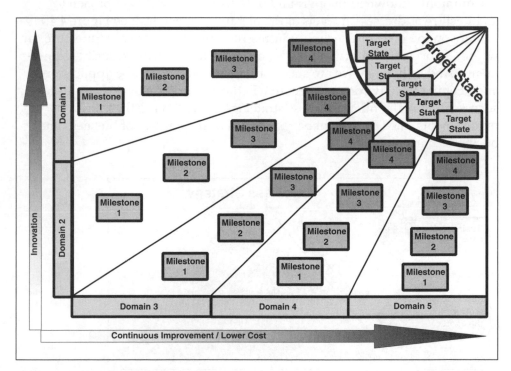

Figure 4.3 Transformation map.

Guide for Developing Open System Environment Profiles [*architectures*]). Since then, our infrastructure methods, as well as complementary methods for developing information systems, security, governance, and business architectures, have been used in hundreds of successful engagements around the world.

The Integrated Architectural Framework (IAF)

The *Integrated Architectural Framework* (IAF) is derived from the collective experience of the Cap Gemini Ernst & Young Group's architects. The essence of the CGE&Y approach to architecture is in its business-driven model. CGE&Y applies technology in harmony with the client business and not as a purpose unto itself.

Business that fails to adequately utilize technology will not achieve its potential. Conversely, technology that is not in alignment with the governing business objectives will fail to adequately support the business. CGE&Y's primary goal with any architecture is to ensure that our client's business and technology are in alignment. Doing this requires an understanding of the business objectives and a proven process from which to derive architecture to support those business objectives.

Commonly known as the pyramid of IT needs, CGE&Y's approach to architecture addresses all levels of the hierarchy of needs. (See Figure 4.4.)

Only on top of a solid technical infrastructure and security architecture can we build successful information systems with their specific security requirements. And only when we are successful in maintaining these applications and keeping the environment secure will the corporation or organization be successful in realizing its business strategies. By achieving these basic requirements, the IT area will be recognized as the enabler of business and business changes.

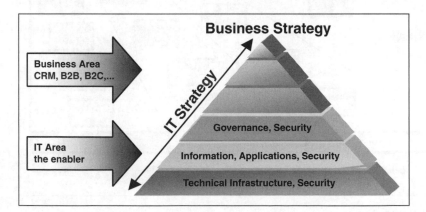

Figure 4.4 The pyramid of needs.

A strong and flexible architecture is the foundation of this success, so let's look at architecture from yet another point of view: how to derive an architecture from business strategies.

The CGE&Y architecture methods define a layered abstraction approach to the complex issue of defining how technology will support business needs. This multiphase abstraction is critical to the process of extracting technology requirements from business strategy. (Refer to Figure 4.5.)

The first phase of this process is called the *contextual phase*. We initiate the project by gathering information to fully support the business case and answer the question, *Why we are doing this?* During this phase, we gather the necessary information to validate the full scope of the project and to generate the detailed project plans used in going forward.

In the *conceptual phase,* we articulate the business strategy in terms that will allow us to determine exactly *what we are going to do* from a business perspective so that the architecture will reflect and support those requirements. During this phase, all distraction factors, including organizational boundaries, political issues, technology biases, and geography, are extracted in order to obtain a simple, clear picture of what the architecture has to achieve.

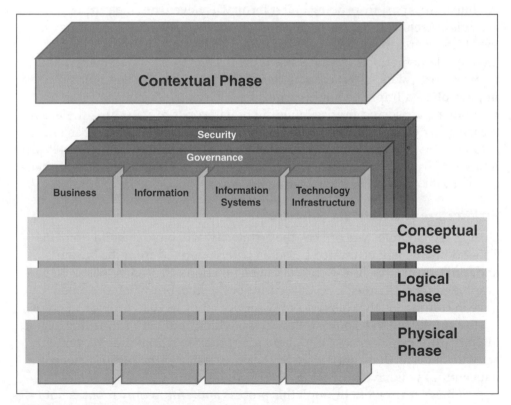

Figure 4.5 Overview and layers of architecture.

In the *logical phase,* we address the question, *How can it be realized?* Scenario planning is utilized in order to determine the best solution and to resolve business imperatives that may conflict (e.g., *We want the lowest-risk solution with the highest flexibility*). At this stage, we begin to reintroduce complexity factors such as geography and organizational boundaries. At the end of the logical phase, the optimal scenario is determined and taken forward into the next phase.

In the *physical phase,* we address the question, *With what can it be created?* At this point, we identify actual products for services that will be utilized to implement the architecture. We also address distribution of data and business logic as well as the physical implementation of platforms (e.g., one or multiple servers? clustering or failover?) in this phase.

The Purpose of Architecture

An architecture describes overarching designs of individual components so that their assembly results in a complete and working product. This design is needed to guide the construction and assembly of components.

Architecture to Support Decisions

Architecture, apart from being an art form, was never the sole purpose of activities. Architecture was, is, and will always be driven by specific needs. In the IT arena architecture will always follow and specify incremental and iterative implementations. Business needs are drivers for speed to market and reduced costs, which start earning their ROI right from the beginning. This implies phased implementation principles such as the famous 80/20 rule. Without a concise and precise blueprint or enterprise architecture, there is no measure in determining which 80 percent to realize and which 20 percent to skip initially. The level of complexity in today's application landscapes is prohibitive of well-educated decisions incorporating the most critical dependencies without an existing enterprise architecture.

Architecture to Accelerate Solutions

During each architecture project, the architect(s) will have to build domains. These domains will be created based on criteria derived from the business and/or IT principles. These criteria can be function oriented, or based on geographies, business units, level of security, or the like. But what all domains have in common is the collection of IT services they have to provide in order to support the business process flow. It is a tedious but very rewarding task to identify duplicate services throughout a corporation's domains. Very often, activities such as these are driven by data ownership principles. Owners of data entities will have to provide all services needed by all other domains.

A CGE&Y team went through this process and within only months not only redesigned the communication systems of a retailer network to a single way of communication but also developed a very strong and complete repository of reusable common components. In order to achieve this objective, the enforcement of project standards and the authorization of any projects by architect(s) became mandatory. Subsequent projects and enhancements experienced a reduction of project durations and costs of up to 50 percent. Clearly, even in an ideal environment such as this, new developments do not simply rearrange existing components for new applications. There will always be some individual coding necessary. But employing reusable components will reduce project durations and costs and, as a side effect, enforce standards. Additional accelerators are *road maps,* describing how to approach the development of a solution, potential hurdles, checklists, and so on, and *patterns,* describing partial solutions to very specific parts of the solution, similar to the plug-and-play approach in other IT industries.

Aspect Areas

Just as a solution is built from multiple components, a balanced architecture must reflect more than just one aspect (e.g., technical) for this solution. An aspect area can be looked at as a view of the solution from one specific vantage point, with one specific objective. Each objective results in a set of related structuring criteria.

The topics addressed in each of the aspect areas differ widely, and the complexity involved necessitates that specialists be engaged for each area (see Figure 4.6). However, there are strong interdependencies between each of the

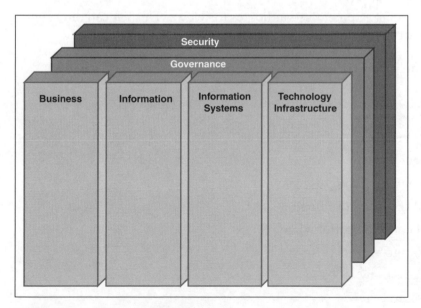

Figure 4.6   Architecture aspect areas.

aspect areas, since the business structure prescribes information structure, which in turn prescribes information systems structure, which then prescribes technology infrastructure. Thus, ideally, all aspect areas are incorporated into the architecture design to ensure its usability as *one system*. Skipping one of the aspect areas implies additional risk, as critical viewpoints have not been considered. Designing solutions based on individual aspect criteria reduces complexity and creates clarity of mind. Only when all aspect areas are over-laid and thereby create the complete solution do we add complexities and reflect a realistic but now balanced view of the solution. Table 4.1 defines the content behind each of these aspect areas.

The last two aspect areas are so critical to the success of an architecture that both aspect areas will have to be reflected in any other area. Here they typi-cally are considered as separate views of the solution within the individual aspect area.

Specialized Views

Developments in business and technology often require special attention in the architecture design. Within the Integrated Architectural Framework, this is translated into the flexibility of integrating any necessary view into the architecture. These views typically do not appear on the horizon all by them-selves. They are carefully selected and derived from recorded business and IT principles. Only a solution that can be measured against evaluation criteria will, in the long term, be successful. The usage of views and viewpoints sup-ports this evaluation process.

Following is a list of potential candidates for views. This list can only sug-gest the breadth of possible views.

- Component view
- Actor view
- Governance view
- Security view
- Data view
- Technology infrastructure view
- External communication view
- Internal communication view
- Data storage view
- Data access view
- Business unit view

Table 4.1   Partial List of Aspect Areas of Architecture

ASPECT AREA	CONTENT
Business	Commercial organization Personnel Administration Processes Finance
Information	Information structure Knowledge management Data warehousing Processing structure
Information systems	Automated IS support IS services Integration of IS services Collaboration between services and components Critical use cases describing functional requirements
Technical infrastructure	Infrastructure support required IT services required Processing platforms and volumes Network diagramming Hardware/software standards and requirements Governance and security services per platform (failover, load balancing, encryption, etc.) Communication methods and standards

Governance	Ongoing support of the different business processes
	Areas of concern related to IS systems:
	    Availability
	    Controllability
	    Performance

Security	Business security needs at the different levels of technology
	  platforms, applications, and network infrastructures
	Security prevention services and elements, such as integrity
	  and confidentiality

A secondary challenge using views is their graphical presentation so that the picture is complete but not confusing.

Levels of Contemplation

CGE&Y's Integrated Architecture Framework (IAF) is the third-generation improvement on the original CGE&Y architectural approach. The IAF differentiates itself by presenting several different aspect areas for architecture.

These aspect areas, while focusing on different elements of the architecture as a whole, still maintain a high-degree of cross-communication and integration, both internally as well as with CGE&Y's estimating, delivery, and quality assurance (QA) methods. The IAF is further differentiated because it is supported by a set of integrated methods that cover the various aspect areas of architecture and drive the entire process from conceptual design to detailed physical design and technology selection.

These methods provide our clients with a high degree of specialization within architecture aspect areas, while still maintaining the business-driven and integrated approach of the original standard (see Figure 4.7).

Recall that the first phase of this process is called the *contextual phase*. The objective of this phase is to prepare everything needed to start the project. In this phase there will be three sets of deliverables:

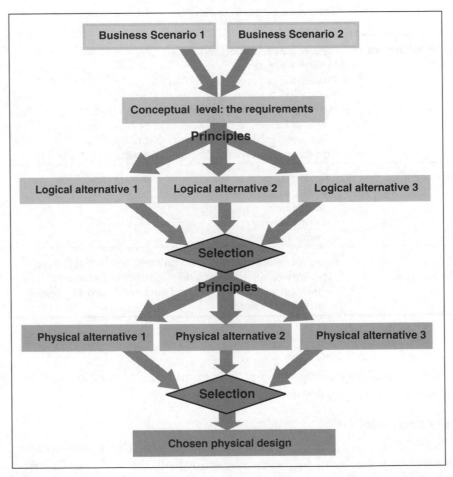

Figure 4.7   A business-driven approach.

1. The *project quality plan*, which will contain all project-related deliverables necessary for such an architecture project.

2. The *internal quality plan*, which will contain all CGE&Y internal management and quality assurance documentation.

3. The *architectural input file*, which will contain all relevant input information for the project that is encountered during this phase.

CGE&Y has set specific objectives to help align us with the needs of the client and the business drivers. We will achieve four objectives during this set of activities:

1. Gather sufficient information regarding the client's business strategy and drivers to identify what will drive, govern, and constrain the architecture.

2. Define the desired end state for the use of technology.

3. Set the scope for the overall architecture and further define the effort required to achieve the desired end state.

4. Establish agreement on the common set of services to be architected and estimated.

The deliverables from this phase will include capturing the strategic vision, business requirements, and project goals. These will drive the new architecture and an information systems (IS) framework for the client's desired environment. Additionally, since the scope of the architecture is not to redefine the information system services needed by the client, but rather to define how they will exist in the new environment defined by the architecture, the existing IS services will be captured and agreed on to facilitate working with a common set across the different streams.

The conceptual phase identifies, structures, and defines the scope of the total architecture. The main objectives of this phase are to:

- Identify all the internal and external elements that must be considered in the architecture. These include external influence such as regulation, emerging industry standards and market pressures, and internal influences such as corporate biases, management directives, and member constraints.

- Define the entities in the architecture that will support the business requirements and the relationships between them.

- Verify that the model supports the business strategy and conforms to the constraints placed on the architecture by both internal and external factors.

The creation of a product follows a specific set of tasks and activities. First, the requirements for the product are gathered and structured. Then the specifications for the product are developed. Finally, the detailed production instructions are designed. Consequently, the conceptual phase in architecture is most concerned with the structuring of the requirements to produce a concept of the architecture.

The aim of the conceptual phase of the architecture is to create the foundation for all architectural decisions. The conceptual architecture must define the business-level requirements that will justify future investments in technology. In order to develop the framework that will define how technology is used to support the business vision, it is important to remember that all aspect areas of the architecture must be addressed.

The scope and specific deliverables of the logical and physical architectures to be conducted will be determined at the conclusion of the conceptual cycle. While the process model remains the same, the content must necessarily differ, depending on the specialized focus areas that are most important to the client's business.

During this phase of the architecture, an additional tool is used to visualize and validate the current progress. Two concepts of the future standard IEEE P1471 have been adopted and incorporated into the CGE&Y methodology. We use the terms *view* and *viewpoint* within the method. The IEEE definitions for these terms are as follows:

Viewpoint.   A pattern or template from which to construct individual views.

View.   A representation of a whole system from the perspective of a related set of concerns.

In other words, we can identify viewpoints and create their views to show the structure from a specific perspective. Examples are an *information view*, an *integration view*, and a *security view*.

The logical phase will use the output of the conceptual phase and address the following:

- All platform services (hardware and operating systems, etc.)

- All networking services (including local area networks [LANs], wide area networks [WANs], and inter/intra/extranets)

- All middleware services (including OLTP monitors, object brokers, and messaging services)

- All common services (e.g., databases, authentication, authorization, and audit)

- All shared services (including directories and common desktop applications)

- Logical application component models

- Component logical life cycle

- The best structuring of shared application services across the enterprise

The main objective of this phase is to translate the conceptual architecture into the (logical) specifications of the architecture. This is accomplished by doing the following:

- Defining the information system components by extracting the requirements

- Projecting the defined information system components onto the desired logical technological framework

- Creating the specifications of the content of the technical components

- Verifying that the proposed architecture meets the requirements

The objectives for the logical phase will be achieved by performing the following four tasks:

1. Creating the logical architecture outline to define the objectives and constraints.

2. Studying alternative scenarios and selecting preferred solution(s).

3. Modeling the (logical) architecture and verifying its feasibility.

4. Making a final selection of viable scenarios and presenting these to the principal client for review and verification.

The outcome of applying CGE&Y's method is a design for all elements of the data processing support and applications infrastructure along with the architecture that supports the critical focus areas of the business. However, the method does not address the detailed functional requirements and specifications of applications that need to be supported by the architecture.

Logical Scenarios

The *logical scenarios* are constructed based on various business drivers, which may have conflicting objectives. Different scenarios are constructed to emphasize particular architectural drivers, which are weighted according to their importance to the client's business. The client and CGE&Y will derive the best scenario from the various options available, and this scenario will be enhanced and detailed so that it can be used as a basis for the physical cycle.

For example, a logical scenario might be driven by least cost. Another might be driven by quickest time to market or by minimum technology risk.

Recall, in the *physical phase,* we address the question, *With what can it be created?* As mentioned previously, here we identify actual products and services that will be utilized to implement the architecture.

Development of the technical architecture in the absence of the business architecture that it is required to support will result in development in a vacuum. This virtually guarantees that the IS framework that is developed will not be supportive of the business, nor will it be able to feed new business opportunities into the organization when those are presented by innovative use of new technology. To enable this alignment and support it is necessary to do the following:

- Define what the as-is technology environment is.
- Define the business requirements that will be supported by the architecture.
- Define the business and technology vision and strategies that will shape the architecture.
- Define the high-level services that are required to support the business requirements.

By bringing the technology and business focus together, we can ensure that technology is utilized within the client's environment in a planned, coherent way that supports the client's business direction. With this understanding, the client will then be armed with sufficient information to clearly evaluate their architectural alternatives.

After all requirements have been gathered in the conceptual phase and they have been translated into the logical architecture design in the logical phase, the detailed physical design of the architecture has to be developed. Each detail of every component in the architecture is designed and described. Verification of the usability of the design is carried out by analyzing the required performance, capacity, and throughput in detail and calculating, or testing via benchmarking and/or prototyping, the physical possibilities of the architecture and comparing the results to the requirements.

The physical phase consists of the same four steps as the logical phase:

1. The phase is prepared by adding the component data flows to the information system components list to create a total list of components that will use computer resources.

2. Alternate scenarios are studied and selected; alternate scenarios will be based on the different technical possibilities.

3. The physical architecture is created and analyzed.

4. The final solution is selected and presented to the client.

Physical Scenarios

The physical scenarios explore the various options available to deliver the services based on the client business constraints and the service levels required of the various components. As an alternative and accelerator to this scenario discussion, CGE&Y offers technology-focused workshops during which the client-specific technical environment will be defined and described. These workshops are highly intensive and require the presence of all decision makers and stakeholders.

Migration Strategy

Consistent with our integrated approach to developing the solution architecture, in the physical phase we include the necessary tasks for designing the migration plan. The migration strategy is concerned with communicating with the business to agree on an outline plan for migrating to the physical architecture. This takes into account ongoing initiatives, business areas that urgently require technology enablement, and interface strategies to allow business operation during the migration period. Using scenarios and views, we can compare the various features of the migration options, evaluate the pros and cons, and then accurately determine the most appropriate approach for moving forward. The objective is to define possible scenarios for implementing the architecture. The scenarios defined here should reflect the business drivers, organizational priorities, and financial considerations of the client. The differences in the approaches of the migration scenarios can have a profound impact on the organization as a whole; therefore, care should be taken to define and validate only appropriate scenarios. Additionally, where possible, the migration strategy includes time lines and cost estimates for the implementation of the architecture.

Security Strategy

The security strategy defines how security will be implemented and managed in the client organization, including the identification of required policies, procedures, and organizational entities.

The aim of the security architecture method is to enable architectures that address business security needs at the different levels of technology platforms, applications, and network infrastructures.

The purpose of the security architecture is to provide a framework to lay out a security architecture that is logically and uniformly derived from business drivers. The emphasis is on what level of security is needed, how to realize it, and with which techniques and products security can be implemented. The objective of our approach is to enable architectures that address business security needs at the different levels of technology platforms, applications, and network infrastructures.

Advantages of this approach are:

- Security expenditures can be justified.

- By clearly defining the objectives of the security environment, the validity of a security budget can be assessed.

- By defining the overall security architecture, the risks of an insecure white spot are minimized. If there are any white spots, at least they are known.

- Having security measures in line with business drivers links business drivers to the security environment. The security environment is therefore easily adaptable to future business changes. When the business changes, then the security can change with it.

- Functionality and products can be decoupled.

The standard approach that has been detailed here—what, how, and with what—translates to the conceptual, logical, and physical cycle of the architecture method.

Governance Strategy

A governance architecture is aimed toward deploying information technology to support the entire business. This overall *enterprise architecture* must provide a structure for guiding the full life cycle of information systems.

Governance is a broad term used to describe more than just the software engineering aspect of the architecture. It should be thought of as the concept that addresses the daily operations, operational impact analysis, budget management, interdepartmental interaction behavior, among other things. By identifying and implementing this type of strategy, the business can manage the enablers such as personnel, capital, housing, and IT on which the business depends. For all these services, the business wishes to control the quality in terms of availability, performance, reliability, continuity, and so on. This strategy defines how technology will be managed at the client, including the identification of required policies, procedures, and organizational entities.

There are four key aspects that need to be implemented in order to realize governance effectively:

1. The business needs must be agreed upon in terms of quality-of-service attributes.

2. The delivery of quality-of-service attributes must be designed and implemented.

3. Control processes need to be implemented to check the agreed quality-of-service attributes against the way they are actually delivered.

4. Control processes need to be implemented to check whether business needs are still correctly represented by the defined quality-of-service attributes.

Applying the Methodologies to Wireless Engagements

In this chapter we've outlined CGE&Y's approach to the project life cycle and the methods and methodologies that are applied during individual parts of the life cycle. How are wireless engagements different from any other project or engagement? Based on our experience:

- Wireless requirements do not differ from former eCommerce requirements.

- Wireless engagements change nothing in the fundamental process of developing the architecture.

- Much of the content creation, preparation, and delivery should be based on existing industry-leading standards.

This does not imply that running a wireless project should be treated as any other project. The biggest challenge is that many of the technologies are new and some of them have not been proven under real-life environments and real-life stress. This results in potentially incomplete standards and products, unexpected efforts in filling shortcomings of standards, increased learning curves by staff, and many unexpected problems during the project life cycle. Being the first to implement a solution has its price, but it also brings with it the rewards of leading the pack.

Before we continue into more of the details of creating a successful and usable architecture that subsequently leads to applications developed according to this blueprint, let's summarize what has to be considered—as an addition—in wireless engagements.

We will have to add many new IS and IT services. Starting at the enterprise level, we have to commit the corporation to being active in wireless system functionality. Drilling down into the project architectures, and specifically the logical and physical definition of services, we must include specific services to address the needs of wireless communication. And let's not close our eyes to the fact that all aspect areas of an architecture are affected by these new services. Not only must we provide the technical services to make wireless communication work, but we must evaluate all the other aspect areas very carefully.

We need to develop our applications within the framework and restrictions of existing industry standards in bandwidth and feature capabilities. Particularly in this newly emerging delivery channel of applications, standards are not yet mature enough to be able to rely heavily on them. This is especially true if we consider international deployments.

If we were successful with our architectural paradigm of separating data and its access layer from business functions from the presentation layer, the actual *testing* of wireless applications could be as simple as modifying existing test scripts to include these new services. The challenge here is capturing the vast majority of different wireless devices and the manufacturer's interpretation of standards. Still, testing wireless applications will not be as simple as rerunning existing tests in a wireless environment.

Deployment and its *preparation* will confront us with new challenges. Will standards restrict our delivery of applications to a limited user community only? Or will we have to develop our applications to multiple standards in order to achieve a higher market penetration? Do we deliver documentation at all? How—within the restrictions of existing standards—can we deliver state-of-the-art training? What is the impact on cost and time of designing really intuitive software? How will we roll out our applications: one service provider at a time? Will we follow a big bang theory? How will we make public that there are new values that our client's corporation has to offer? How thick will wireless devices be? Do we need to follow synchronize (e.g., similar to PALM devices) or download policies? In other words, will the application be real-time wireless or require a synchronization with another system?

We will have to upgrade versions on a regular basis. How will we manage the possible configurations? How will we measure the response times our users experience? How will we recognize that users might abandon the features we offer if they encounter problems?

As we can see, there are some very specific questions and answers that surround the implementation of wireless features and values. Subsequent chapters will demonstrate solutions to these and a lot more questions.

Chapter 5 will begin to solve the technical issues of architecting a wireless solution so that many of these problems we highlighted will be addressed right from the beginning: the idea and blueprint of the solution.

An Overview of Wireless Architecture

In most system development and integration efforts, there is some existing technology investment that you are trying to reuse. When the World Wide Web first gained prominence in the mid 1990s, we saw quite a few deployments that did not try to reuse existing technology investments. However, people soon realized that the value of a channel to a customer, supplier, or any other business partner lies in the savings you gain from automating functionality. In most cases, this involves integration to an enterprise resource planning (ERP) system, a billing system, or the like.

We are seeing a similar evolution in the wireless world. As a result, one of the most important points we want to make in this book is how wireless technology will effect your existing technology investment and the architecture thinking behind it. This chapter will give you a general understanding of these effects and will set the stage for the case studies in Part 2 and beyond. The case studies will describe many of the points made in this chapter in much more detail.

We will describe some of these points in the context of the Integrated Architecture Framework (IAF), which we first introduced in Chapter 4. To recap, the IAF is a methodology used to define architecture. You may ask, What good does this do me if I am not a Cap Gemini Ernst & Young employee? The IAF defines the scope that we cover and provides the direction for this book. However, the content of the book is described in general architecture terms and does not require knowledge of the methodology.

Figure 5.1 defines the scope of this book. We define architecture in four phases:

1. *Contextual architecture:* In the contextual stage, we define the scope of the architecture and make sure that all the information we need is provided.

2. *Conceptual architecture:* We define what needs to be done in this stage.

3. *Logical architecture:* Now that we know what needs to be done, we define how it is done.

4. *Physical architecture:* Now that we understand how we are going to do it, we need to determine with what we will do it. This means that we make product choices and define the details for servers, and so on.

Figure 5.1 also shows four aspect areas that are covered in each of these phases, as follows:

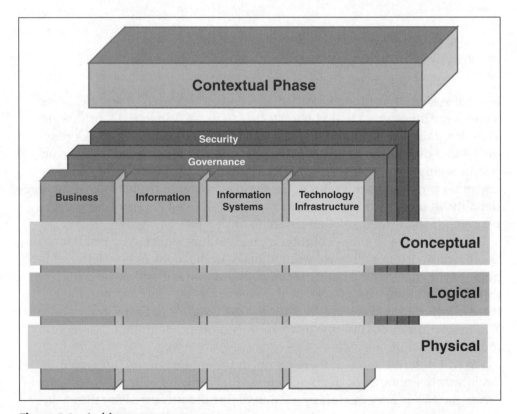

Figure 5.1   Architecture scope.

1. *Business:* Defines the business case and associated business processes.

2. *Information:* Defines all aspects of data ownership, storage, and movement.

3. *Information systems:* Defines all aspects of the applications used to build the system.

4. *Technology infrastructure:* Defines all aspects of the hardware and networks needed to run the system.

Additionally, there are two special aspect areas:

1. Security: Defines all aspects of network and information security.

2. Governance: Defines all aspects of what you need to run the system.

Given the nature of this book, we will not cover contextual architecture here. Instead, we will begin with the conceptual architecture.

Principles of Conceptual Architecture

The conceptual level answers the question, *What are we trying to do?* During this phase, all distraction factors, including organizational boundaries, political issues, technology biases, and geography are extracted in order to obtain a simple, clear picture of what the architecture has to achieve. The conceptual level defines the services that are required and what is required from each service. A service, such as email or Web access, is a structured representation of requirements that forms the highest level of abstraction in architecture. In addition to describing services, the conceptual level also provides a model that structures the active form of cooperation between the services.

The advantage of service-based architectures is that they are easily derived from business requirements and are recognizable for clients. Services are loosely coupled by nature because a service is a distinct entity. Loose coupling enables architects to flexibly compose an overall structure, and new technologies are allowing dynamic real-time collaboration between services. This development has allowed designers to move from silo applications to loosely coupled and integrated systems.

When applied to wireless applications, the conceptual level of an integrated architecture will show the set of wired and wireless services, and how they interact. There are generally two scenarios for wireless applications:

An extension of an existing application to include a wireless channel.   An example of this is providing wireless access to ordering products in the B2B exchange, described in Part 2 of this book. These applications are generally deployed to enhance the current business offerings.

An application whose only use is wireless. One example of this is a location service for a taxi company that enables operators to constantly know where their taxis are. These applications are generally deployed to expand the scope of a business.

An example of a conceptual view of wireless access to an enterprise can be seen in Figure 5.2.

The reason for creating a conceptual picture is to show what we are trying to accomplish in such a way that nontechnical personnel and executives can easily understand and buy into it. The conceptual artifacts (typically pictures) offer no information about how to make the pieces interact but do describe the set of services and interactions between them. Finally, the conceptual architecture also defines the scope of the work. This is done in terms of the aspect areas described in the introduction.

An aspect area looks at one system from a specific standpoint. The topics addressed in each of the aspect areas may differ, but there are strong interdependencies between each of the aspect areas. The next section will review the aspect areas in further detail, and will show how they relate to wireless application architecture.

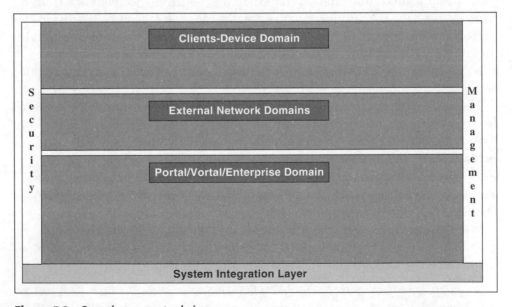

Figure 5.2 Sample conceptual view.

Conceptual Business Architecture

Every architecture must be business driven. If you are building a system without a business case, you will likely have to make drastic business changes to it when it is used in a business scenario. As a result, the needs of the business define the choices made by the architecture team. Several business documents are required to accomplish the goal of making an architecture business driven.

IT strategy.   An IT strategy influences the architecture choices and products that are used to implement it. For instance, it may be inappropriate to recommend using an IBM relational database management system (RDBMS) if Oracle is already the corporate standard.

Business requirements.   The business plan for what is being architected must be known to make the appropriate decisions.

Environment.   An architecture must be completed with knowledge of the business environment. For instance, we may define an architecture to provide mobile services for both a health care provider and a car manufacturer, but the results may be very different due to future business objectives and organizational structure.

These tenets of enterprise architecture signify the importance of having a clearly defined business architecture. The business architecture is elaborated in the business aspect area, a set of business system requirements that warrant the support of information technology (IT) for its delivery. Well-defined business architecture provides an understanding of the structure and required quality of service (QoS) of the enterprise. It gives a framework for deducing requirements for other aspect areas and, in some cases, how the business itself can be improved. The business aspect area allows a company to align its business principles with its IT services.

One of the difficulties in describing business architecture in general is that the content is driven by the business needs and business principles. Therefore, detailed descriptions of business architecture will come in subsequent chapters in this book, when we design an architecture to solve the needs of a specific industry. However, there are business needs for wireless architectures that apply horizontally across many channels. In a recent survey (CGE&Y Mobile eBusiness B2B Survey 2000/2001), many companies asserted that they are currently focusing on the increased efficiency of internal processes with their wireless efforts. They identified the following set of focus areas:

- Field service automation
- Sales force automation

- Mobile office
- Mobile payment
- Logistics
- Telematics
- Financial transactions

To provide these services, a business architecture will have to answer questions such as:

1. What effect will our wireless offer have on our place in the market?
2. Where on the wireless technology adoption spectrum are we most comfortable: early adopter, wait-and-see, or late adopter?
3. What processes will have to be changed or created to meet our wireless needs?
4. What effect will wireless have on our organization?
5. What effect will wireless have on our personnel?
6. What will we have to do to administer our wireless endeavor?
7. What will be the financial impact of wireless?
8. When future wireless technology improvements become available, will our business be prepared to leverage the additional functional capabilities?

We will focus on four of these areas in the case studies provided in this book:

- Business-to-business (B2B) wireless exchange, with a focus on supply chain automation
- Customer relationship management (CRM), with a focus on sales force automation
- Financial services, with a focus on financial transactions
- Mobile office

However, every company will have to answer this type of question for itself based on its business direction.

Once the business architecture has been defined, the next step is to determine the data needs of the defined business services.

Conceptual Information Architecture

The information aspect of the conceptual architecture describes the data flows between the processes defined in the business architecture. It deals with issues such as:

Standards and policies.   This defines the standards and policies for information architecture.

Capabilities.   This defines what the information architecture should be able to do.

Technology directions.   This defines which technologies have been chosen already.

Distribution strategy.   This describes how data can be distributed.

Constraints.   This describes some of that factors that may prevent us from making certain decisions. An example of a constraint is certain data being located in a legacy database that can be accessed only via screen scraping.

The information aspect straddles the business and its systems; the business requirements are an input for generating the information requirements, which, in turn, are an input for the information systems (IS) requirements.

Quite often, information is divided into *domains* or *zones*, allowing actors to have different levels of access. *Actors* are defined as user groups or systems that initiate actions in a system. An example of information zones is shown in Figure 5.3.

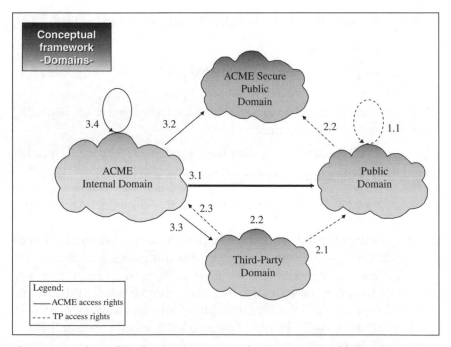

Figure 5.3   The public domain is an example of a conceptual information zone.

The model allows actors to have access to the data necessary to meet their business needs in the system, but prevents them from having access to other data. The information domains must incorporate information of certain types, for example:

- Public information is placed in the general public domain (the Internet).
- Business-to-business appointments are placed in an ACME secure public domain (an ACME extranet).
- ACME internal business information is placed in the ACME internal domain (the ACME intranet).
- Third-party internal business information is placed in the internal domain of the third party (the third-party intranet)

The connections between the domains support the following access rights:

1. The general public has access to:
   a. The general public domain (unrestricted/restricted update)

2. Third parties have access to:
   a. The general public domain (unrestricted read, unrestricted/restricted update)
   b. One or more ACME secure public domains (restricted read, restricted update)
   c. The ACME internal domain (restricted read, restricted update)

3. ACME staff has access to:
   a. The general public domain (unrestricted read, restricted update/content management)
   b. One or more ACME secure public domains (restricted read, restricted update/content management)
   c. One or more TP internal domains (restricted read, restricted update)
   d. The ACME internal domain (restricted read, restricted update)

Access rights 1.a and 3.d are related to the respective domains. Access right 2.a is outside ACME control.

This example shows a typical set of information domains. When applied to the wireless space, there *should* be no impact on the information model. Our process for creating architecture has the business driving the information, not the devices used. However, the state of wireless technology forces the tail to wag the dog, so to speak: Until the functionality of wireless devices matures, its limited information capabilities must be considered when creating the information architecture.

Conceptual Information Systems Architecture

The information systems aspect of the conceptual architecture defines the structure of information systems required by the communication structure. It is based upon the IS principles of the organization, such as buy versus build policies and requirements for degree of systems automation. This conceptual view defines the IS services provided at a very high level, using the artifacts produced in the conceptual information aspect area for input. In addition to describing services, it also describes the interfaces for accessing those services.

The conceptual IS aspect may or may not change for wireless-enabled applications, depending on the wireless services desired. For example:

- Lack of wireless bandwidth could prevent some IS services, such as streaming video, from being offered to wireless clients.

- Inability to guarantee nonrepudiation could force some security-conscious transactions to be conducted by only wired clients.

- Concerns about data privacy could prevent some services from being provided for wireless clients.

Barring any of these types of concerns, however, the conceptual view of the client should not matter to the systems providing services.

These services are being taken to a whole new level with modern Web services toolkits. Services are aggregations and collaborations of components to provide value to a business need, and Web services allow this to happen dynamically. They make the IS value Web a reality, where components can be used effectively in ways not envisioned by their creators. With these dynamic collaborations come increased automation of services, increasing efficiency, and an improvement in the quality of service that can be provided.

Modern technologies allow customization by client channel. Every request to a Web server contains information about the class of client making the request. As a result, Web servers can customize responses for those clients. For example, a Web server receives a request from a Wireless Markup Language (WML) capable WAP phone, applies the same business logic to formulate a response as it would from any other client, then takes the content generated by that business object and formats it in WML before sending the response.

Another enhancement to the user experience is the "myWebPage" concept of content customized by the individual user for that particular user. The user specifies the content he or she finds most useful from the total set of content available, and the selections are remembered for future interactions with that customer. It is amazing how many users find their experience enhanced significantly just by seeing the "Welcome, [your name here]" content so typical today.

These are examples of the kinds of services created in information systems with wireless channels. If we recall the areas being focused on by businesses for wireless integration (field service automation, sales force automation, mobile office, mobile payment, logistics, telematics, and financial transactions), the important artifact to generate in this part of the architecture is what IS services are being provided by which objects, as specified by the business architecture.

Conceptual Technology Infrastructure Architecture

The technology infrastructure (TI) aspect of the conceptual architecture describes the design of an integrated set of hardware, systems software, and network components that fully support the business requirements. It provides an end-to-end technology framework in which services are created to meet the business needs. At the conceptual level, TI is also concerned with the technology principles to be used when creating and deploying the application.

At the conceptual level, TI is vendor and package neutral, specifying the classes of services to be provided, not tying the architecture to any particular way in which the technology might be deployed. For example, a conceptual TI for a bank might specify that the accounts database be connected to the Web application server via a local area network (LAN), but not specify things like "using a star topology 100-megabit Ethernet."

Unlike the aforementioned aspect areas, the content of TI is different for wireless applications since it specifies the systems used, albeit in very general terms. A detailed description of the technologies is found in Chapter 3, so we will not duplicate the effort of that chapter here. What is salient for this architectural view is the use of those technologies in accordance with the principles of the organization.

Again, let us take the bank example. This time, the desire is to provide users accessing the system via WAP phones access to their account information. One of the technology principles of the firm has to do with usability—the mean response time for any request can be no longer than 8 seconds. The bank commissioned a usability study, which concluded that after 8 seconds, the user experience suffered tremendously. This has impact for both the IS and TI architectures, the former being pressured to provide an efficient means of arriving at a response and the latter tasked with pushing that response within the allotted time. What is important from the TI perspective is that the architecture be created with these principles as inputs.

Conceptual Security Architecture

The security specialized aspect area defines how security will be implemented and managed in the client, including the identification of required policies, procedures, and organizational entities. The aim of the security architecture method is to enable architectures that address business security needs at the different levels of technology platforms, applications, and network infrastructures.

The purpose of the security architecture is to provide a framework to lay out a security architecture that is logically and uniformly derived from business drivers. The emphasis is on what level of security is needed, how to realize it, and with which techniques and products security can be implemented. The aim is to enable architectures that address business security needs at the different levels of technology platforms, applications, and network infrastructures.

Advantages of this approach are that:

- All security expenditures are justified by going through this approach.
- By clearly defining the objectives of the security environment, the validity of a security budget can be assessed.
- By defining the overall security architecture, the risks of an insecure white spot are minimized. If there are any white spots, at least they are known.
- Security measures in line with business drivers link those drivers to the security environment. The security environment is therefore easily adaptable to future business changes. When the business changes, then the security can change with it. The products you are choosing do not define functionality. Instead, the functionality is defined before the products are chosen.

The conceptual level defines the business needs and links them to security services for various actor types and security areas.

Conceptual Governance Architecture

The governance specialized aspect area determines how the customer wants to approach the management, maintenance, and exploitation of its information technology and information systems. This should include items such as configuration management, problem management, change management, availability management, software distribution and control, capacity and performance management, contingency management, cost management, service-level management, operations management, and security management. A governance architecture is aimed toward deploying information technology to support the entire business. This overall *enterprise architecture* has to provide a structure for guiding the full life cycle of information systems.

Governance is a broad term used to describe more than just the software engineering aspect of the architecture. It should be thought of as the concept that addresses several issues, including the following:

- Daily operations of all systems
- Operational impact analysis of changes to the system
- Operations budget management
- Interaction with other departments in the enterprise

By identifying and implementing this type of strategy, the business can manage the enablers such as personnel, capital, housing, and IT, on which it depends. For all these services, the business wishes to control the quality in terms of availability, performance, reliability, continuity, and so on. This strategy defines how technology will be managed at the client, including the identification of required policies, procedures, and organizational entities.

There are four key aspects that need to be implemented in order to realize governance effectively:

1. The business needs must be agreed upon in terms of quality-of-service attributes. These attributes include scheduled downtimes, response time for unexpected downtimes, and so on. All of these attributes are defined in a service level agreement (SLA).

2. The delivery of quality-of-service attributes must be designed and implemented. This means that we now have to design how the scheduled downtimes can be kept to the agreed minimum, how the response time to unexpected downtimes can be kept at the required level, and so on.

3. Control processes need to be implemented to check the agreed quality-of-service attributes against the way they are actually delivered.

4. Control processes need to be implemented to monitor whether the defined quality of service still correctly represents business needs attributes.

Our method focuses on the governance of information services, communication services, related technology, and organization and processes (IT). A governance architecture defines the organizational and supporting IT structures needed to operate the system. In addition, it provides the specifications for organization processes, as well as the governing and governed technologies based upon business needs.

To deliver a governance architecture, the standard approach that has been detailed here—what, how, and with what—is followed. The conceptual cycle addresses the following:

Quality-of-service parameters required for each business process.   More specifically, what are the most important aspects of the system that need to be taken care of during its operations? Is it important to have a high level of availability? A high level of flexibility?

Overview of governance responsibilities.   Who has the responsibility for certain parts of the operations, such as the failover solution?

Overview of governance influence and control.   What is the scope of the governance solution?

Overview of IT dependency.   What IT systems are needed to accomplish good governance and governance procedures?

Summary of Conceptual Architecture

Upon completing the conceptual architecture, the stakeholders should have a clear understanding of what the system is going to accomplish. There should be artifacts specifying the business requirements, the information requirements, the IS services to be provided, and the technical framework in which the solution will be developed. At this point, each aspect area of the architecture will be revisited in greater detail in the logical phase.

Principles of Logical Architecture

The conceptual architecture gives us a clear idea of what needs to be accomplished and the scope of the effort. The logical architecture takes the conceptual architecture as input and answers the question, *How are we going to do it?* As with the conceptual architecture, the logical architecture also focuses on the four architecture aspect areas: business, information, information systems, and technology infrastructure. We will also focus on the two special aspect areas: security and governance.

Let's start by understanding some of the key points that architecture has to provide.

The Big Picture

A good architecture has several key characteristics, as illustrated in Figure 5.4.

An enterprisewide architecture may have several key characteristics:

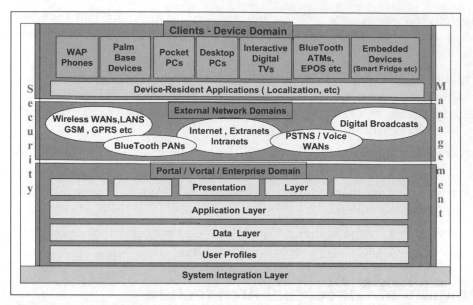

Figure 5.4 Architecture key characteristics.

Performance.   The architecture must support building a system that performs according to the user expectations.

Stability.   Any system built on the architecture must be stable. This is an interesting aspect in wireless application development since many of the technologies in that market space are cutting edge. The difficulty is that you give up a lot of exciting functionality if you give up some of the more cutting-edge products. This book will not focus on products, but any architect using this book must be aware of this trade-off and have a clear understanding of the implications of the choices he or she makes.

Flexibility.   The architecture should be flexible enough to include new components and channels without having to redo everything you have done already. Requirements always change and the architecture must be able to accommodate this.

Scalability.   Most large-scale systems are implemented to support a certain amount of user transactions. An architecture is designed to allow this implementation to be scaled without reimplementing everything. For instance, if we suddenly realize that we have to support an additional 100,000 users, we should be able to add some servers and distribute some application components as opposed to redeveloping my applications and throwing away my hardware.

Reliability.   A system must be running to service transactions and to keep customers happy. The architecture must support this principle.

Logical Business Architecture

This book will focus on four areas where companies are currently looking to improve their operations by using wireless technologies:

- B2B wireless exchange, with a focus on supply chain automation
- CRM, with a focus on sales force automation
- Financial services, with a focus on financial transactions
- Mobile office

These are areas where we will see widespread wireless development in the next two to three years. As Figure 5.5 shows, the business applications of wireless technologies will become more interactive with the increased network capabilities.

We are now in the stage where most companies are porting some of their existing capabilities to the wireless Internet. As Figure 5.6 illustrates, we expect to see mobility enabling a more dynamic business model with the deployment of 2.5G and 3G environments. Several processes that are traditionally static will be available via wireless channels.

What is the impact of these new services on the architecture? We have to add new components to the system. We need to identify the business components first and then go into information systems, technical, security, and governance components. Before that, we must identify the information model and where the data resides in the system.

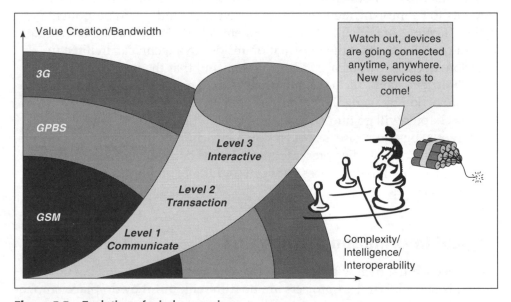

Figure 5.5 Evolution of wireless services.

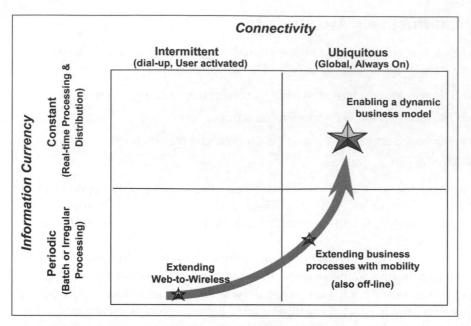

Figure 5.6 Evolution in wireless business applications.

Let's start with the business components. Some business components that may exist in a mobile office environment include locating a person, calling in sick, and having wireless network access.

The business components require new data and information systems/components to be implemented. For example, we may find an IS component called "Person" and an interface component called "Retrieve name." We can assume in this case that the information resides in a technical architecture component called "Database" in our system and that the interface component "Retrieve name" gets the data from there. The user can log in via a wireless connection by using the "Wireless login" component.

This chapter will go into no further detail on these concepts, but you will see how the business case is broken down into different types of components in each of the case studies presented later in this book. The rest of this chapter will focus on the differences between traditional architectures and wireless architectures.

Logical Information Architecture

All projects must deal with where data comes from, who owns it, and how it is stored. Wireless projects are no different, but they do place emphasis on different aspects of these issues. Some of the main differences are:

Profiling.   Though an important aspect of most Web applications, profiling will become even more important in the wireless world. Most current wireless applications are marketed for people on the run. By definition, this group of people will not want to spend a lot of time looking for what they need. This will push content and service providers to make their services more customer focused. In the future, many applications that are now static will become wireless. For instance, a company may want an up-to-the-minute sales update. All sales representatives may have access to an application that provides them with a quick way of doing this.

Data ownership.   This is really no different from what we have experienced with Web development projects. At some point, it became practical and cost effective to integrate existing systems and information. This is no small challenge and brings up many issues about who owns what data. The solution is different for every project.

Data movement.   In the first stages of wireless development, customers are not always connected. As a result, the user behavior is similar to what was experienced on the Web: Customers dial in, find what they are looking for, and leave. This will change as we move into the 2.5G and 3G environments. Customers will be online as long as their phones are on and, with some technologies, can also always be found. This will change not only user behavior, but also the behavior of information providers. The information flow will be much more constant, though perhaps it will appear in smaller chunks, and will result in systems seeing much higher concurrent user volumes.

This does not affect the structure of the logical architecture, but it will affect the scaling requirements of the solution. A truly wireless architecture must consider the increased storage and network requirements.

Logical Information Systems Architecture

The logical architecture defined several logical application layers:

- **Presentation layer.**   The presentation layer is responsible for presenting the user-requested data in the correct format for the correct device. This layer should be the only layer that has any knowledge of the environment that the user has.

- **Application layer.**   The application layer houses all the business logic and system application components. It should not know what device the user is using to look at the result of a query.

■ **Data layer.** The data layer houses all information that is specific to the system you are building. It has no knowledge of business logic or the device the user is requesting the information from.

■ **Personalization layer.** The personalization layer is responsible for housing all business logic to apply personalization parameters, such as preferred language, home location, address, and preferred screen layout.

■ **Integration layer.** The integration layer is responsible for integrating external parties into the system. For instance, if an ERP system has been integrated into the system, this layer to the data layer may replicate data. Alternatively, this layer may have the transaction logic to allow the system to form synchronous requests to the ERP system. There are several ways to integrate and the case studies presented later in this book will illustrate some of these.

These layers are no different in the wireless world.

Figure 5.7 shows several points that must be carefully considered when building a wireless system.

Channel integration. In most applications, wireless will be one way to access the system. For instance, if I want to look at the sales data for my company, I would want to be able to do it through my wireless handheld device, my phone, the Web, and maybe through an interactive voice response (IVR) system. Wireless forces the issue of multichannel applications by its popularity and ubiquity.

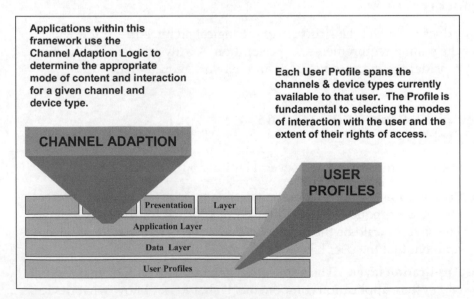

Figure 5.7 Wireless application architecture concerns.

Personalization.   As previously discussed, personalization will now encompass determining the device and interface of the user and adjusting the content accordingly. I don't want to have to tell my sales application that I am now using my phone to access information—I want it to know and to give me the result I desire. This takes personalization to a new level.

Back-end integration.   Web development went through several stages. First, companies wanted to distribute information. Then companies wanted to interact with the users. Now everyone wants users and systems to interact. If you look at this from an integration perspective, you can see that this evolution shows an increasing level of systems integration. Wireless development will go through the same stages and will force companies to further integrate their back ends. Another point to note is that many companies want to integrate their wireless applications with their existing Web applications. These have suddenly become legacy systems!

Where will the logical components reside for the four business areas on which we focus in this book (B2B wireless exchange, CRM, financial services, and the mobile office)? In the green-field scenario, where we have a perfect architecture, all the business logic components will reside in the application layer, the presentation logic will reside in the presentation layer, and the all systems associated with these applications will be integrated in the application layer. Unfortunately, green-field scenarios are rare. The following sections will describe several examples of real applications that have been modified to have a wireless channel.

Logical Technology Infrastructure Architecture

Wireless does involve some new technology components. In the early stages of wireless development, much emphasis has been placed on the network technology that will deliver high-speed wireless access to anything, from your handheld device to your refrigerator. When looking at the systems side of a wireless application and actually providing these services, we see many of the usual components. However, the technological emphasis has once again changed.

We have already written about the always-on aspects of wireless technology and the resulting higher concurrency of users. Let's talk about this in more detail. The largest Web-based systems of today can support maybe 100,000 concurrent users. If we look at the projected number of mobile phones in 2004 by market analysts, we see projections of hundreds of millions of users. Presumably, many of these users will have 3G phones in 2004 that will always be on. Let's say that 1 percent of these users choose the ACME portal as their entry point to the wireless Internet and that these are now always online. This implies that we now have 2.5 million concurrent users. Given that we have not yet included any other access devices into this calculation, you can easily see where this is leading: We will see orders-of-magnitude more concurrent usage than before. Our technology infrastructure will have to scale to this. More specifically, some of these issues are as follows:

Component deployment. You may be forced to deploy components across multiple locations.

Network bandwidth. Even low usage requires a huge amount of bandwidth for this many concurrent users. This not only includes Internet connectivity, but also internal bandwidth.

Legacy system integration. Chances are that many legacy systems in production today will have problems handling the volumes of tomorrow. Not only will the technology infrastructure of the new components you are deploying be bigger, but you will have to take a close look at the legacy systems you are integrating to make sure that they can handle the volumes you are targeting.

Despite this, the situation may not be as difficult as it seems. Though we may have a large number of concurrent users, many of these will not be active. An interesting question to resolve is how to handle the users that are not active in the systems we build.

Most sites today don't have to worry too much about having implementations that span multiple sites or the level of disaster recovery to keep this kind of a site running. Future sites may well have to deal with these issues, however. When building a new site or adding a wireless channel to an existing one, the volumetrics you are looking at should be clearly understood because this will impact the architecture and the cost of the site tremendously.

Logical Security Architecture

Customers must have trust in your site to use it. This is especially true if they are conducting transactions with you. Referring to the evolution of the Web once again, many users, to date, don't like conducting their financial transactions over the Web. Wireless transactions will pose similar challenges for companies trying to develop transactional sites, and security solutions must be good to ease the fears.

Security solutions can be broken down into the following three areas:

1. *System security.* The security of the system that handles transactions and distributes information. This is what companies are building.

2. *Transfer security.* This is the security of the information traveling between the system and the device from which you are accessing your information.

3. *Device security.* This is the security of the device you are accessing the information from.

This book is primarily concerned with the first two areas. Figure 5.8 shows an example infrastructure solution for protecting your system.

Please note the following features:

- Firewalls protect the interface to any external and internal network.

- All applications reside in demilitarized zone (DMZ) 2, while all legacy systems reside in the corporate network.

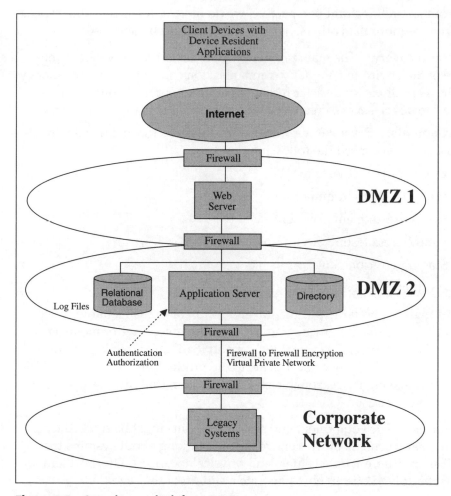

Figure 5.8   Generic security infrastructure.

- Firewall-to-firewall encryption is used between all systems in the DMZs and the legacy environment.

- The application server contains security components that handle authentication, authorization, integrity, and reputability.

- The applications log all activity on the system and provide system administrators a way to track this activity.

In all, the security solution is a combination of application components and technical infrastructure components. Security will be described in further detail in Chapter 6.

Logical Governance Architecture

The same concept as before applies to governance: Though not radically different from an architectural perspective, we do have to focus on some aspects of governance more than others. Several topics should be discussed:

Disaster recovery.   The main problem with disaster recovery here is the size of the environment. Many companies don't want to spend the money to have a full-blown disaster recovery environment for a solution of this size. However, several creative solutions are available to cut the cost.

Environments.   Several environments are required to run any site. In most cases, you need the following:

- Development environment

- Integration environment

- System testing environment

- Load/stress testing environment

- Staging environment

- Production environment

We spoke mainly about the production environment in the technology section, but the issue of configuring and maintaining these environments is no small one. For instance, the system testing and load/stress testing environments should be relatively close in scale to the production environment. How is this accomplished at a reasonable cost for a system that must support 2.5 million concurrent users?

Versioning.   The component and system versions must be tracked carefully in any system. However, we are now talking about systems that will span different data centers and physical locations. This problem must be solved effectively to operate a system of this size.

Subsequent chapters will illustrate the solutions to some of these issues.

Summary of Logical Architecture

Upon completion of the logical architecture, you should have a pretty good idea about how to implement wireless services. The next step is to select products and to specify exactly how to do it. We will not spend much time in this book describing the physical architecture aspects. This chapter does not cover physical architecture at all since we are just talking about a general case, but the case studies included in this book will at least touch on it. The case studies will also cover the logical architecture in much more detail.

Summary

As you can see, wireless business ideas and their implementation will have a wide-ranging impact on our existing architecture thinking. Though nothing is fundamentally different, the focus of our efforts will shift to pay more attention to several key areas:

- Profiling
- Data ownership
- Data movement
- Channel integration
- Legacy system integration
- Component deployment
- Network bandwidth

The case studies provided in the following parts of this book will illustrate some of these key characteristics in much more detail. Part 2 will cover B2B wireless exchange, with a focus on supply chain automation; Part 3 will cover CRM, with a focus on sales force automation; Part 4 will cover financial services, with a focus on financial transactions; and Part 5 will cover the mobile office.

Security for Mobile Systems

This chapter will describe the security of mobile systems from the user's and the network points of view. Today's mobile phone users, devices, and portals trust the cloak of invisibility provided by the current 2G systems. Yet the impending evolution from second-generation to data-capable, always-on third-generation mobile systems will introduce to the world of mobile communications the threats affecting today's wired Internet devices. From the technical perspective, the new security requirements are similar to the security needs of today's corporate intranet environments.

This chapter will focus on security issues affecting mobile technologies such as GSM, CDMA, GPRS, CDMA2000, WCDMA, WAP, wireless LAN, wireless Web, wireless portal, and IP-based mobile services. The chapter will also address the security issues of air interfaces, transports, networks, applications, protocols, and handheld devices. In addition, we will identify some of the threats confronting the mobile subscriber, including level of protection, charging, privacy, terminal security, and lawful interception. Furthermore, this chapter will study the security infrastructure and feature migration from existing mobile systems to third-generation technologies. Finally, the chapter will evaluate the security challenges associated with mCommerce, as well as evaluating which security platform, authentication, and access management technologies offer the best protection against fraud.

Introduction

The telecommunications industry is ushering in a new era of mobile Internet commerce, commonly referred to as *mCommerce*. This rapidly emerging field includes everything from simple access to enterprise portals to complex pay-for-use services in both B2B and B2C environments. In the early instances of wireless Internet services, security was not a critical concern, as most services were low-risk applications such as access to news, weather reports, or directions, and the user paid for these services, via either a flat rate or per-minute charges. The existing levels of network security were sufficient to mitigate the low-level risks that operators faced in these environments.

However, the mobile evolution is in full swing. Today more than half the enterprise workforce is mobile. Recent trends indicate that companies buy more laptops than desktops for their employees. Employees now store the bulk of their business communications on personal digital assistants (PDAs). This statistic is troubling because most companies do not have policies in place to govern the use of such devices, have no control of the information contained on them, and for the most part don't know the potential impact to their organization should the information contained in these devices become compromised.

The tools carried by the mobile workforce are no longer restricted to housing address books, email, and short notes. The new devices coming onto the market are now reaching into the networks of many organizations to perform business-to-consumer or consumer-to-consumer (i.e., person-to-person) transactions. Security measures need to be implemented not only to protect the data stored on mobile devices, but also to shield personal information, passwords, access codes, and financial information.

The benefits of mobility are too tantalizing to resist. The mobile workforce is here to stay, and the general sentiment is that it is the employer's responsibility to provide the mobile workforce with the necessary tools and know-how to protect the information employees carry. Yet the corporate world must also make sure that its mobile workers have the tools they need to be productive. And that means fast, open access to the corporate network, freedom from the restrictions typically imposed by time and distance, access to the latest technology, and availability of technical support to use the technology quickly and efficiently.

Fast, open access to the corporate network is not as simple as it sounds. With today's presentation files, video and audio clips, and email attachments, fast access implies big pipes. Broadband access, through digital subscriber lines (DSLs) or cable modems, is becoming an accepted way of accessing the corporate network from a tethered device, usually a desktop PC. And the assumption is that such access is relatively safe, since the user must authenticate to the network before gaining access.

In addition, freedom from restrictions implies the ability—and the permission—to store, manipulate, and transmit sensitive corporate information while outside the network or the physical plant. This generally means that employees will have easy Internet access, including the ability to roam outside the normal zones, and carry that Internet access with them wherever they may be. But once the information has traveled outside the firewall, it is vulnerable. Any employee who deals with such data becomes a potential weak link in the security chain. To empower employees, organizations must arm them with the knowledge, the procedures, and the tools that enable them to safely deal with sensitive information while enjoying the advantages of mobility.

Judging recent advances in cell phone sophistication, WAP-enabled phones will soon become a critical component of the executive's toolkit. Organizations cannot risk continuing to treat these mobile devices as nonchalantly as many have for a long time. While empowering the workforce, a company needs to ensure that it has carefully deliberated the role such devices will play in dealing with confidential information, and ensure that policies and procedures are built around their use.

Empowerment means that your employees do not need to become experts in technology. If they have to worry about updating virus signature files or remembering to run update programs, or if they have their work interrupted by seemingly senseless tasks, they will find ways to avoid doing such tasks.

However, as more and more sophisticated services and applications are being rolled out, security will become a more pressing concern. While network operators may not have an incentive to increase the level of security, content providers and enterprises, which are experiencing greater risks, will certainly desire enhanced security measures. This chapter will discuss some of the issues surrounding security and explore prospective solutions to those problems.

Background

Mobile commerce, or mCommerce, is conducted over highly integrated and heterogeneous networks, including private intranets or the public Internet. Today's wireless devices support a variety of interfaces—HTML, CHTML, HDML, WML, VOXML, and others. Worldwide, there are numerous interest groups, committees, Web sites, and companies overseeing the development of wireless commerce and in particular the protocols used for this service. The point we are trying to make is that the industry is advancing and developing at breakneck speeds, as are security applications, policies, and procedures.

Although the adoption of wireless commerce is showing an acceleration similar to the penetration of its counterpart and predecessor, wireline eCommerce, these technologies are inherently different. In essence, mCommerce is eCommerce on steroids, not bound to a physical location, available from anywhere at any time. And although mobile commerce is performed on devices that typically have relatively low communication speeds and limited computational resources, these technology limitations will soon be overcome, further adding to the medium's renegade status. It is these differences that provide an ideal hunting ground in which to establish stringent security measures to protect the medium. However, simply identifying these opportunities is unfortunately not enough.

Whereas the security practices of eCommerce have been well established and many standards have been adopted, de facto and otherwise, mCommerce security has yet to catch up.

Obviously, the greatest concern at the device level is the threat of viruses or unauthorized programs downloaded to the end user's equipment. In the following section, we will discuss the specific components required of a secure mobile commerce system.

Requirements of a Secure mCommerce System

To provide secure transactions in any medium, there are four requirements that must be met:

- **Confidentiality**
- **Authentication**
- **Integrity**
- **Nonrepudiation**

Security as a Service

Some of the best practices today for delivering a rock-solid system are to provide everyone in the organization with security as a service. It does not matter whether these managed security services are delivered by the internal IT department or an external, third-party service provider. The end user should never have to worry about security, wonder how it works or whether it is working, or be required to take extra steps to ensure that policies are being followed.

Security services may comprise a collection of various products, including firewalls, virtual private networks (VPNs), antivirus software, authentication tokens, and most probably some consulting services built around the products. Currently there is a growing demand for managed extranet services with related certificate authority (CA) and public-key infrastructure (PKI) services.

In addition to offering a comprehensive portfolio of services, a third-party provider can also cut infrastructure costs, much in the same way that corporate customers are using VPNs to lower their data communications bills. This approach works especially well for operators who are challengers in a particular market; these organizations can start to serve their customers using VPN links over the local Internet, and build the expensive, dedicated overlay network when, and only when, a business case can be established for bringing the technology in-house.

For end users, security as a service draws a picture of security being automatic, invisible, reliable, always on, and up to date. *Automatic* means the corporate system will handle all security requirements for end users and will do so in accordance with corporate policies. End users do not have to download the latest antivirus software, scan their files, or encrypt them each and every time; the system does that for them automatically. Further, the system will work even when end users are accessing corporate data through portable devices while on the road or from a hotel room.

Invisible means the service is transparent to end users. They will not know the system is implementing security policies, scanning files, downloading the latest versions of antivirus software, or encrypting files. And it means they won't know the system is working even when they're accessing corporate files through portable devices on the go.

Reliable means just that. End users won't have to deal with corrupt files or worry about losing time or money because hackers have crashed the corporate system, for instance. It means they can access the system off site or while on the road over portable devices and be assured that the system is fully operational at any given moment.

Always on means the security system never sleeps and provides $24 \times 7 \times 365$ protection.

Up to date means the system consistently protects users against the latest viruses or other security threats.

For administrators, security as a service has a slightly different connotation. It entails policy-based management, centralized management of a widely distributed user base, and instant alerts.

Policy-based management is crucial. Administering security on an ad hoc basis is ineffective and potentially opens the system to security breaches. Policy-based management, as the name suggests, lets administrators work out a set of guidelines covering all devices enterprisewide and ensures that the security service is applied uniformly throughout the organization.

Centralized management of a widely distributed user base is the only effective way to implement policy-based management. By centrally locating the administration of all security services, administrators can more effectively supervise the system and its maintenance while preserving scarce resources.

Instant alerts are crucial to corporate security administrators. The faster they can respond to security threats, the less time there is to damage the corporate system, and the greater the chance of tracking and catching the perpetrators.

Current mCommerce Security Practices

The Wireless Application Protocol (WAP) represents an important step forward in the wireless revolution. It is an open, global standard for communication between digital mobile handsets and devices and the Internet or other value-added service. WAP-based technology enables the design of advanced, interactive, and real-time mobile services, such as mobile banking and electronic commerce. The WAP specification enables solutions from various suppliers to work consistently for end users on the digital networks. More than 500 members have joined the WAP Forum to promote this standard and the accompanying interoperability.

WAP Security Model

The Wireless Application Protocol takes a client/server approach. It incorporates a relatively simple microbrowser into the mobile phone, requiring only limited resources on the wireless device. This makes WAP suitable for thin clients and early smart phones, as the intelligence is placed in the WAP gateways. Microbrowser-based services and applications reside temporarily on servers, not permanently on the phone. Thus, WAP is aimed at turning a mass-market mobile phone into a network-based smart phone. As a representative on the board of the WAP Forum commented, "The philosophy behind Wireless Application Protocol's approach is to utilize as few resources as possible on the handheld device and compensate for the constraints of the device by enriching the functionality of the network."

The WAP is a layered protocol stack that contains an application protocol, a session protocol, a transaction protocol, a security protocol, and a datagram protocol. This stack isolates the application from the bearer when used as a transport service. WAP enables a flexible security infrastructure that focuses on providing connection security between a WAP client and server. WAP can provide end-to-end security between WAP protocol endpoints. Actually, the endpoints for the WAP security layer are the mobile terminal and the WAP gateway. When the WAP gateway makes the request to the origin server, it will use the Secure Sockets Layer (SSL) below HTTP to secure the request. This means that the data is decrypted and again encrypted at the WAP gateway. If a browser and origin server desire end-to-end security, they must communicate directly using the WAP protocols.

The complete secure connection between the client and the service can be achieved in two different ways. First, the safest method for the service provider is to place a WAP gateway in its own network. Then the whole connection between the client and the server can be trusted because the decryption will not take place until the transmission has reached the service provider's network rather than the mobile operator's network. When placing the WAP gateway outside the mobile operator's network, a remote access server is needed. The origin server could be created to include the functionality of the WAP gateway. This provides the highest security solution available.

The service and content providers can also trust the mobile operators' gateway and use virtual private networks to connect their servers to the WAP gateway. But then they do not have the possibility of managing and controlling the parameters used by the Wireless Transport Layer Security (WTLS) at the WAP gateway.

WAP-compliant phones use the in-built microbrowser to:

- Make a request in Wireless Markup Language (WML), a language derived from HTML especially for wireless network characteristics.

- This request is passed to a WAP gateway that then retrieves the information from an Internet server either in standard HTML format or preferably directly prepared for wireless terminals using WML. If the content being retrieved is in HTML format, a filter in the WAP gateway may try to translate it into WML. A WML scripting language is available to format data such as calendar entries and electronic business cards for direct incorporation into the client device.

- The requested information is then sent from the WAP gateway to the WAP client, using whatever mobile network bearer service is available and most appropriate.

The negotiating parties are able to decide the security features that they want to utilize during the connection. According to the security requirements, the applications enable and disable the WTLS features. For instance, privacy may be left out if the network already provides this service at a lower layer. The connection between two terminals can also be secured by the WTLS.

WAP is envisioned as a comprehensive and scalable protocol designed for use with any mobile phone from those with a one-line display to a smart phone or any existing or planned wireless service such as the Short Message Service (SMS), Circuit Switched Data (CSD), Unstructured Supplementary Services Data (USSD), and General Packet Radio Service (GPRS).

Indeed, the importance of WAP can be found in the fact that it provides an evolutionary path for application developers and network operators to offer their services on different network types, bearers, and terminal capabilities.

The design of the WAP standard separates the application elements from the bearer being used. This helps in the migration of some applications from SMS or CSD to GPRS, for example. WAP supports the following:

- Any mobile network standard such as Code Division Multiple Access (CDMA), Global System for Mobiles (GSM), or Universal Mobile Telephone System (UMTS). WAP has been designed to work with all cellular standards and is supported by major worldwide wireless leaders.

- Multiple input terminals such as keypads, keyboards, touch screens, and styluses.

WAP has a layered architecture as shown in Table 6.1. Additionally, Figure 6.1 illustrates how WAP fits into a satellite-based wireless infrastructure.

Table 6.1 WAP's Layered Architecture

Wireless Application Environment (WAE)
Wireless Session Protocol (WSP)
Wireless Transaction Protocol (WTP)
Wireless Transport Layer Security (WTLS)
Wireless Datagram Protocol (WDP)
Bearers, e.g., data, SMS, USSD

The Devices and the Technology

There are presently a number of different types of networks in operation that support the wireless Internet. These networks reflect several different technologies: ARDIS from Motorola, MOBITEX, Circuit Switched Data (CSD) for digital cellular systems, Cellular Digital Packet Data (CDPD), and General Packet Radio Service (GPRS).

Although the network operator has a whole host of security concerns, ranging from unauthorized network access to the use of services without a subscription, for mCommerce the real concern is authentication. Authentication is the starting point for any mCommerce transaction and should be successfully executed prior to allowing access to wireless Internet applications or services.

ARDIS and MOBITEX are both two-way paging mechanisms. Although two-way paging does account for a portion of wireless data service, it currently is not a significant part of mCommerce and does not pose the same risks as purchasing goods and services or managing personal finances.

In a CSD model, the authentication is mostly the same as that which exists for voice communications. The strength and efficacy of this authentication vary across technologies (from 56 to 64 bit), but is generally not as sturdy as the standard for eCommerce transactions (128 bit). Also, the authentication performed by the network entails only the verification of the device—it does not ensure the identity of the actual user. This suggests that for enterprise data, or mCommerce transactions, a second layer of authentication is required prior to allowing access to sensitive resources.

In the CDPD systems, the authentication processes are executed by the underlying cellular network, either via an analog (Advanced Mobile Phone Service, or AMPS) or digital (D-AMPS) system. In the case of AMPS, we are dealing with a very low strength authentication. In the case of D-AMPS, the Cellular Authentication and Voice Encryption Algorithm (CAVE) is employed, which utilizes a 64-bit cipher. In either case, it will still be necessary to augment the network authentication techniques to truly secure mCommerce activities.

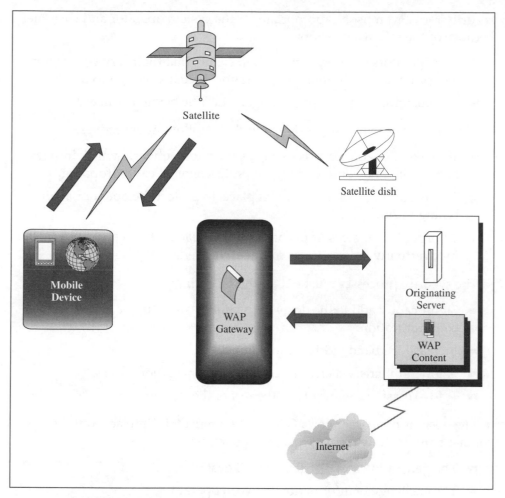

Figure 6.1   WAP environment.

As next-generation cellular systems emerge, GPRS will be one of the first architectures deployed. These new architectures will likely involve an enhanced authentication, authorization, and accounting (AAA) element that should include greatly improved authentication measures. However, as mentioned before, although AAA elements will likely provide the operator protection against unauthorized terminal access, they will still require reinforcement for commerce transactions and private data access.

Global System for Mobile Communications (GSM)

Existing cellular systems have a number of potential weaknesses, and those weaknesses have been considered in the security requirements for GSM. The security for GSM must be appropriate for the system operator and customer because of the following factors:

- The operators of the system wish to ensure that their services are not compromised, including the generation of customer invoices.
- The customer requires privacy against traffic being overheard.

The countermeasures are designed with the following objectives:

- To make the radio path as secure as the fixed network, which implies anonymity and confidentiality to protect against eavesdropping.
- To have strong authentication in place, to protect the operator against billing fraud.
- To prevent operators from compromising each other's security, whether inadvertently or because of competitive pressures.

The security processes must not do the following:

- Significantly add to the delay of the initial call setup or subsequent communication.
- Increase the bandwidth of the channel.
- Allow for increased error rates or error propagation.
- Add excessive complexity to the rest of the system.

The designs of an operator's GSM system must take into account the environment and include secure procedures, such as:

- The generation and distribution of keys
- The exchange of information between operators
- The ability to ensure confidentiality of the algorithms

The security services provided by GSM are as follows:

Anonymity.   To prevent users from easily being identified when using the system.

Authentication.   To allow operators to know who is using the system for billing purposes.

Signaling protection.   To protect sensitive information such as telephone numbers on the signaling channel.

User data protection.   To protect users who are passing data over the radio path.

Anonymity is provided by using temporary identifiers. When a user first switches on a cell phone, his or her real identity is used. Immediately, a temporary identifier is issued. From then on, the temporary identifier takes over. Only by tracking the user is it possible to determine the temporary identity being used. *Authentication* is used to identify the user (or holder of a smart card) to the network operator. Authentication uses a technique usually referred to as *challenge and response.* The process is illustrated in Figure 6.2. First, a random signal, or *challenge,* is issued by the network and sent to the mobile device. Upon receipt, the device encrypts the challenge using the random number R as the input (plaintext) to the encryption. Next, the device uses a secret key unique to the mobile, Ki, transforms this into a *signed response* (SRES) (ciphertext), and sends it back to the network.

The operator, having access to the key, can now verify that the response to the challenge is correct.

Eavesdropping on the radio channel reveals no useful information, as the next time the device is turned on, a new random challenge will be used.

User data protection and *signaling protection* continue where authentication left off. The response generated by the mobile device is passed through an algorithm (A8) by both the mobile device and the network to obtain the key Kc used for encrypting the signaling and messages to provide privacy (A5 series algorithms).

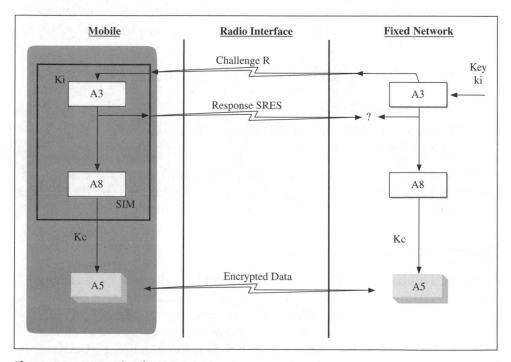

Figure 6.2   Encryption for GSM.

Implementation and Roaming

The authentication algorithm *A3* is an operator option, and it is implemented within the smart card, known as the *subscriber interface module* (SIM). GSM allows triplets of challenges (*R*), responses (SRES), and communication keys (*Kc*) to be sent between operators over the connecting networks, so that the operators may interwork without revealing the authentication algorithms and mobile keys (*Ki*) to each other.

The A5 series algorithms are contained within the mobile equipment, as they have to be sufficiently fast. There are two defined algorithms used in GSM known as *A5/1* and *A5/2*. The enhanced Phase 1 specifications developed by the European Telecommunications Standard Institute (ETSI) allow for interworking between mobiles containing A5/1, A5/2, and unencrypted networks. These algorithms can all be built using a few thousand transistors, and they usually occupy only a small area of a chip within the mobile device.

Worldwide Use of the Algorithms

As you've seen, there are three levels of security provided by GSM: unencrypted, using the *A5/1* algorithm, or using the *A5/2* algorithm to secure the data. The reason for three standards is that GSM was designed for Western Europe, and export regulations did not allow the use of the original technology outside Europe. Using algorithms in the network operator's infrastructure is regulated by the GSM Memorandum of Understanding (MoU), described as follows:

- The present A5/1 algorithm can only be used by the 43 countries that are members of the European Conference of Postal and Telecommunications Administrations (CEPT).

- The algorithm A5/2 is intended for operators in countries that do not fall into the above category.

Export controls on mobile devices are minimal, and the next generation of mobile devices will support A5/1, A5/2, and no encryption. The protocols to support the various forms of A5 (up to seven) are available in GSM.

Protocol Installed Base Issues

Compared with the installed base of SMS-compliant phones, the relative number of handsets supporting WAP is tiny. None of the existing GSM bearers for WAP, including the SMS, USSD, and CSD, are optimized for WAP.

USSD is a means of transmitting information or instructions over a GSM network. USSD has some similarities with SMS since both use the GSM network's signaling path. Unlike SMS, however, USSD is not a store-and-forward service. It is session-oriented in such a way that when a user accesses a USSD service, a session is established and the radio connection stays open until the user, application, or time-out releases it. This has more in common with CSD than SMS. USSD text messages can be up to 182 characters in length.

USSD has some advantages and disadvantages as a tool for deploying services on mobile networks:

- Turnaround response times for interactive applications are shorter for USSD than for SMS because of the session-based feature of USSD, and because it is *not* a store-and-forward service. According to Nokia, USSD can be up to seven times faster than SMS to carry out the same two-way transaction.

- Users do not need to access any particular phone menu to access services with USSD. Instead, they can enter the USSD command direct from the initial mobile phone screen.

- Because USSD commands are routed back to the mobile network's home location register (HLR), services based on USSD work just as well and in exactly the same way when users are roaming.

- USSD works on all existing GSM mobile phones.

- Both the SIM Application Toolkit and the Wireless Application Protocol support USSD.

- USSD Stage 2 has been incorporated into the GSM standard. Whereas USSD was previously a one-way technology useful for administrative purposes such as service access, Stage 2 is more advanced and interactive. By sending in a USSD2 command, the user can receive an information services menu. As such, USSD Stage 2 provides WAP-like features on *existing* phones.

- USSD strings are typically complicated for the user to remember, involving the use of the * and # characters to denote a string's start and finish. However, USSD strings for regularly used services can be stored in the phonebook, reducing the need to remember and reenter them. As such, USSD could be an ideal technology for WAP on GSM networks.

Code Division Multiple Access (CDMA)

The most important use of spread-spectrum techniques in the commercial world is in multiuser communications. Spreading the signal of multiple users with a unique spreading waveform assigned to each user can allow simultaneous access to a shared communication channel. This technique is called Code Division Multiple Access (CDMA) and forms the basis of the IS-95-A standard.

The IS-95-A design provides superior multiple-access capabilities, in addition to displaying a number of other desirable attributes for wireless cellular service. Some of these attributes include optimum subscriber station power management, universal frequency reuse, soft handoff, and the enablement of the use of optimum receiver structures for time-varying multipath fading channels.

The IS-95-A standard covers both analog and digital signaling formats, providing specifications for both modes of operation. The analog portion of the specification describes how the system must operate in order to be compatible with the existing Advanced Mobile Phone Service (AMPS) equipment.

The digital IS-95-A standard employs *direct-sequence* (DS) spectrum spreading by multiplying the user's narrowband waveform by a wideband signal. This wideband signal is generated by spreading *codes* consisting of sequences of 64 *chips*, associated with each symbol interval. The entire sequence of chips is used to modulate the carrier during each symbol period, resulting in a widened or *spread* spectrum. The underlying information sequence is mapped into the chip sequence in different ways in the forward vis-à-vis the reverse channel. In both cases, the characteristics of the spread-spectrum signal provide some important advantages:

- The spreading sequences are chosen so that multiple subscribers can access the channel at the same time over the same frequency band. The spreading sequences are chosen so that signals corresponding to other subscribers are exactly or substantially uncorrelated with a given subscriber, after appropriate processing at the receiver.

- The autocorrelation functions associated with wideband spread-spectrum signals are much narrower than those associated with the underlying information-bearing signal, thus enabling much finer delay resolution for multipath signals. Since the sets of spreading waveforms employed are independent of the data, accurate gleaning and combining of such multiple signals is possible at the receiver.

- CDMA systems can operate with a much lower carrier-to-interference (C/I) ratio, allowing significant immunity to various types of interfering signals when operating at a given carrier power.

When considered purely as a multiple-access technique over Additive White Gaussian Noise (AWGN) channels, CDMA will actually support fewer users within the same bandwidth as Time Division Multiple Access (TDMA) or Frequency Division Multiple Access (FDMA) techniques. However, certain aspects of the mobile cellular environment allow CDMA to provide vastly increased efficiency over the other two techniques:

- Universal frequency reuse is possible, since users occupy a common spectrum.

- Transmitter energy scattered over multiple paths can be consumed more efficiently at the receiver. The spread-spectrum waveform enables the constructive combining of multipath signal components, as described previously.

- Soft handoff between cells is possible, allowing more efficient use of base station power. This also allows continuous coverage, with the potential elimination of dropped calls between cells.

- Since interference of all types appears noise-like at the base station receiver, it is possible to perform accurate interference power estimation and closed-loop subscriber power control. This allows subscribers to transmit at minimal power levels, minimizing their interference power on other subscribers as seen at the base station. This effectively eliminates the common near-far problem traditionally associated with DS spread-spectrum systems.

In addition to CDMA spread-spectrum modulation techniques, the IS-95-A standard employs other sophisticated signal processing techniques. Powerful convolutional coding techniques are employed, in order to embed controlled redundancy into the transmitted symbol stream. This redundancy is exploited at the receiver to allow accurate data bit estimates to be made, based on observations of very noisy received symbols. Channel symbol interleaving is also employed, allowing the occurrence of low-reliability symbols to be randomized at the decoder input. This allows the effective use of the large class of random error-control codes.

Wideband Code Division Multiple Access (WCDMA)

Wideband Code Division Multiple Access (WCDMA) is a refinement of CDMA technology that could raise data transmission rates up to 2 Mbps. WCDMA spreads the chips of the wireless signal over a much wider band of frequencies than CDMA does. This is the air interface technology selected by the major Japanese mobile communications operators, and in January 1998 by ETSI, for wideband wireless access to support third-generation services. This technology will permit very high-speed multimedia services such as full-motion video, rich content Internet access, and videoconferencing.

Vendor Solutions That Provide Components of a Secure mCommerce System

Most of the players in the wireless security market are the established Internet security vendors, including VeriSign (www.verisign.com), Entrust (www.entrust.com), Baltimore (www.baltimore.com), Certicom (www.certicom.com), Diversinet (www.dvnet.com), and iD2 (www.id2tech.com).

VeriSign and Entrust, both leaders in wireline Internet security, came a little late into the mobile market. Companies such as Baltimore and Certicom were the first to implement WTLS 1.1 with their WAP security toolkits. VeriSign and Entrust joined the WAP bandwagon from the implementation of WTLS 1.2 onward. On the other hand, vendors including Diversinet focus primarily on providing security solutions almost exclusively for wireless and mobile commerce applications. Diversinet's solution does not follow the conventional approach of deploying certification authority (CA) and some other features of PKI infrastructure. Instead, Diversinet's product uniquely provides digital permits that can carry application-specific information such as rights and privileges to authorize users.

Various vendors are at different stages of PKI implementation. The following section examines the major vendors' offerings (see Figure 6.3).

VeriSign

VeriSign is a complete solution provider, offering a full suite of applications. The company delivers a secure mCommerce environment via the following services and applications:

- WTLS server/gateway certificate service
- Short-lived certificate service for WTLS server/gateway

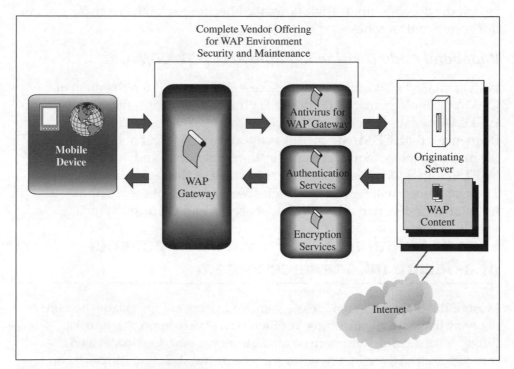

Figure 6.3 Basic vendor offering.

- VeriSign-managed PKI services for wireless clients
- Wireless platform PKI enablement toolkit
- Go Secure! packaged services for wireless eCommerce

The VeriSign server/gateway certificate service is available in two formats: X.509 and the WTLS mini-certificate. VeriSign employs both Rural Statistical Technologies (RSA) and Elliptic Curve Cryptography (ECC) technologies for certificate cryptography.

PKI services for clients include digitally signed transactions and authentication. The wireless client contains a private key and digital signing logic, but does not need to store or process a certificate for the corresponding public key. The certificate can be stored in a directory or other form of repository within the wired infrastructure, and can be picked up and used by application servers whenever a user's digital signature needs to be verified. With this approach, there is no concern about the size of the certificate. Therefore, standard X.509-format certificates, as are used in the wired Internet world, can also be used for client key pairs in the wireless environment. The advantage of this approach is that servers can use precisely the same functions to verify digital signatures from wireless clients as they use for verifying digital signatures from wired Internet clients. Wireless client certificates will be issued through VeriSign-managed certificate issuance services or related VeriSign service provider products, and delivered to repositories for storage and use in the wired infrastructure. Certificates may also be delivered to and stored on client devices or SIM cards. VeriSign is a leading provider of Internet trust services and was the first to provide Wireless Short-Lived Server Certificate service as part of a planned wireless application trial. The Motorola WAP Server Solution is an open standards-based solution and was the first WAP server on the market to support short-lived certificates.

Baltimore

Baltimore has made its Telepathy WST suite fully compatible with its PKI product suite and UniCERT's architecture. However, UniCERT has only been extended to produce WAP digital certificates (WTLS certificates). Currently, Baltimore does not release information about the client certificates. Other products for PKI implementation include the following.

Baltimore Telepathy WAP Security Gateway (*WSG*) provides an end-to-end security solution for content providers who need complete unbroken security from the mobile phone all the way to their Web/WAP servers. The offering provides WTLS authentication, confidentiality, and integrity as a stand-alone application independent of the WAP gateway/server. It can be configured to redirect WAP data traffic to any WAP server or gateway.

Baltimore UniCERT has been extended to produce WAP digital certificates (WTLS certificates). *Baltimore Telepathy WAP Certificate Authority (WCA)* extends the UniCERT plug-in modules already available to now include wireless and wired world users.

The *Baltimore Telepathy PKI Registration System (PRS)* provides mobile device users with their digital identity in a wireless network. It provides the technology to bind mobile device users to their digital certificates stored in PKIs. It enables users to authenticate themselves and participate in mobile commerce. This system can be used for both WAP and GSM/SIM phones.

The *Baltimore Telepathy PKI Validation System (PVS)* is a unique digital certificate retrieval and validation system, with minimal bandwidth and storage requirements, specifically designed for the constrained wireless environment. The system retrieves and validates using certificate identifiers rather than the complete digital certificate, allowing the user access to multiple certificates without the need to store them locally.

The powerful *Baltimore Telepathy Digital Signature Toolkit (DST)* developer toolkit allows content providers to build systems to process wireless digital signatures. It is a tool for anyone creating a mobile commerce solution that requires digital signatures to be used for authentication, confidentiality, integrity, and nonrepudiation. WAP 1.2 specifies digital signatures in WML. The toolkit implements that specification and allows content providers to receive signatures sent from a mobile device and verify their validity using a PKI.

Certicom

Certicom implements PKI through its MobileTrust product suite. MobileTrust tools, products, and services are designed to provide an integrated ECC-based PKI. MobileTrust-managed certificate services consist of the following tools and products.

The *MobileTrust Certificate Authority (CA)* provides certificate services to both servers and mobile clients. The CA is hosted in a highly secure and reliable facility to ensure guaranteed service. It uses ECC and provides X.509v3 certificates. The system uses certificate revocation lists (CRLs) for revocation.

The *MobileTrust Registration Authority (RA)* platform enables an enterprise or service operator to administer certificates for large deployments. The RA subsystem interfaces with principals requesting certificates, with sources and processes authorized for approving certificate requests, and with designated certificate authorities for final request policy processing, certificate signing, and publishing. The RA calls on the MobileTrust CA to issue certificates. It has the following features:

- 100 percent java-based application.

- Interfaces with supporting database and directory components through standard Java Database Connectivity (JDBC) and Java Naming and Directory interface (JNCI) APIs.

- Flexible policy support defined in Java Beans Graphical and menu-driven administrative consoles to configure operational policy and modify or approve certificate requests.

MobileTrust PKI client software performs key generation, certificate requests, client certificate management, and trusted root management on mobile devices. Original equipment manufacturer (OEM) developers get a complete and consistent interface for integrating MobileTrust PKI services into their applications.

Future Development

The WAP model does not support certificate revocation as a result of decisions reducing the size of the certificate. WAP compensates by keeping the certificate validity period short, which requires that the certificate must be renewed or refreshed frequently. A short-lived certificate service is an excellent solution for certificate revocation. In the event a server is compromised or decommissioned, users cannot unwittingly continue to execute what appear to be valid, secured transactions. WAP server/gateway certificates are relied upon and processed by the mobile client devices, which do not have the local resources or the communication bandwidth to implement revocation methods used in the wired world such as certificate revocation lists (CRLs) or the online certificate status protocol (OCSP). With the short-lived certification approach, a server or gateway is authenticated once in a long-term credentials period—typically one year—with the expectation that the one server/gateway key pair will be used throughout that period. However, instead of issuing a one-year-validity certificate, the certification authority issues a new short-lived certificate for the public key, with a lifetime of, say, 25 hours, every day throughout that year. The server or gateway picks up its short-lived certificate daily and uses that certificate for client sessions established that day. If the certification authority wishes to revoke the server or gateway (e.g., due to compromise of its private key), it simply ceases issuing further short-lived certificates. Clients will no longer be presented with a currently valid certificate, and so will cease to consider the server authenticated.

The SIM Toolkit is an ETSI standard for value-added services and eCommerce transactions over GSM phones. The SIM Toolkit is programmed into the special GSM SIM card, which enables the SIM card to drive the GSM handset interface, builds up an interactive exchange between a network application and the end user, and accesses or controls access to the network. For the first time the SIM card has a proactive role in the handset. This means that the SIM initiates commands independently of the handset and the network.

Bluetooth

Bluetooth is a short-range wireless technology that connects electronic devices, including cell phones, printers, digital cameras, and handheld computers. Bluetooth is designed to exchange data at speeds up to 720 Kbps and at ranges up to 10 meters.

The Bluetooth protocols are intended for rapidly developing applications using the Bluetooth technology. The lower layers of the Bluetooth protocol stack are designed to provide a flexible base for further protocol development. Other protocols, such as RFCOMM, were adopted from existing ones that have been modified slightly for the purposes of Bluetooth. The upper-layer protocols are used without modifications. In this way, existing applications may be reused to work with the Bluetooth technology, ensuring interoperability.

The technology of Bluetooth has significant positive and negative implications. Mobile users will have the ability to access the Internet and service providers by their proximity alone. Interfaces will extend beyond that of human–computer to include autonomous device-to-device communication. Merely walking by a particular cluster of appliances could have your Bluetooth device probed for information that its owner may not want to disclose. Similarly, walking near a vendor or advertiser will open a connection to your device, which these parties could use to push desired or unsolicited information your way. You might say that the problem could be eliminated were the user required to enter a personal identification number (PIN) to authorize the exchange. Yet this solution poses a challenge that compromises the system's core concept and functionality; it would be immensely irritating for a user to repeatedly have to enter the PIN, using one or possibly multiple Bluetooth-enabled devices. Further, the user may already be out of range before even getting the opportunity to enter the code.

Still, Bluetooth poses a considerable security risk, as users can be completely blind to what is happening to their mobile devices as they travel from place to place. Stealth viruses can exploit this new medium and behave in the same manner as an airborne virus, jumping from device to device. In the past, users infected their computers with viruses by sharing information through the physical medium of a floppy disk or by downloading an infected file from the Internet. With Bluetooth, you just need to walk by someone. Picking up a mobile device virus from a stranger may become easier than contracting the common cold.

Although the industry is not fully developed to provide the utopian electronic community where consumers or businesspeople roam freely and share data, users will quickly embrace the technology for its merits of freedom. The implications to mCommerce transactions using this medium are staggering. If successful technologies such as Bluetooth would significantly propel and enhance all forms of mCommerce, especially in the consumer-to-consumer market. Accordingly, security risks will increase at the same rate and must be met with strict resolve, antivirus technology, and personal security policies.

In addition to catching a mobile virus, Bluetooth technology allows users to be tracked by their PINs. Mapping a PIN to a user's identity opens the door for a whole new list of potential violations against personal freedom. Tracking the whereabouts of a user can provide a corporation with insightful information about consumer habits. Similarly, the technology opens channels for government to keep tabs on the citizenry. The threat of such information falling into the wrong hands and being misused increases manyfold over previous threats associated with conventional Internet access devices. While security practices are under development, manufacturers focus on educating the users of their products in an effort to create awareness and to take precautions to protect their information and to keep it secret.

Wireless LANs

Wireless connections to workstations may be utilized wherever high mobility of computing services is required in and around the premises of an organization. As you would expect, leveraging wireless local area networks brings with it a set of security issues. For example, outside users may gain access to the network by purchasing a wireless network interface card (NIC) and situating themselves within proximity of an access unit.

To address this threat, the new series of wireless LAN products combine mobility and flexibility that users need with the throughput and security that system managers demand. Whether it's a cell phone, a PDA, or an 802.11b LAN, security can be a major sticking point for any user thinking of moving to wireless technology.

3Com, a large corporation that manufactures network components, aims to change that. The company now offers a new product that provides 802.11b, 11-Mbps wireless LAN users with VPN-like data encryption and security. 3Com's Superstack II Router 400—scheduled for release in the Fall of 2001—will provide secure, encrypted network access in a wireless LAN, supporting up to 256 wireless clients per router.

Security will always be a problem when you have signals traveling through the air, as Figure 6.4 illustrates. Most of the products available for wireless LAN security have involved using the Layer 2 access control method and Layer 2 data encrypting. However, this approach really manages the access control problem at the wrong layer. The protocols at this level are designed for configuring and tracking devices—and not the users. In an

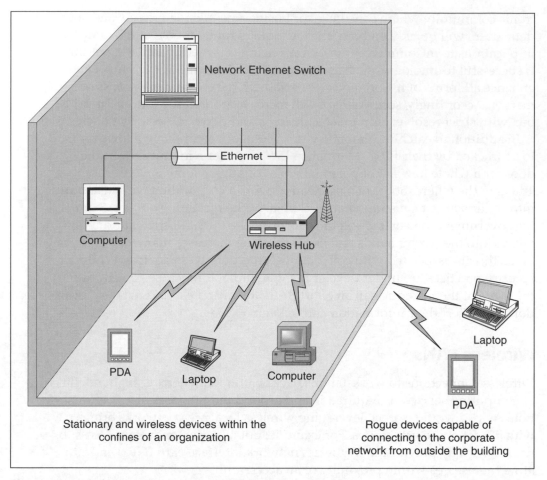

Network Ethernet Switch

Ethernet

Computer

Wireless Hub

PDA

Laptop

Computer

Laptop

PDA

Stationary and wireless devices within the
confines of an organization

Rogue devices capable of
connecting to the corporate
network from outside the building

Figure 6.4 Wireless LAN intrusion possibilities.

unprotected wireless LAN environment, an intruder need only position him-
or herself outside the building at a location where they can intercept a signal
and access your network at their leisure.

Wired Equivalent Privacy (WEP)

The 802.11 standard describes the communication that occurs in wireless
local area networks. The Wired Equivalent Privacy (WEP) algorithm is
intended to protect wireless communication from eavesdropping. A secondary function of WEP is to prevent unauthorized access to a wireless network; this function is not an explicit goal in the 802.11 standard, but it is
frequently considered to be a feature of WEP.

WEP relies on a secret key that is shared between a mobile station and an
access point/base station. The secret key is used to encrypt packets before
they are transmitted, and an integrity check is used to ensure that packets are

not modified in transit. The standard does not define how the shared key is established. In practice, most installations use a single key that is shared between all mobile stations and access points. More sophisticated key management techniques can be used to help defend from their decryption.

Cisco Systems, Microsoft, and other organizations use a combination of the Extensible Authentication Protocol (EAP, an extension to Remote Access Dial-In User Service, or RADIUS) that can enable wireless client adapters to communicate with RADIUS servers and IEEE 802.1X, a proposed standard for controlled port access.

Authentication is based on username and password, and each user gets a unique, session-based encryption key. When the security solution depicted in Figure 6.5 is in place, a wireless client that associates with an access point is refused access to the network until the user completes a network logon. When the user enters a username and password into a network logon dialog box or its equivalent, the client and a RADIUS server (or other authentication server) perform a mutual authentication, with the client authenticated by the supplied username and password. The RADIUS server and client then derive a client-specific WEP key to be used by the client for the current logon session. All sensitive information, such as the password, is protected from passive monitoring and other methods of intercept. Nothing is transmitted over the air in the clear.

The sequence of events is as follows:

1. A wireless client associates with an access point.

2. The access point blocks all attempts by the client to gain access to network resources until the client logs on to the network.

3. The user on the client supplies a username and password in a network logon dialog box or its equivalent.

4. Using 802.1X and EAP, the wireless client and a RADIUS server on the wired LAN perform a mutual authentication through the access point. One of several authentication methods or types can be used. With the Cisco authentication type, the RADIUS server sends an authentication challenge to the client. The client uses a one-way hash of the user-supplied password to fashion a response to the challenge and sends that response to the RADIUS server. Using information from its user database, the RADIUS server creates its own response and compares that to the response from the client. Once the RADIUS server authenticates the client, the process repeats in reverse, enabling the client to authenticate the RADIUS server.

5. When mutual authentication is successfully completed, the RADIUS server and the client determine a WEP key that is distinct to the client and provides the client with the appropriate level of network access, thereby approximating the level of security inherent in a wired switched segment to the individual desktop. The client loads this key and prepares to use it for the logon session.

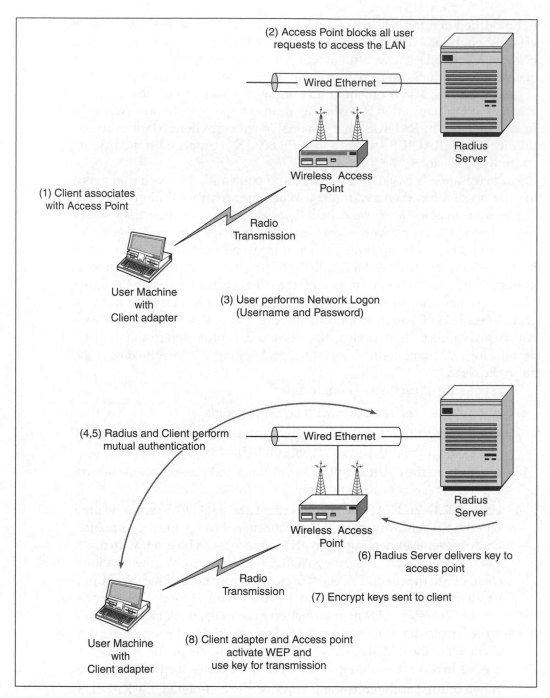

(2) Access Point blocks all user requests to access the LAN

Wired Ethernet

Radius Server

Wireless Access Point

(1) Client associates with Access Point

Radio Transmission

User Machine with Client adapter

(3) User performs Network Logon (Username and Password)

(4,5) Radius and Client perform mutual authentication

Wired Ethernet

Radius Server

Wireless Access Point

(6) Radius Server delivers key to access point

Radio Transmission

(7) Encrypt keys sent to client

User Machine with Client adapter

(8) Client adapter and Access point activate WEP and use key for transmission

Figure 6.5 Cisco security solution.

6. The RADIUS server sends the WEP key, called a *session key*, over the wired LAN to the access point.

7. The access point encrypts its broadcast key with the session key and sends the encrypted key to the client, which uses the session key to decrypt it.

8. The client and access point activate WEP and use the session and broadcast WEP keys for all communications during the remainder of the session.

Standard Keys and IP-Based Mobile Services

Leading wireless service providers and equipment suppliers at the 2000 GSM World Congress in Cannes, France, announced the formation of an industry group aimed at speeding up the adoption of open, mobile Internet standards. The new industry group, called the Mobile Wireless Internet Forum (MWIF), boasts among its 45 founding members many of the world's leading computer, networking, and telecom companies, including Alcatel SA, Cisco, Compaq, Fujitsu, IBM, L.M. Ericsson, Motorola, Nokia, Samsung Electronics, Sprint PCS, and Vodafone AirTouch PLC.

MWIF was formed to provide a venue for service providers and equipment suppliers to jointly identify and resolve issues surrounding development of key specifications, and to enable early implementation of wireless networks based on IP, according to a statement issued by the founding members. Such specifications will be aimed at creating common infrastructure and services, as well as enabling seamless integration between mobile telephony and new IP-based services, including data, voice, video, and multimedia applications, according to the statement.

Many of the new offerings are aimed at bridging the gap between current second-generation GSM networks and future, fully integrated third-generation technologies. The first 3G networks are scheduled to launch late in 2001, with Japan's NTT Mobile Communications Network first out of the gate.

Although still in its infancy, packet-based communications, including IP telephony, could make up as much as half of the traffic on the world's advanced wireless networks by 2004, predicted a leading mobile device manufacturer detailing the company's new concept for IP-based radio access networks.

The Solution

There is a strong perception by many organizations that computer- and network-security products are magic prevention technologies. Firewalls are perceived to prevent unauthorized traffic from entering an internal network; authentication mechanisms are perceived to prevent unauthorized people from logging onto computers; encryption products are perceived to prevent unauthorized people from reading files or other information.

However, security is not a product; security is a process. Prevention technologies are good, but prevention is only one part of security. Effective security management must be an integrated set of tools and associated processes to police valuable corporate resources. Business practices surrounding mCommerce are in their infancy compared to previous methods of conducting electronic business. Companies should develop security methodologies that become second nature to their employees, being incorporated into daily activities. Proprietary information no longer resides on stationary computers behind the protected walls of a corporation. Instead, the information now travels with employees everywhere they go.

There are two dimensions that need to be considered when discussing an integrated security system: (1) distribution, and (2) management (see Figure 6.6). *Distribution* defines how far and how widely security is managed. A common approach in the past was to secure the perimeter. The alternative was to rely on individual products to secure each device—often depending on the end user to install and run the products. Given the increasing proliferation of wireless devices and their ability to store information, it is imperative that corporate security policies be implemented to the fullest extent and that the security measures' distribution extends to every device at multiple levels throughout the organization.

Management defines how security is controlled and implemented—in a fragmented way, managing every device separately, or in a centralized fashion, using policy-based management.

As mentioned earlier in this chapter, security must be based on corporate policies, and these policies must be enforced on every remote and local device at all times. Policies are set once and then distributed everywhere. To accomplish this, we need a centralized management scheme, but with the added complexity of wide distribution. Not only must the policies be formulated and communicated, they must be enforced, automatically and transparently. Furthermore, we need to ensure that the various software and hardware components used to enforce these policies are aware of one another and work well together. Consider encryption and malicious code detection software. If encryption and antivirus products were not aware of each other, the antivirus product might neglect to inspect encrypted files and leave the malicious code undetected.

As you can see, the various aspects of data security need to be integrated into one functional package. This is an extremely complex, difficult, and costly process. Many companies have adopted enterprise network management frameworks such as Tivoli TME, Computer Associates CA-Unicenter, or

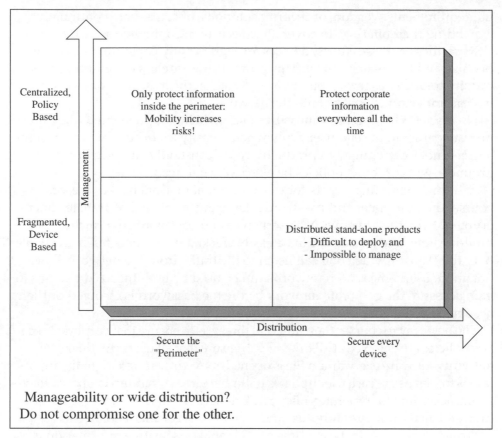

Figure 6.6   Management versus distribution.

HP OpenView as the management platform of choice. Integration of the security solution requires integration with these frameworks to leverage the company's previous investments and to provide an integrated, cohesive view of the network state.

Summary

mCommerce must develop a security system that clients trust to provide the same level of comfort as obtained in a face-to-face transaction. At this early stage of development regarding mCommerce security offerings, it is unlikely that a one-size-fits-all solution will emerge any time soon. The merchants and enterprises considering mobile Internet-based services or applications will need to leverage a slew of tools to build a solution that matches their individual requirements. Vendors of security solutions, of course, are continually expanding their offerings to cover all aspects of mCommerce security.

Regardless of these efforts, devices are significantly constrained by memory size and processing power. This prompts the wireless development community to utilize as few resources as possible on the handheld device and instead move functionality onto the network.

No matter where security measures end up being implemented, they must be transparent to the end user. Only when security is effectively hidden from a user—not interfering with productivity yet constantly running in the background—will corporate policies be effectively enforced.

Extremely tempting targets for the malevolent internal hacker are, of course, the usernames and passwords that many applications transfer unencrypted. Even the Windows NT usernames and passwords that include some built-in protective measures can be easily cracked using tools freely available over the Web, provided one can listen to the traffic from the network. Once obtained, usernames and passwords can be used by unauthorized persons to gain access to the corporate network before the password is changed or the account is disabled.

The best way to counter these internal threats is to ensure that corporate security policies are enforced at all times. *First*, ensure that all users run their antivirus software every time a file is opened, moved, or stored. Equally important is to make sure each user updates his or her virus signature database often—again, according to corporate policy. Since new viruses and malicious code are invented all the time, an antivirus product is only as good as it is current.

Second, make sure that each time confidential or sensitive information is transmitted across the corporate LAN, it is encrypted. Don't assume that, simply because a person works for your company, he or she should be trusted with company secrets. Because no access control system is foolproof, employees without a need to know could easily intercept sensitive information in transit.

Third, encrypt data whenever it is stored. The fact is that most corporate users should encrypt most of their data most of the time. Encrypting files considerably improves security in Windows NT 4.0 and Windows 95/98 environments. The encryption and decryption process shouldn't require any additional steps by the user, but the process needs to be invoked on the fly whenever a document is opened, saved, or moved after the valid user is authenticated.

Keep in mind that many applications also store open documents, backup documents, and other potentially sensitive data files as temporary files. Some of these files can remain on the hard disk for a long time. On-the-fly encryption ensures that the temporary files are also protected and nobody else can gain access to them. In addition, should the computer accidentally run out of power, the data remains encrypted on the hard drive, as only the unencrypted copy is stored in the computer's memory.

Fourth, protect sensitive information from department to department. It is often relatively easy to determine which departments within your organization should *not* have access to information from other departments. Examples include human resources (to which almost no one should have access) and finance (to which only certain individuals should have access). By implementing a distributed firewall between departments, enabling or disabling access can be made a straightforward process.

PART

Two

Architecting a Wireless Exchange

An Overview of Wireless Exchanges

The first vertical channel for which we will create an architecture that entails a wireless component is an exchange. We will explore the different varieties of exchanges and how wireless technology can be used in each. With this overview as a background, we will explore one exchange in great detail: a B2B exchange in the metals industry called the MetX Exchange. We will look at the business case of the exchange as it exists today, see how wireless fits into the overall exchange architecture, and determine the impact of adding a wireless dimension. The case study will be covered in subsequent chapters in Part 2; this chapter will provide the context. So let's start by defining a wireless exchange.

What Is a Wireless Exchange?

To answer this question, we will first look at exchanges in general. For thousands of years, people have congregated to trade, or exchange, goods and services, either for money or for other goods and services. Exchanges have three constituents:

1. Supply-side exchangers (sellers in a commercial exchange).
2. Demand-side exchangers (buyers in a commercial exchange).

3. Service providers (those that facilitate the interactions of exchangers and exchangees).

The exchange process was basically unchanged from ancient times until the mid-1800s, when mass production radically altered how companies were able to bring goods to market. The traditional channels through which trade was conducted began to evolve, and whole new trading processes emerged. Markets moved beyond the local scope, encompassing larger and larger geographic areas.

Technology has continued to affect exchanges just as dramatically in modern times, often extending what can be traded. Information, for example, has now become a commodity to be exchanged just like time-share rental properties. Technology has also given rise to the connected society, in which trade can be conducted electronically. The central meeting place, no longer a physical venue, but a virtual *marketplace*, provides participants access from any part of the connected world. Just as mass production opened up the supply side of the exchange in the middle of the nineteenth century, the connected economy has opened up the demand side in the late twentieth century.

Now the concept of an electronic exchange is being taken one step farther, providing access via a wireless channel where participants are no longer tethered to a workstation. Imagine lining up a birdie putt on the fifteenth green, when your cell phone chirps, giving you a message that the concert tickets your spouse has been asking you to get for months are now available on your favorite ticket exchange Web site. Not only are they available, but they are also within a price range you chose. You quickly put in a buy order before they are taken—long before most other exchange participants even know they exist. Sound too good to be true? Sound like nothing but market hype? Well, the technology has arrived, and that's one of the goals of this book: creating road maps to possible futures.

Different Varieties of Wireless Exchanges

The drivers for rendering an exchange wireless are the same drivers for making it electronic: to facilitate the meeting of geographically dispersed members in the virtual market space. Members use an exchange when they derive value from it, and most electronic exchanges can increase that value by extending their infrastructure to include a wireless dimension. There are several kinds of exchanges:

- Consumer-to-consumer (C2C)
- Business-to-consumer (B2C)

- Business-to-employee (B2E)
- Government-to-citizen (G2C)
- Business-to-business (B2B)

The creation of each kind of exchange has its own set of difficulties, and each provides unique benefits. Let's take a look at the advantages and challenges particular to each variety of exchange.

C2C Exchanges

Internet connectivity has allowed consumer-to-consumer (C2C) exchanges to flourish on a global scale. In the past it was a challenge to exchange goods and services, even when transactions were conducted locally. The virtual exchange provides a means by which private individuals can act as either the *exchanger* or the *exchangee*, and the set of exchangeable commodities is only limited by the imagination. For example, one of the Web's most popular C2C exchange sites, eBay.com, conducts an auction-style consumer exchange. It has traded everything from a celebrity's half-eaten breakfast to *Star Trek* collectibles.

As has been mentioned, information can be exchanged as a commodity, just like goods and services. For many C2C exchanges, information exchange is free. Take SomeWhereNear.com, for example, which provides information on things to do in London. Suppose you are out one night in the city and suddenly have the urge to eat Thai food (it could happen). You can pull out your WAP-enabled phone, go to SomeWhereNear.com, and not only learn about the existence of several Thai restaurants in the vicinity, but also obtain evaluations of the quality and quantity of food, the price, the ambiance, and the overall experience as expressed by patrons who had previously dined at the establishment and then submitted their reviews to the site. Assuming you trust the opinions of others, the service can enhance your dining experience, and the wireless aspect gives you the freedom to be impulsive yet informed.

When designing and implementing wireless C2C exchanges, the primary concern is scalability. As mentioned earlier in this book, the penetration of wireless devices is about to explode. As a result, the potential number of concurrent users will explode as well. Our ticket buyer on the golf course used wireless access as a competitive advantage; this will certainly be true, but only on a tactical level. In the future, more and more users of the exchange will learn about and want the wireless benefit. Wireless access will cease to differentiate your use from everyone else's. The architect for the exchange has to take this into consideration, because soon everyone who wants those concert tickets will be accessing the system concurrently. Assuming the offering is a very popular one, there could be a staggering number of concurrent users trying to secure admission, and a strategy must be in place to handle the spikes.

B2C Exchanges

Historically one of the most prevalent kinds of exchanges, business-to-consumer (B2C) exchanges link businesses with their customers, and vice versa. Our previous example of buying concert tickets between putts is also an example of a B2C wireless exchange. Think about what would be required to make this example of contact on the golf course possible: a prearranged agreement between the consumer and a business with the capacity to delivering the ticket offer. The consumer would have to identify a set of commodities he or she deemed desirable, the price limit he or she would be willing to pay for them, and a means by which the consumer could be notified of the commodities' availability. The business would have to provide a mechanism for capturing that consumer's information, an intelligent agent that watches the exchange for the desired commodities to become available, and the infrastructure to reach the consumer when the specified parameters are met.

Any customer-facing business can use the customer relationship management (CRM) aspect of a B2C exchange to enhance its offerings. For example, a gas station can use new data mining techniques to identify high-volume consumers who use their brand infrequently. They can then target that sector of the market, polling its members to determine how the gas station could modify its value proposition to win them as regular customers. The station may offer incentives, such as a point system for use of the service, enabling high-volume customers to earn rewards for purchasing the particular brand. This *value exchange* means that the company invests more in a particular segment of the market, and as a result obtains a return on investment (ROI) that is much higher as a result of the increased sales volume. It is a win-win exchange, because the business is able to increase revenue, while consumers are rewarded for doing something they have to do anyway.

Adding a wireless dimension can further enhance the value offered in this scenario. Since the gas station has identified me as a person who prefers convenience over cost, it can provide me with an offering that suits my tastes. For example, the station could provide me with a Bluetooth-enabled device that identifies me as I pull in. The clerk at the full-service pump knows to fill up my tank with premium gas and charge it to my Visa card, so I don't even have to roll down my window. This scenario illustrates the ultimate in convenience. In exchange for my loyalty as a customer, I receive the two things I value most: convenience and points redeemable for discounted or free merchandise for every dollar I spend. All enabled, facilitated, and fortified by a wireless service offering.

B2E Exchanges

Business-to-employee (B2E) exchanges are used by companies to allocate internal resources. For example, Cap Gemini Ernst & Young uses a B2E exchange to meet its very dynamic teaming requirements to staff engagements. A project manager identifies his or her staffing needs, and then posts them on the exchange. Consultants who currently are not on an assignment or are soon rolling off a project will go to that exchange and specify parameters, such as geography, skill-set requirements, and employment level, to identify a need for which the employees are best suited.

Working smoothly in the traditional environment, this functionality could be enhanced even further with a wireless offering. Suppose a client is having performance problems with their Web application. CGE&Y is brought in to assess and resolve the issue. The consultancy applies a set of metrics and determines that the bottleneck is being caused by the company's SQL server database. As time is of the essence, the bottleneck has to be removed by Friday, yet the determination is made on Wednesday afternoon. To make matters worse, nobody on the current assessment team is an SQL server expert and thus capable of fixing the problem on the spot. What is needed immediately is a mechanism to identify and reach out to the consulting firm's SQL server experts. With the wireless B2E exchange in place, the request can be posted onto the database as an ASAP requirement. An intelligent software—based agent receives the request, searches the employee skill-set database for those in the firm identified as SQL server experts, cross-checks those employees with the employee utilization database to see who is available, and then cross-checks that set to see who is within a reasonable geographic proximity to the client. The intelligent agent now has its set of employees that match the urgent need.

Now comes the interesting part: how to contact the set of candidates in a way that will minimize response time. The intelligent agent knows the preferred contact method for each identified consultant, which, given the mobile nature of today's workforce, will most likely be wireless. Reaching the consultant could be something as simple as sending a message to a pager or to a more functionally rich device, such as a wireless-enabled PDA. The messages are sent to the available SQL server experts, and a consultant immediately joins the team and saves the day. The client is happy, because their problem was solved. The SQL server expert is happy, because he or she was staffed on a project that leveraged his or her particular skill set. The project manager is happy, because the client issue was resolved, and the team was assembled in a rapid, cost-effective manner. All this is possible with the integration of a wireless channel into the corporate infrastructure.

G2C Exchanges

Another interesting exchange enabled by technology is a government-to-citizen (G2C) exchange. The purpose of such an exchange could range from providing citizens with a simple discussion forum to enhanced governmental services. An example of the latter is an alert system conducted by the park service. It could be configurable by the users—park visitors—to provide as much or as little interaction as they see fit.

The system would work as follows: Visitors to a national park would be issued handheld wireless devices, and they then travel to whatever parts of the park they wish to visit. Their handheld devices would be connected from any park location, and park officials could use them to notify the visitors of a dangerous situation, such as a forest fire. Such a device would be invaluable for coordinating the evacuation of visitors and the placement of rescue workers.

The same wireless device could be used by park visitors to notify park officials about dangerous situations, such as rabid animals or forest fires. But it could do more, if the visitors wish. It could become a virtual tour guide, with built-in location-based services. Suppose you take your family to a wirelessly enabled national park. You stand before a beautiful rock formation, and your wireless device/tour guide describes the composition of the igneous rock outcropping currently being admired. It could tell you how the formation was created and what significant historical events took place right where you are standing.

The device could also be used to enable chats among visitors. There could be public access discussions, such as identifying the best spot to view gardenias today, or person-to-person chats to coordinate a rendezvous for dinner (you know how mad parents get when you're late for dinner). The wireless enablement would allow the freedom to roam with the security that help and/or immediate contact with your party is literally in the palm of your hand.

But if it talked too much, or if you just wanted a map with an always-accurate "you are here" marker, you could set the device to provide only those services you desire. There would be an easy way to set the level of device interaction, ranging from "notify me only if I'm in immediate danger" to "tell me everything that ever happened everywhere in this park," as suits the users' tastes. The flexibility would allow the wireless park to always enhance the visitor's experience.

While these applications provide real value to all parties involved, one can expect to encounter a Luddite-like resistance to such an intrusion of technology into the escape from modernity that the parks provide. There may also be a Big Brother type of fear that the government would misuse or even abuse its newly gained capability of communicating with its citizens anytime, anywhere. These concerns are quite valid and will need to be overcome should such a system be created.

The goal of this example, of course, was not to debate the moral and social issues around government intrusion into private lives, but rather to show how a wireless exchange could improve the ability of park rangers to monitor their lands and enhance their offerings. In the end, it is always in the hands of the providers and users of a technology to ensure a responsible way of leveraging its capabilities.

B2B Exchanges

The last category of exchange we will discuss is the business-to-business (B2B) exchange. All businesses either buy materials from other businesses, sell materials to other businesses, or do both. The challenge is to make the process as cost effective as possible. As a buyer, the goal is to obtain the materials needed as quickly and cheaply as possible. As a seller, the goal is to find the highest price, then deliver the goods as cheaply as possible by maximizing the supply chain efficiency. As the connected society grows and begins to mature, businesses are becoming more comfortable leveraging technology to effectively connect to suppliers and business customers. Members derive value from the electronic B2B exchange by the economies of scale created.

If you recall, in Chapter 5 we identified a set of areas in which today's companies are focusing their wireless efforts. We believe the following subset of those focus areas will give value to wirelessly enabling B2B exchanges and their corresponding supply chains:

- Field service automation
- Sales force automation
- Mobile office
- Logistics
- Telematics

These focus areas will be applied to three parts of a B2B exchange in which we believe wireless can have the most impact:

1. Making real-time demand of goods and services.
2. Making real-time offerings to supply goods and services.
3. Supply chain visibility and improvements.

The case study to be detailed in Chapters 8 through 11 will focus on wireless enhancements used to improve supply chain efficiency and give all the

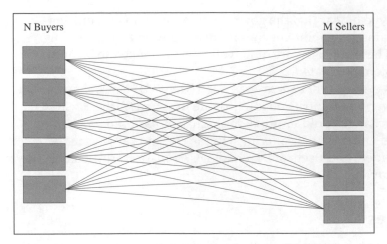

Figure 7.1 Direct connections between buyers and sellers.

stakeholders visibility to the process. We will now focus on using wireless technology to actually buy and sell goods and services in a B2B exchange.

We will begin discussing the MetX Exchange, where metals companies perform contract-based buying and selling. The process is optimized by having one location to find trading partners, as opposed to maintaining individual relationships. Without the MetX exchange, N sellers would have to establish relationships with M buyers, meaning $N \times M$ relationships have to be established and maintained. With the exchange, $N + M$ relationships exist, with each participant having to maintain only one relationship. In reality, fewer relationships would exist, because the set of buyers and the set of sellers on the exchange would not be disjoint. These relationships, and how they can be simplified with the introduction of an exchange, are depicted in Figures 7.1 and 7.2.

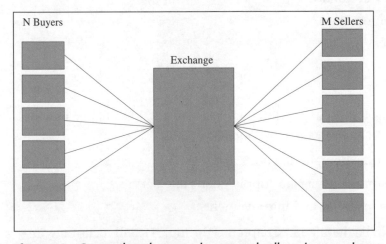

Figure 7.2 Connections between buyers and sellers via an exchange.

Even if no other advantages are obvious, membership reduces the complexity and cost of data integration, because data-level integration has to occur with only one entity: the exchange. Yet there are additional benefits. Now that members have easy access to their trading partners, they will benefit from the increased volume of transactions enabled and facilitated by the system. By taking these advantages to the next level, introducing a wireless dimension, both buyers and sellers derive additional value, as illustrated in the sidebar examples.

FACILITATING INTER-ENTERPRISE TRANSACTIONS

Buy-Side Value

Suppose you are a sales representative for a metals distributor and you visit your best customer, Acme Pens. As you arrive, you notice the mood is tense. Scott McLean, the president of Acme Pens, tells you that, due to a miscalculation by his soon-to-be ex–purchasing manager, he has only enough of the solvent used for making pens to last that day. If more solvent doesn't arrive by that afternoon, his employees will be playing touch football in the parking lot tomorrow. Leaping to his aid, you pull out your wireless device and visit the MetX Exchange, your wirelessly enabled metals exchange Web site. You are able to save the day by finding the solvent and a carrier who will deliver it that afternoon. Needless to say, Mr. McLean is very grateful for your timely assistance.

Sell-Side Value

Suppose you are the owner of a bakery, and you receive an order for 1,000 blueberry muffins. You diligently create the muffins, but when the customer arrives they tell you there has been an error: They really only meant to order 100 muffins. What are you to do with the other 900 muffins, which will only be useful as paperweights by 2 P.M.? You pull out your wireless device and visit BreakfastFoodsExchange.com, your B2B exchange, and post the muffins. It turns out that a convention center two blocks away has a broken oven, so they need food delivered fast. They buy the muffins—and at a higher price than you would have received from the original customer. These two examples demonstrate how the old adage is as true in business as it is in sports: Speed kills. There is competitive advantage from exchange membership, enhanced by using wireless technology.

Summary

This chapter gave a brief overview of exchanges and examples of how wireless could enhance their offerings. The different kinds of exchanges and the issues they raise are illustrated in Figure 7.3.

Exchanges exist to facilitate the offering and acquisition of commodities, and members use an exchange when, and only when, they perceive value in its use. Many exchanges are commercially based, but some exist to provide a forum in which ideas and opinions can be expressed.

Advances in production technologies and methodologies allowed suppliers to offer their commodities on a global scale, and the advent of electronic exchanges in our connected economy now allows both sides to participate globally. Exchanges benefiting from electronic channels almost always enhance their value proposition from the implementation of a wireless dimension, as their members, leveraging the wireless channel, have a competitive advantage over those tethered to a wired infrastructure.

In the subsequent chapters of Part 2 we will explore a specific B2B exchange, the MetX Exchange, to demonstrate the specific value a wireless channel can provide. We will begin by exploring the business case, and we will look at the complexity and volume of data a wireless channel requires, the information systems needed to handle that data, and the technical infra-

	Comments	Primary Exchanger Concern	Primary Exchangee Concern
B2E	Connects businesses and their employees	Up-to-date and accurate data is key to maximize resource utilization	Need to keep personal data accurate and current to be useful
C2C	Connects consumers and other consumers	Scalability – possibly millions of concurrent users	Scalability – possibly millions of concurrent users
G2C	Connects governments and their citizens	Integration with infrastructure is key	Scalability – possibly millions of concurrent users
B2C	Connects businesses and their consumers	Integration with infrastructure is key	Scalability – possibly millions of concurrent users
B2B	Connects businesses and their business partners	Integration with infrastructure is key	Integration with infrastructure is key

Figure 7.3 Different kinds of exchanges and their concerns.

STREAMLINING INTERNAL INFRASTRUCTURE

Buy-Side Value

You work for the Acme Corporation, which has a large and active metals division. In fact, you are the purchasing manager for that metals division, and have been for several years. When you assumed the post, business took place at a much slower pace than today. As competition increased, so did the pace of business. When Acme looked for ways to trim operating expenses, you chose to enroll in the MetX Exchange, the aforementioned metals exchange Web site. At the time, the iterations between purchase requests took four to five days. As competition has increased, pressure is mounting to shorten those iteration times. Your task today is to find 10,000 liters of helium to be used for a new line of Acme balloons (MetX also sells alternate items, such as helium gas). You check the exchange and find several offers for helium. Which one do you want? Can you get it elsewhere cheaper? Who should make that assessment? You quickly realize that the process is more complicated than you originally realized. You need to contact someone on the technical side to determine what grade of helium is used in balloons. You need to contact someone on the financial side of the project to see what the budget is for helium. Having these questions answered in a timely fashion is critical to successfully releasing the product on time, and decreasing the response time is necessary in order to compete successfully.

It could take days to get responses to voice mail or traditional email notes. So what's the answer? Wireless infrastructure. You send an email to the technician in charge of inert gases, who is traveling across the country at the time. At his layover in Chicago, he is able to check his mail on his wireless-enabled PDA, cutting several hours to a day off the response time that would be required if he didn't have wireless access. He responds that you obviously don't know much on the subject, because helium is helium, and helium goes into balloons. Even though the answer was filled with barbs and sarcasm, it was a response, and it was received more quickly than would have been possible without wireless connectivity.

You similarly reach out to the financial guru. Good news and bad news: The good news is that she is at her desk and can give the matter attention immediately. The bad news is that she does not have the budget broken down by individual materials, so she does not know how much money can be allocated for helium. She needs to schedule a meeting with several members of the financial team—something that ordinarily takes days or weeks to accomplish. But with the wirelessly connected enterprise, time is found in everyone's schedule for the next morning, and meeting notifications are sent out. Since everyone has wireless access, they all receive the meeting

(Continues)

(*Continued*)

notification in time to prepare for the conference. At noon the following day, an exact figure for the limit on helium spending is determined and conveyed to you, the purchasing manager.

The wireless infrastructure has given Acme the competitive advantage it has been seeking; coupling that infrastructure with the MetX membership makes Acme's purchasing process as efficient as any in the business. The challenge was met—the same depth of analysis was provided by the purchasing team to determine the relative merits of an offer for goods in a shorter time span. The balloons are manufactured on time and within budget.

Sell-Side Value

On the sell side, the process is essentially the same as on the buy side, only the content differs. Acme must perform an analysis to determine the highest price they can fetch for the goods they have to sell. If an offer to buy our goods was rejected, do we wait for another buyer? Do we restructure our offering to make it more attractive? Can we afford to do so? Who should make that assessment? The same advantages the wireless infrastructure provides on the buy side apply to the sell side as well. The point is to get the most efficient access to the right people to obtain the knowledge and/or decisions necessary to conduct business.

structure that can make it all possible. By analyzing these four aspects, we can develop four distinct but highly interrelated architectures, each useful to a particular constituency responsible for turning the idea of a wireless exchange into a reality. In addition, the following discussion will provide a clear understanding of the costs and risks associated with the endeavor, maximizing the possibility for success.

CHAPTER 8

Conceptual Business Architecture

Introduction

This chapter looks at a case study developed for a wireless metals exchange. The goal is to provide you with insights regarding business-level requirements—in other words, the *what* aspects of the application's architecture design. We will define some of the business components: What services are required of the exchange, and what architecture model best structures the cooperation between these services.

The exchange is designed to connect participating buyers and sellers of metals, letting each member streamline their business operations and increase profitability. Built on an open eCommerce platform, the exchange provides a neutral, industry-endorsed system to deliver process improvements and increases to the bottom line. Service offerings integral to the exchange include enterprise resource planning (ERP) connectivity, catalog management, contract management, order management, and transportation management, as well as logistics and supply chain optimization. It should be noted that mobility plays an important role in all these service offerings— much the way it does throughout the traditional business enterprise. We will look at the functional requirements of a wireless application supporting the

NOTE ON CLIENT NAMES

Due to nondisclosure agreements, client names are kept confidential. For this reason, we have used fictional client names throughout this book. In addition, the subject matter of the exchange has been altered slightly.

metals exchange in the *global visibility* portion of supply chain optimization. Business components that are part of this particular application include the following:

- Creating orders from electronic data interchange (EDI), using templates, or manually
- Using a core carrier or acquiring a sponsor for an open bid
- Scheduling orders and shipments
- Supply chain tracking, that is, gathering source data
- Monitoring and/or managing nonconformance reports
- Ongoing standard and customized management activities and reports, such as vendor management inventory (VMI), shipping documentation, and settlement
- Collaborative feedback loops for decisions, inputs, and actions when necessary

Supply Chain Overview

The metals industry has built a complex global supply chain that links its participants in an effort to streamline the movement of material and information from the supplier to the end customer, with the objectives of maximizing customer service and asset utilization while minimizing total delivered cost and inventory investment. Tightly integrating business processes that entail the making, moving, buying, and selling of products or services can result in gains for all participants in the network. However, the success of the integration effort across the entire chain is only as strong as the chain's weakest link.

One of the areas that continue to challenge today's supply chain participants in the metals industry includes the integration of transaction processing systems. Whereas the streamlining of physical processes, such as the procurement of iron ore and scrap metal, or the shipment of semi-finished goods from an integrated steel plant or a mini-mill to the distributor, has been progressing steadily over recent years, transactions are frequently conducted in a manual fashion. The lack of automation in transaction processing can result in bottleneck-like inefficiencies that prevent the supply chain members from fully reaping the benefits from improving the logistics within their industry.

An electronic exchange promises to be an ideal solution to remedy the negative side effects of a transaction system characterized by redundancies and errors associated with a manual information interchange between buyers and sellers in a complex trading network.

The optimization of logistics and business processes through the exchange network offers the potential of significant savings while retaining the high level of reliability and service that the metals industry expects and delivers. Supply chain optimization eliminates the need for redundant systems, resulting in reduced costs and actually strengthening the relationship between buyers and suppliers. The exchange helps its participants tap into the benefits of a standardized, automated information exchange and provides both parties with the following benefits:

- A seamless single point of connectivity to back-end information systems

- The ability to automatically route and track transactions

- The tools to translate transactions into a readily acceptable format

- The infrastructure to develop and maintain all trading partner information per facility across the globe

- A state-of-the-art, automated product cross-reference table engine

Buyers realize new savings through the following:

- A single point of access to all metal suppliers

- An industry-endorsed network standardized for many-to-many connections

- The most secure, efficient, and cost-effective way to transact the repeat order processing for ongoing contract buying of metals (collapsing the loop cycle to near real time)

- The ability to streamline contract negotiation, approvals, and management

- Enterprise resource planning (ERP) integration to facilitate global inventory reduction between them and their trading partners

- Sharing of demand and production planning information with suppliers

- Transportation optimization through carrier integration

- Lower life cycle connection costs versus single point-to-point connections

- New opportunities to outsource noncore competencies through the exchange network

Sellers realize closer customer relationships and savings through the following:

- Linking their supply chain directly with the supply chains of multiple customers, through one industry connection

- Improved speed, cost, and accuracy of order entry and fulfillment for repeat contract customers

- Use of the exchange as an electronic front end for buyers requesting Internet-based ordering

- The ability to focus on core value propositions such as differentiating product supply, maintaining inventory, and providing customer service

- Lower connection costs through an industry-endorsed system standardized for many-to-many connections

- ERP integration to facilitate global inventory reduction

- Reduced risk of stock-outs and better on-time delivery

- Lowered freight rates resulting from more efficient use of total transportation systems

- Improved ability to deal with logistics crossing country borders

- Closely linked demand and production planning with customers

- Increased reach through the exchange catalog system

World-Class Supply Chain Excellence

Let's take a high-level view of the key features found in the supply chain function of the metals exchange before taking a closer look at the conceptual functions of our wireless global visibility application. Some key characteristics required of the exchange include:

- *Speed* in terms of flexibility, adaptability, and real-time processing.

- *Connectivity* through a standardized technology platform that enables a many-to-many relationship capability.

- *Collaboration* in terms of real-time information sharing between clients, suppliers, and partners in the areas of forecasting, replenishment, vendor management inventory, and transportation. This is accomplished through standardized and aligned inter-company processes and practices.

- *Optimization* in the aforementioned collaborative areas through the use of tools and technologies that store and analyze data, and identify improvements and cost efficiencies leveraged across the exchange participants.

- *Visibility* that provides product availability, reduced inventory levels that remove cost inefficiencies while improving flexibility and customer service, and the global tracking and tracing of shipment and order status.

Global Visibility

Speed, connectivity, collaboration, and optimization are all important factors, yet it is *visibility* that is the most critical element of the supply chain as it is provides insight into the various stages of the process. The Web-based metals exchange is a complex undertaking; not only does it promise to improve the status quo by more closely aligning the network's participants, but it also offers each participant the opportunity to expand its reach. Because the exchange represents a scalable platform, participating suppliers will be able to service additional participating buyers, while participating buyers will be able to forge relationships with additional participating sellers.

Participants in the metals exchange recognize that:

- The majority of information needed to optimize their operations resides outside their enterprise.

- Zero-latency interactions with real-time alert capabilities is required between companies and their trading partners. This is necessary in order to achieve a high-velocity, tightly synchronized business process with the flexibility to accommodate changing conditions and business objectives.

- There is a need for increased agility in meeting more dynamic customer demands. This requires shortening the time-to-plan horizon while lowering safety stock. Without visibility of real-time information on what is happening across the *complete* supply chain, order fulfillment in this environment is difficult and costly.

One of the chief functional needs in the area of global visibility is a system to support dynamic transaction relationships, one that operates at any time and from anywhere and enables *on-the-fly* reconfiguration of supply chain fulfillment activities. In other words, wireless mobility.

Figure 8.1 illustrates the conceptual architectural framework of the exchange's wireless visibility application. Figure 8.1 also shows the inputs provided by various actors and the outputs generated by the logistical operations of the system.

Our wireless global visibility application focuses primarily on the transportation management portion of supply chain operations. The application interfaces with other components for status reports, alerts, and the input of change notifications in both automated and manual ways. These components include:

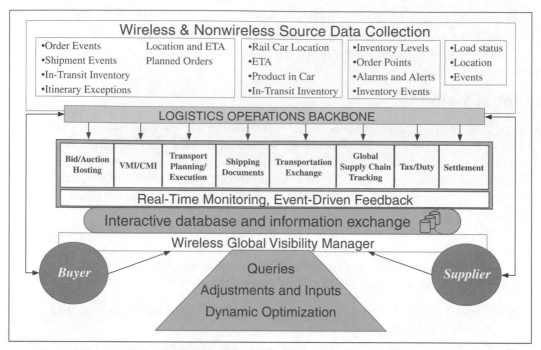

Figure 8.1 Wireless visibility application

- Order creation and bidding
- Vendor management inventory (VMI)
- Transportation planning and execution
- Transportation exchange (carriers and shippers)
- Global supply chain tracking
- Shipping documents
- Tax/duty and settlement

WHAT'S AN ACTOR?

Actors, at least from the architectural viewpoint, are components of the system that interact with each other in some way. Actors can be people, other systems, other business processes, or any other subsystem that is interactive with the system as a whole. This concept is similar to that of the Rational Unified Process approach or other object-oriented methodologies.

The order creation component is a key activity within the supply chain process. Our interest here lies especially in the up-front parameters defined in a negotiated contract as they relate to supply chain deliverables such as quantity, security, costs, and delivery dates. These parameters determine what materials and inventory management orders are placed, and they determine carriers and shipper bids. In addition, we need to look at the order process as a feedback point from the supply chain. An order delivery date might need to be renegotiated, an existing order might be modified, or an order might be canceled altogether and created anew.

Besides the fundamental functionality of order creation/modification and bidding, the exchange offers vendor management inventory (VMI) services. The VMI feature monitors inventory at the customer's site and proactively replenishes the site based on preestablished rules. This allows the buyer to outsource raw material planning and procurement activities at multiple levels up and down the supply chain. Direct results of this strategy include increased inventory turns, the elimination of stock-outs and spot shortages, the reduction of inventory stockpiling, and faster response time in relation to the company's customers and its trading partners, as well as improvements in customer satisfaction.

The Transportation management (planning and execution) system of the exchange, illustrated in Figure 8.2, interfaces with a member's order manage-

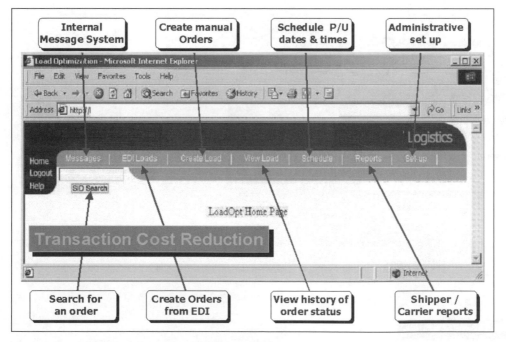

Figure 8.2 Transportation management technology.

ment systems to plan, optimize, and direct its transportation initiatives. This provides cross-company synergies, opening savings opportunities no individual member would be able to exploit on its own, that is, outside of the network. The transportation management system improves the utilization of assets through the identification of consolidation and backhaul opportunities, resulting in reduced operating expenses that can be passed through to customers. In addition, the order fulfillment service offers members of the exchange the benefits of an integrated solution, thus freeing the members from expending the time and effort required to build an application on their own. The service enables increased information flow internally and externally between carriers, customers, and the marketplace. It facilitates what-if scenarios, and it provides flexible freight payment options, including pre- and post-audit, payment, and self-invoicing. Finally, the system allows for enhanced carrier management and control, resulting in improved carrier commitment.

Transportation Exchange

The transportation exchange component encompasses a variety of online trading mechanisms that consolidate and execute in the highly fragmented transportation marketplace (see Figure 8.3). It is a real-time, rules-based, and

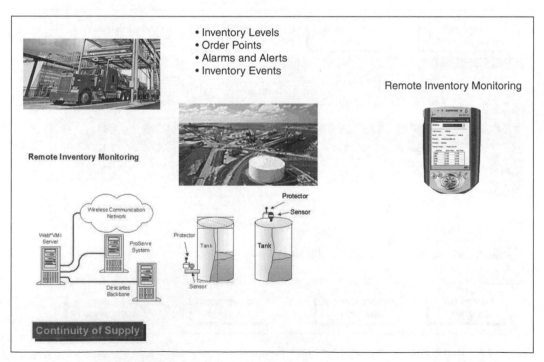

Figure 8.3 Transportation exchange.

automated trading platform that brings together a user-defined community of shippers and carriers into a central marketplace.

This component offers a value-added service that provides a 99-percent placement guarantee for all shipments run through the exchange. The system is able to achieve such a high level of placement by intelligently allocating shipments to a member's private or public contracted carriers, to a shipper's core carrier, or through alternate placement channels that adhere to member-specified requirements. The exchange system supports managing carrier compliance, exception intervention, gathering status updates, and executing shipping orders.

Shipping Documents, Tax/Duty, and Settlement

The online freight settlement service provides the shipper of merchandise and the carrier automated reconciliation of their respective bills of lading and invoices, with payment to carriers for service within three days, minimum, by the exchange partner's bank (see Figure 8.4). The automated and collaborative freight settlement service efficiently settles freight and resolves freight discrepancies, with the bulk of any manual processing being handled by the exchange—thus reducing in-house-related workloads and expenses. The exchange's order fulfillment service offers clients the benefits of an integrated solution, again avoiding the cost and effort associated with creating their own solutions. Typical benefits for the buyer include:

- Seamless interfaces with rating engines and enterprise accounting systems

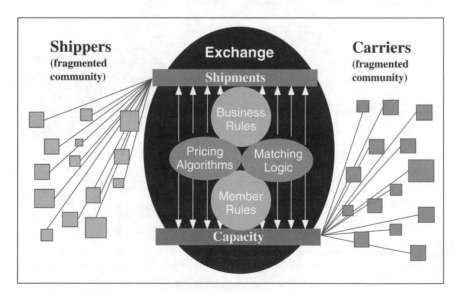

Figure 8.4 Shipping documents: track enabled entities and composition.

- Increased process efficiency
- Fast, accurate payments for better vendor relations
- Detailed monthly statements
- Online tools for quick claims and dispute resolution
- Avoidance of costly, cumbersome post-audits

Benefits for carriers include:

- Increased cash flow and profit margins, with fast, accurate payment
- Payments as quickly as 24 hours after service completion
- Reduced responsibilities for collecting accounts receivable
- Overall cost reductions

Global Supply Chain Tracking

Global supply chain tracking is the capability to track a shipment throughout the order fulfillment process and life cycle of the shipment, including red flag—that is, exception-reporting—capability (see Figure 8.5). It provides visibility across all modes of transportation and all regions, enabling clients to avoid the costs and complexities of connecting to multiple providers. The sys-

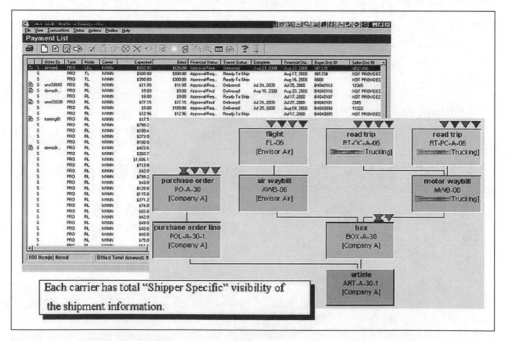

Figure 8.5 Supply chain tracking: managing a track-enabled entity.

tem prepares in advance for shipping exceptions and delays through electronic, real-time alerts, reducing expediting effort. It improves real-time visibility, especially in multimode shipments, and customer service through proactive response exceptions.

Wireless Global Visibility Application

The wireless global visibility application helps users monitor and control their customer fulfillment networks by tracking the order flows through the transportation exchange network and managing information reporting. We are concerned with three primary aspects of the application: source data collection, real-time monitoring and event-driven feedback, and dynamic optimization.

Figure 8.6 illustrates the dimensions of the wireless application.

The following conceptual business functions direct the architecture for this application:

- Enterprise systems and data-capture devices collect data throughout the fulfillment network, ensuring a high level of data quality. This information is then made available via message-based transfer mechanisms to support the users and systems in charge of fulfillment-related processes (source data collection).

- Users must be able to access this data with a thin-client decision support tool, browser, or Java applet to obtain unified views of the supply chain, independent of the original data source (real-time monitoring).

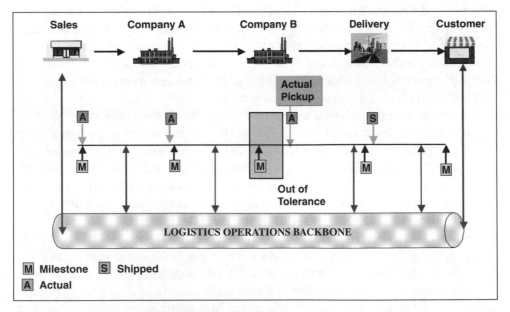

Figure 8.6   Wireless user interface into visibility.

- Alerts are automatically reported to individual users, user groups, and information systems when planning, scheduling, and execution data diverge from defined tolerance levels (event-driven feedback).

- Wireless Internet and intranet VPN functionality is available to access the data and send alerts and notifications to users in specific user groups (real-time monitoring and event-driven feedback).

- Quick responses to new enterprise and inter-enterprise information exchange requirements are enabled via configurable and open software instruments (dynamic optimization).

Our wireless application consists primarily of a thin client or Java applets and a centralized *visibility manager*. The visibility manager offers an unrestrained view of the fulfillment cycle that allows personnel to monitor, control, and report on logistics activities taking place throughout the customer fulfillment network. Specifically, the visibility manager allows users to:

- Monitor the activities taking place in the supply chain.

- Receive proactive alerts and notifications when problems and delays are expected, and retrieve additional information quickly and easily to determine the impact of the disturbance on other orders.

- Quickly investigate the progress of an order, investigate its details, and review related dispatched SKUs, palletized loads, containers, flights, ocean voyages, and road trips.

Users gain from an intuitive functionality that can create personalized reports on demand. Additional functionality allows users to further analyze these reports in common office applications. In addition, reports can be sent via email to support customer and trading partner service exchanges. Links can be created that allow users of the reports to review the source information and to investigate additional details of the business items. Reporting can be generated on an ad hoc or scheduled basis.

Automatic exception-based events and notifications are essential aspects of supply chain visibility (see Figure 8.7). Exceptions and alerts allow users with responsibilities for delivering on commitments to improve their customer service levels. Users are notified as soon as irregularities and predefined notification milestones occur. For example, functionality is available to generate alerts when specific information is not reported in time. Recipients have context-sensitive access to the present status of the reported problems that are embedded in the alert and notification messages themselves.

The *global visibility manager* is used to define personalized criteria that proactively alert users and systems about departures from the plan. It is here that users can also define their own notification criteria and direct alert messages for follow-up action to coworkers, trading partners, or customers.

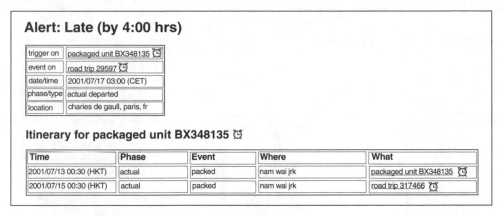

Figure 8.7   Alerts and exception based event monitors.

Global Visibility Clients

Many users require easy access at any time and from anywhere to subsections of supply chain information that relate to them. These users might be members of the finance staff or sales department at the organization, its customers, the shipping company, or the organization's trading partners.

Global visibility clients provide a customized view of logistics activities meeting the specific information needs in customer fulfillment. Client interfaces are user-configurable to provide functionality tailored to support specific tasks (see Figure 8.8).

The following clients are used:

Monitoring.   The most basic interface available to the user, the monitor, offers direct, real-time tracking and tracing of orders and shipments.

Viewing.   The viewer is a Java-based browser that offers authorized users direct, real-time status updates on order and shipment information. Users can customize their browser desktops with sets of predefined searches and search result formats based on their specific information requirements.

Managing.   Here the browser supports the management planning and execution portion of the application. Users are able to provide quick responses to alerts, missed milestones, and new enterprise and inter-enterprise information exchange.

Communicating.   This portion of the client is used to communicate with other collaborators, clients, and suppliers who have joint investment, ownership, or input into various aspects of the visibility process.

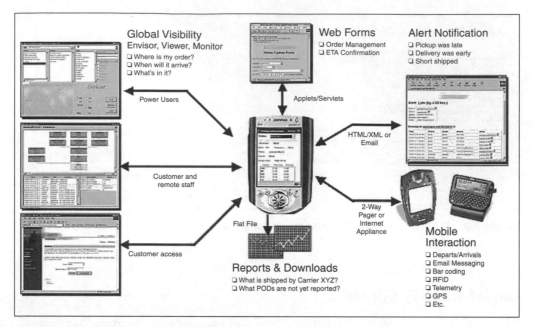

Figure 8.8 Wireless global visibility clients.

Use-Case Example

Let's look at a use-case scenario from the wireless metals exchange to follow the interaction and logic flow between several of the functional components we have discussed. We'll take a remote telemetry reading as the input for the use case.

A seller on the exchange needs to maintain a certain amount of liquid nitrogen in inventory to be able to quickly respond to the organization's customers' dynamic production demands. However, due to several constraints it is not cost effective for this company to maintain large amounts of liquid nitrogen on hand, as tying up resources in inventory beyond what is the absolute minimum to satisfy demand would prevent the organization from competing effectively in its market space.

To allow the organization to keep inventory at minimum levels, a remote monitoring device constantly measures the available volume of gas in real time. This information is relayed over a wireless link to the visibility manager. The company has also set up links into its ERP back-office order management system that controls current production volumes for this particular product based on incoming orders received from customers, the remote sales force, field technicians, and the metal exchange itself. Thus, the visibility manager tracks the current production targets and the amount of liquid nitrogen in the tanks, and projects supply targets based upon current volumes and incoming sales.

When the visibility manager *sees* that current levels will not meet demand expected within the following 24 hours, an order is placed on the exchange with a previously negotiated supplier. If the supplier is not able to fulfill the request, the order is placed on the exchange as a bid. In this illustrative case, the contracted seller was not able to meet the organization's demand. However, there was a major supplier of liquid nitrogen which not only maintained its own fleet, but whose drivers regularly reported positions, delivery confirmation, and deviations from schedule back to the supplier's ERP system. In addition, this supplier had several remote sales representatives in the field who were wirelessly tied into the exchange. The supplier's sales reps monitored the exchange for bids that originated in their geographic location regularly, in other words, throughout the day. When the rep closest to the organization noticed the open bid, he searched his company's deliveries in the area and realized that a partial shipment was en route to the organization's holding tanks. Since this shipment of the liquid nitrogen was already paid for by contractual penalties, the sales rep was able to easily undercut other bidders and win the business. All that remained was to enter the order into the system, and the exchange automatically prepared the required documents and routed them to the appropriate points.

The liquid nitrogen supplier was able to maintain its sellable inventory without interruption to fulfilling demand, while the sales representative at the third-party supplier was able to book an additional sale and develop a relationship with a new customer. All this was made possible by the wirelessly enhanced exchange.

Summary

This section described the conceptual business functions of the application, a case study developed from an engagement to build a large global exchange for the metals industry. We looked at the functional requirements of a wireless application that supports the metals exchange in the global visibility portion of supply chain optimization. One of the major functional requirements in the area of global visibility is a system to support dynamic transaction relationships—a system that enables *on-the-fly* reconfiguration of supply chain fulfillment activities at any time and from anywhere. We were concerned with three primary aspects of the application:

1. Source data collection.
2. Real-time monitoring and event-driven feedback.
3. Dynamic optimization.

We found that our process would have to include:

- Data capture mechanisms that are distributed throughout the fulfill-ment network and that ensure a high level of data quality
- Thin-client decision support tools
- Proactive alert notifications
- Wireless Internet VPN capabilities
- Real-time access to enterprise- and exchange-based information
- Configurable and distributable reporting mechanisms

Our wireless application consisted primarily of a thin client or Java applets and a centralized visibility manager. The customizable visibility manager offered an unrestrained view of the fulfillment cycle, thus providing personnel to monitor, control, and report on logistics activities taking place throughout the customer fulfillment network. Mobile users interface with the manager through a variety of channels or views, provided by remotely enabled visibility clients that provide insight into logistics activities meeting the user's specific information requirements. Client interfaces are user-configurable to provide functionality tailored to support specific tasks.

The wireless global visibility application helps users monitor and control their customer fulfillment networks by tracking the order flows through the transportation exchange network and managing information reporting at *any time* and from *any place*. Results are demonstrated in increased inventory turns, eliminated stock-outs and spot shortages, reduced inventory stockpiling, faster service response time, improved customer service levels, and faster response time when dealing with trading partners.

Now that we have created the business architecture, we are one step closer to our goal of a well-architected solution. In Chapter 9, this business architecture will be used as an input to create the second architecture our framework entails: the data architecture. We will see how the business architecture becomes an input to determine the data needs that our business processes require, and how those needs are viewed from a data perspective. The data architecture will bridge the gap between the business needs and the information systems (applications) used to fulfill them. Let's now look at the data architecture of our case study, the MetX Exchange.

Information Architecture for a Wireless Exchange

Our framework for architecting solutions dictates the derivation of data requirements from the business needs. In this way, we ensure that the business dictates the use of technology, and technology is not wedged into business processes for technology's sake. To make this happen, we create the first of our four architectures, the business architecture, as we did in Chapter 8.

From that starting point, the second architecture created is the data architecture, which we will cover in this chapter. It frames the business needs in terms of the data they require. Once the data architecture is completed, we can then determine which information systems are best suited to meet those data needs. That is called the *information systems* architecture. Finally, we decide upon the technical infrastructure in which those information systems will operate, known as the *technical infrastructure* architecture, our fourth and final architecture. These architectures are distinct, but interrelated, and every one leads us toward our ultimate goal: a flexible system that works today and is positioned to still work tomorrow. Right now, we are ready to discuss the data requirements implicitly specified by the business drivers outlined in Chapter 8.

Overview

The overall goal of our application is to move materials from their source to their destination. The data architecture facilitates our reaching this goal by defining the data to be exchanged between business processes. Figure 9.1 demonstrates a high level view of how the various stakeholders participate in this data communication process. The major benefits of creating a data architecture are to uncover the interdependencies between business processes and to create a clean definition of what information is exchanged. The data architecture will serve as a contract between the business components, creating a level of trust in the services provided. To quickly reiterate, the business processes outlined in Chapter 8 include:

- Bid/auction hosting
- Vendor management inventory (VMI)
- Transport planning and execution
- Shipping documents
- Transportation exchange
- Global supply chain tracking
- Tax/duty and settlement

We also discussed in Chapter 8 that the following three aspects of the wireless application would provide the most value:

1. Source data collection.
2. Real-time monitoring and event-driven feedback.
3. Dynamic supply chain optimization.

In addition to the aforementioned business processes, we also use the MetX information principles as inputs to create the data architecture. To illustrate,

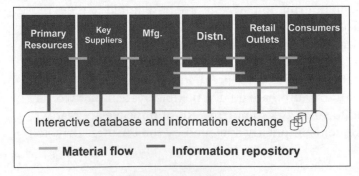

Figure 9.1 Inter-enterprise value chains.

the following principles have been identified as pertinent to source data collection, real-time monitoring and event-driven feedback, and dynamic supply chain optimization:

- The majority of information required by an enterprise to optimize its operations resides outside the enterprise.

- Companies and their trading partners require zero-latency interactions with real-time alert capabilities. This is necessary in order to achieve high-velocity, tightly synchronized business processes with the flexibility to accommodate changing conditions and business objectives.

- Companies have the need for increased agility in meeting more dynamic customer demands. This requires shortening the time-to-plan horizon while lowering safety stock. Without real-time information on what is happening across the *complete* supply chain, order fulfillment in this environment is difficult and costly.

These principles need to be considered when creating any part of the general data architecture. In addition, in order for each of the specific application components to be wirelessly enabled, it will have its own data principles. The data architecture for each application component, therefore, combines general information principles and those specific to that component.

Information Requirements for Source Data Collection

- Enterprise systems and data capture devices collect data throughout the fulfillment network, ensuring a high level of data quality.

- This information is then made available to support the users and systems in charge of fulfillment-related processes. The data is transferred via message-based mechanisms or real-time feed, based on system requirements.

Information Requirements for Real-Time Monitoring and Event-Driven Feedback

- Views of this information give a unified picture of the supply chain, independent of the original data source.

- Users access these views via different channels, including thin-client decision support tools, browsers, or other applications.

- Alerts are automatically reported to individual users, user groups, and information systems when planning, scheduling, and execution data diverge from defined tolerance levels.

- Internet and intranet virtual private network (VPN) functionality is available to access the data and send alerts and notifications to users in specific user groups.

Information Requirements for Dynamic Supply Chain Optimization

- Quick responses to new enterprise and inter-enterprise information exchange requirements are enabled via configurable and open software instruments.

We have discussed business processes and information principles as two inputs into the creation of the data architecture. We will now introduce the third: the business flow, describing interactions between the business processes. Understanding the business flow will require us to look at the business processes more closely. From that analysis, we can derive the data requirements of the general system. Figure 9.2 contains a very high-level view of the processes of the MetX Exchange, a view of how the fourth-party logistics (4PL) provider interacts with the supply chain/order management and fulfillment space, and a rudimentary representation of the data flow.

We will now discuss the components of the process flow, and how adding a wireless infrastructure will enhance the source data collection, real-time monitoring and event-driven feedback, and dynamic supply chain optimization capabilities of those components.

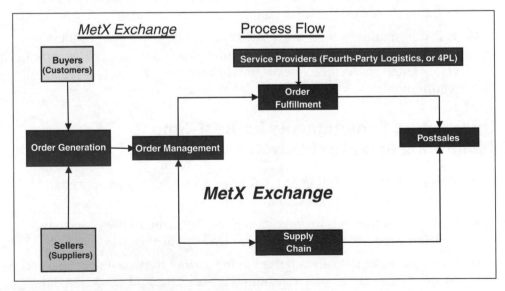

Figure 9.2 MetX Exchange high-level processes.

The Information Architecture for Order Generation

Examples of wireless order generation have been given in both Chapter 7 (Acme Pens and the blueberry muffin baker) and Chapter 8 (the liquid nitrogen distributors). The examples show how wireless orders can come from either the supply side (the seller) or the demand side (the buyer). Regardless of whether the request comes from an individual using a Web browser, a WAP phone, a wireless PDA, or some business logic module generating an order, the data requirements are more or less the same. Therefore, we can expand upon the liquid nitrogen distribution use-case example from Chapter 8 as an illustration of the data architecture needs for order generation, and feel comfortable that we are doing the subject justice.

VMI Capability

The example from Chapter 8 is a VMI system connected to the enterprise via a wireless monitoring device. Figure 9.3 depicts how such a VMI system fits into the MetX Exchange framework.

Remember our architectural philosophy that we are trying to first answer the question, *What are we trying to do?* In short, we are trying to keep the minimum amount of liquid nitrogen on hand to maximize profitability. Too much, and we waste money on excess inventory. Too little, and we lose the opportunity to fulfill a customer's request. The next question we ask is, *How can we do that?* Well, we need to keep a close eye on our liquid nitrogen stock, our production levels, and incoming orders from customers. Finally, we ask, *With what can we accomplish our goals?* The answer here is that we need a system capable of receiving inventory levels, current production, and demand inputs, all in real time. Furthermore, the system needs to be able to coordinate the three inputs to maximize profitability.

Since our goal is to demonstrate how adding wireless capabilities enhances the business offering, we will go by the assumption that our liquid nitrogen producer has an ERP system in place with the ability to perform the aforementioned resource coordination. What it needs is a cost-effective means of receiving the three inputs in real time and the flexibility to respond to them. Clearly, this system can benefit in all three of our wireless focus areas: source data collection, real-time monitoring and event-driven feedback, and dynamic supply chain optimization.

Let's first focus on the wireless source data collection capabilities of the VMI system. As described in the use-case example in Chapter 8, a wireless link connects the visibility manager to a remote sensor at the storage facility. The sensor reports liquid nitrogen levels more or less in real time. Since the

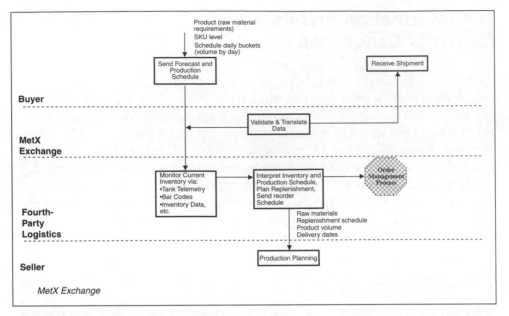

Figure 9.3 Order generation process: VMI capability.

telemetry sent to the visibility manager is digital, it is a collection of samples rather than a true real-time reading (a car with a digital speedometer samples in such a fashion).

The impact on the data architecture is straightforward: This kind of system requires a large volume of simple pieces of data, since its polling loop is negligible. There is nothing wrong with this kind of system, which is, in fact, in accordance with the previously stated information principles of zero latency and the specific source data collection requirements. However, there may be security issues because a competitor could learn of this telemetry flying freely over the airwaves and not be able to resist listening in. The competition could use this tactical information as a negotiating tool, should a glut or shortage be reported by the wireless system. Therefore, the system needs a mechanism for encrypting the data.

Data being transferred from the remote sensor would be stored locally in the visibility manager for reporting purposes. Because the number of transmissions is high, the time-out for a lost or incomplete message would be relatively short, since the next message is right behind it. Lost messages would be tolerable, as long as they were intermittent; too many consecutive lost messages could indicate a problem with the device or with the receiver.

The logic for handling the telemetry consists of a business object that receives the telemetry, checks it against predefined limits, generates an alert should the data be out of tolerance, then forwards the data to the persistent storage engine. An out-of-tolerance event could be too much or too little liquid nitrogen, or too long a time since telemetry was received from the wireless device.

To prevent swamping with old telemetry from the device, the visibility manager must have a strategy for limiting the volume of data stored persistently. It would probably include limits on the amount of time the telemetry is stored (data more than a week old would be purged, for instance). These limits could be tuned as data reporting requirements and database hardware or software limitations dictate. Should this persistent data store be lost, the consequences would be small. Events requiring user intervention are reported before the data enters this store, so the only loss is the ability to compile some telemetry reports. The alerts would not be affected.

The data feed is an example of a cost-effective means of wireless data collection providing value. It allows the visibility manager to perform real-time monitoring and event-driven feedback. What is needed to make this part of the system useful is a set of business logic rules defining scope of tolerance and actions to perform when out of tolerance. Since the liquid nitrogen production process is so dynamic, having wireless alerts when an out-of-tolerance event occurs is essential. What is the use of having subsecond turnaround time on fault detection and reaction when it will result in the creation of an event such as an email, which may go unread for several hours?

The Information Architecture for Order Management

Once the demand- and supply-side participants come into contact with one another, they need a mechanism to come to an agreement on an order. Quite often, this is done through an auction-style exchange. Wireless bidding and countering at auction-type exchanges has been discussed in Chapter 7. Once an agreement is reached, there are a number of steps to be performed before the order fulfillment process can begin. These are illustrated in Figure 9.4.

Bid/Auction Hosting

We already elaborated on the advantages of wireless auctions earlier in this book. However, they need to be revisited at a lower level to understand their data requirements. In order to truly derive the tactical speed advantage of wireless auction participation, traditional Web-based application strategies will not be effective. The majority of Web-based applications today have one common characteristic: They are client initiated. A Web server *responds* to a client request; it never initiates the interaction. As a result, Web clients pull from the server; the server never pushes to the client.

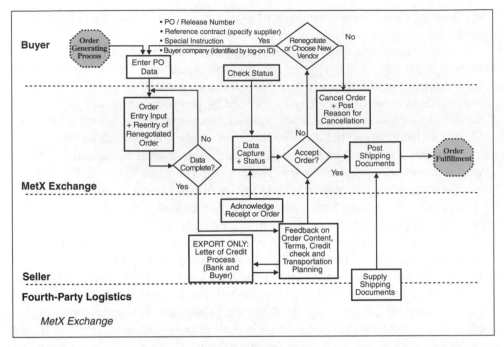

Figure 9.4 An overview of the order management process.

Therefore, an alternative model must be used for a wireless-enabled auction to be truly useful. Data must be able to be pushed to the wireless client device to perform the highly synchronous activities required to participate in the auction process. The data model is typically not very complex, but it is the strategy for pushing the data that involves new technologies.

The Information Architecture for Order Fulfillment

The piece of the supply chain puzzle that is probably best positioned to benefit from wireless enablement is order fulfillment. Figure 9.5 gives an overview of the fulfillment process of the MetX Exchange. Studying the illustration, you will quickly see that there are many parts that would benefit from a wireless dimension.

The overriding characteristic of the wireless-enabled fulfillment process is its ability to respond very rapidly to unexpected change. This section will explain how the process flow and data interchange, if managed properly, can reap large efficiencies made possible by wireless access.

The work flow through the order fulfillment process primarily involves fourth-party logistics processes. The buyer and seller are the first and second parties; the MetX Exchange is the third party; any service providers acting as

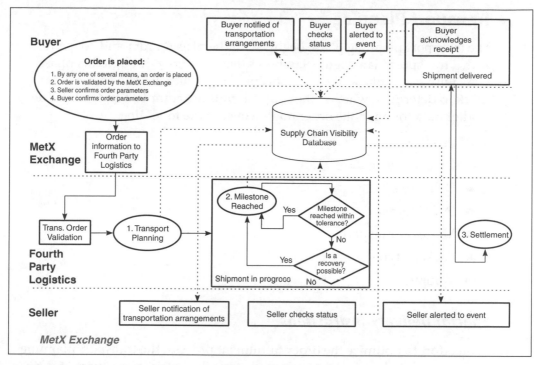

Figure 9.5   An overview of the order fulfillment process.

facilitators for the transaction are fourth parties. The need for fourth parties is the cause of the information principle that states that the majority of information required by an enterprise to optimize its operations resides outside the enterprise. As a result, the MetX Exchange has an entire system with its own architecture dedicated to fourth-party logistics (4PL).

The fulfillment process begins when the order management process is finalized. The MetX Exchange system passes the order to its 4PL system, which then checks the order for accuracy. If the order passes these data validity and completeness tests, it goes to the transport planning phase, then through the transport arrangement phase, before the shipment finally is transported and delivered. We will now look at each one of these phases in more detail.

NOTE

Because we live in the twenty-first century, and everything that has to do with computers begins with an *e*, fourth-party logistics is sometimes referred to as *e4PL*.

Transport Planning

The transport planning phase begins with the placed order and concludes with a transport plan. Figure 9.6 shows the work flow through this phase.

As with every phase, the transport planning phase begins with a data check to determine whether the source inputs are complete and valid. The kind of data found in the order might concern the following

- Product
- Origin/destination
- Delivery schedule
- INCO terms
- Mode of transport
- Package type
- Volume

Special Delivery Instructions

In an effort to optimize the order fulfillment process, the transport planning phase seeks to identify the best way to move the order from source to desti-

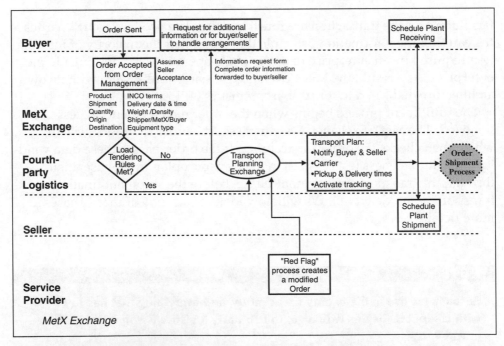

Figure 9.6 Transport planning phase of the order fulfillment process.

nation. There are two criteria for determining this *best* path: time and money. Both play into the decision-making process for every delivery, but their level of importance can vary. For example, the receiver of a donor heart would probably be willing to pay more for cross-country transport of the vital organ than if it were a consignment of cornflakes. The logic to make that determination is data driven and housed within the transport planning exchange.

Transport Planning—Transport Exchange

The transport exchange takes an order, analyzes its parameters, and determines the best means to move the order to its destination. At its center lies the transportation planning module, which uses the data parameters to optimize the delivery mode most suitable for the shipment needs and to find available resources on the exchange that match these shipment criteria. The transport planning module contains the logic that allows it to understand where the shipment falls on the continuum of being time critical versus being cost critical. The exchange is depicted in Figure 9.7.

One of the primary missions of the transport exchange is to find the most cost-effective carrier willing and able to make the delivery, given the other shipment constraints. There are two paths to identifying that carrier: prenegotiated contracted rates and real-time negotiated agreements. Typically, the buyer, the seller, and the MetX Exchange itself all have contract rates for certain carriers making specific deliveries. The transportation planning module requires visibility to all these rates in order to find the most appropriate means of delivery. This information is located within a data warehouse that either is part of the exchange or is outsourced. If an acceptable contract rate does not exist, then the vendor exchange will be used to find a carrier.

For a vendor to be selected to transport the shipment, the company must be both cost efficient and available. Just as the liquid nitrogen company was able to find a supplier in real time via the exchange, the 4PL transportation planning system has a vendor exchange for carriers. This is an example of the dynamic supply chain optimization wireless enhancement, made possible when field personnel (for example, truckers) have a mechanism to report their location and availability. But for the exchange to be useful as a carrier exchange, the data provided by participants must be current and accurate. Thus, the zero-latency information principle of the MetX Exchange certainly applies to the 4PL vendor exchange.

Membership in the vendor exchange is a two-way commitment, providing value to all participants. Exchange members derive value from having access to a large number of potential carriers, the most cost effective of which is automatically selected by the transportation planning module. Carriers derive value from having access to a large pool of potential work. In return,

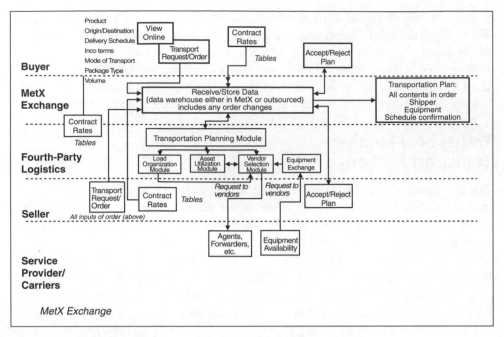

Figure 9.7 The flow of the transport planning exchange.

the carriers must consent to a minimum set of data input requirements provided to the exchange. Since carrier employees are mobile by the very nature of their work, having bidirectional wireless access is essential for them to participate in the dynamic, efficient supply chain network.

The wireless access need not be very sophisticated; a truck driver with a cell phone meets all the requirements for participation. The important thing is to have the reporting process in place and to ensure that the process is being used consistently to maximize data accuracy. The system will work for a large shipping company managing large fleets of ground-, air-, and sea-based transport equipment just as well as it does for a trucker working on his or her own. However, because the complexity of managing resource availability is strictly located on the supply side, the large shipping company is required to manage its own ERP system.

The output of the work performed by the transportation planning module is the transport plan, consisting of the following:

- All the contents of the order
- The carrier's identity
- The equipment to be used
- The pickup schedule
- The drop-off schedule

The transport plan is then sent to the buyer and seller for approval, to coordinate the pickup with the seller's shipping department, and to coordinate the delivery with the buyer's receiving department. Once those approvals are received, the shipment is ready to be released. The carrier picks up the shipment at the prescribed time, and shipment tracking is activated.

Although there are efficiencies gained already in the initial transport planning phase, even more benefit will be derived should the process be forced to return to the planning phase. Such a return may occur as the result of a breakdown in the supply chain sequence of events, as we will explore later in this chapter. At any future point of the work flow, an out-of-tolerance event deemed to be nonrecoverable would force the shipment back into the transport planning phase, where real-time supply chain optimization can reap significant benefits.

Global Tracking

Carriers that participate as members in the exchange are required to submit data inputs meeting or exceeding the requirements for the transport planning component. They are also required to provide real-time visibility during the transport process itself. Again, this visibility could be as simple as a trucker submitting information via his or her cell phone, or it could be as sophisticated as a fleet of trucks outfitted with GPS tracking devices providing real-time telemetry. Obviously, buyers or sellers with a need for the more advanced visibility will pay a premium for that service. Figure 9.5 introduced the flow during transport; now Figure 9.8 shows that flow in more detail.

This kind of tracking is common practice for carriers such as Federal Express, UPS, and other commercial shippers, which have been providing Web-based visibility to the shipment process for a long time. Why is the 4PL transportation process any different? There are two answers to this question.

First, the tracking at FedEx and UPS is largely an *intra*-enterprise process, while tracking at 4PL providers takes place at the *inter*-enterprise level. At the large shipping companies, data never leaves the confines of the company. On the other hand, the 4PL carriers need the infrastructure of the exchange to provide a context in which they can report their data to all other companies participating in the system. Enforcing this requirement is in keeping with the exchange information principle of providing real-time information on what is happening across the *complete* supply chain.

Second, the exchange provides a context for those requiring visibility to the shipping process (i.e., the buyer and the seller) to view the data they need (see Figure 9.9).

Let's look at the data required to make this tracking feature work. First, the carrier is responsible for ensuring that the progress reports reach the MetX

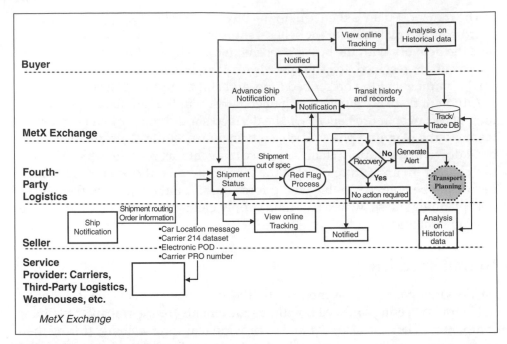

Figure 9.8 Tracking deliveries.

supply chain visibility database (introduced in Figure 9.5). This could consist of an operator at the MetX site data entering phoned-in status reports—the carrier would simply give a shipment ID and the milestone reached. Or it could be a truck with a GPS locator and a wireless transceiver sending out telemetry to a wireless network. The data sent would be the same as on the cell phone call, just done more frequently. If the progress reports do not arrive in the visibility database within the prescribed allowances, an alert is generated. This requires an intelligent agent, since systems normally respond to events, not their absence. Without the context of the exchange, buyers and sellers would have to rely on whatever tracking visibility the various carriers chose to provide. Thus, the exchange ensures that a minimum set of visibility data is provided in real time. Every carrier involved with the exchange will work under a service level agreement (SLA) that specifies the set of milestones, or expected events, at specific points in time, to which the carrier will provide visibility. There will also be specific tolerance levels of deviation from these milestones and remedies should those tolerance levels be exceeded.

Figure 9.10 illustrates a typical shipment work flow and its associated milestones. It also depicts an out-of-tolerance or out-of-spec, event.

If we refer to Figure 9.8, we see the prescribed steps:

1. The process is red-flagged.

2. Notification is sent to the buyer and seller of the out-of-spec event.

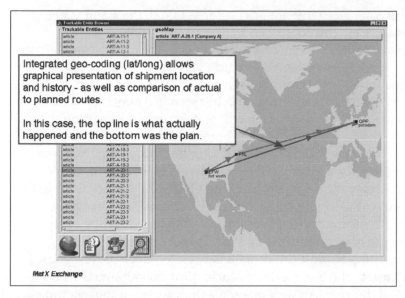

Figure 9.9 Providing clients with real-time visibility into the shipping process.

3. A determination is made about whether the problem is recoverable or nonrecoverable.

The *red-flagged* process may be a simple matter of time slippage, from which recovery is possible. The truck driver may have hit unexpected traffic congestion or blown out a tire and had it fixed as he waited on the side of the

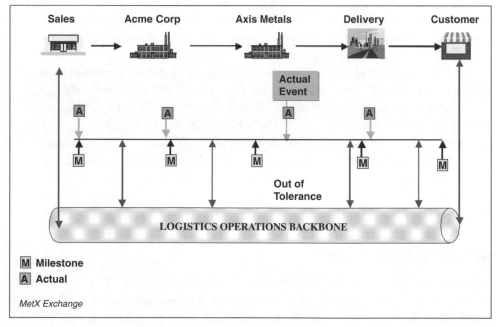

Figure 9.10 Out-of-tolerance event handling.

road. Should this tolerance fault not be considered fatal, the event is logged and the delivery process can continue.

On the other hand, the slippage might be nonrecoverable. This might occur if the driver had an accident where his vehicle was rendered inoperable. Alternatively, the time slippage might be unacceptable to the buyer, if the company's time constraint is absolute. Another scenario may include the milestone being out of spec for a reason completely unrelated to time schedules, as would be the case if part of the shipment were missing or damaged, for instance. If this happens, the red-flagged process is deemed unrecoverable, and two activities are initiated: First, an alert is generated, and, second, the shipment reverts to the transport planning phase of the work flow. In this scenario the transport planning exchange really demonstrates its value. Referring to Figure 9.6, the red-flag alert process creates a modified shipping request and then places the request into the transport planning exchange. The modified shipping request can contain parameters such as new source location (in the case of the broken-down truck), time constraints (in the case of the time-critical delivery), or new materials (in the case of the incomplete shipment). Since the event is recorded in real time, this new order with its modified parameters hits the exchange almost immediately.

Let's tie in the liquid nitrogen example from Chapter 8: The enterprise determines that it needs a shipment of liquid nitrogen in its New York office by 6:00 P.M. that day. The order goes through the MetX Exchange, and a supplier is found in Baltimore. A trucker with availability is free to make the run at the right price and therefore is chosen by the transport planning exchange process. Unfortunately, the truck carrying the nitrogen breaks down on I-95 just north of Philadelphia at 1:00 P.M. The truck driver uses the wireless reporting capabilities to notify the exchange of the predicament, and the exchange responds by promptly generating a red-flag event.

At 11:00 A.M. that same day, another participating carrier reported himself free in downtown Philadelphia, having just completed a delivery. He is just finishing his second cheese steak sandwich when his cell phone rings, offering him the job. He accepts, meets the other truck on the highway by 2:00 P.M., and is able to make it to New York by the 6:00 P.M. deadline. This certainly could not happen without the exchange linking buyers and sellers efficiently, the transport planning exchange linking carriers and shipments efficiently, and the tracking and tracing infrastructure providing real-time access to shipment status.

Shipping Documents

Additional benefits provided by the technologically improved business processes are better shipping documents. Many companies have realized that eradicating traditionally paper-based activities had the following positive impacts on their organizations:

- Data is entered only once, eliminating the duplication of effort.

- Data is more accurate, as data capture occurs at the data point's origin. In other words, there is no transcribing necessary, eliminating a data entry clerk guessing about the handwriting of colleagues.

- A well-written data entry application will perform data validity and completeness checks at the time of entry, so the potential for incorrect entries is reduced and errors of omission are eliminated.

- Since the shipping documents exist in the system as soon as they are created, they can be distributed more rapidly.

- Changes to the shipping documents can be distributed in real time, without the hassle of having old versions lying around.

These benefits have been driving the paperless initiative for years. Couple that with wireless clients, and the benefits increase dramatically for a mobile workforce. More accurate shipping documents mean fewer misconfigured shipments, reducing errors and increasing profitability. Figure 9.11 shows the shipping documents needed for the MetX Exchange order fulfillment process.

Suppose the liquid nitrogen suppliers need to ship their product to Montreal from New York City. They quickly find a carrier that is on his way right on schedule. He is just about to cross the border into Canada when he realizes he does not have all the documentation necessary for the international shipment. Ordinarily, this could cause a delay of several hours, as the docu-

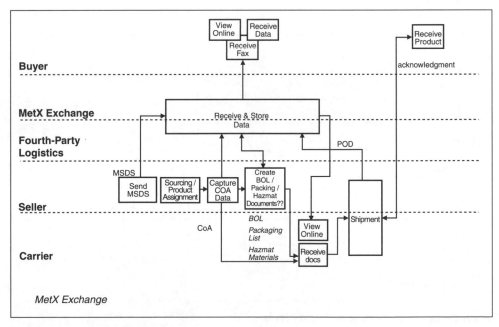

Figure 9.11 Shipping documents.

ments try to catch up to the shipment and appease the feisty Canadian border patrol. But, because our carrier is wirelessly enabled, he is able to request the missing documentation from his handheld device and receive the necessary materials long before he reaches the border. Once again, the schedule is saved by wireless connectivity permeating the entire supply chain.

Post-sales Activities

Once the shipment is delivered, there are still several tasks that need to be performed. The shipment must be verified as complete and intact, and payment must be made. Figure 9.12 depicts the steps involved in the settlement process.

Another significant impact that wireless connectivity can have on the entire supply chain process is the facilitation of the settlement process.

Settlement

Wireless document management is useful for more than the shipping documents. It can be used to streamline and automate the processes depicted in Figure 9.12, allowing the shipper to generate a valid freight bill. Having this bill validated and paid immediately would be of immense value to the seller

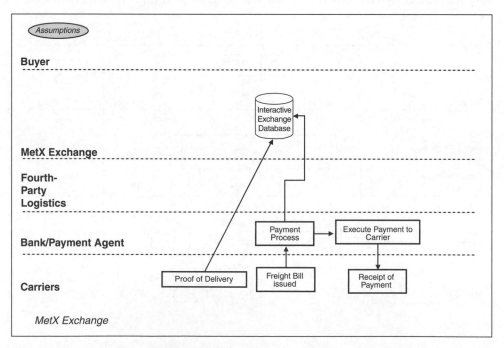

Figure 9.12   Post-sales activities: settlement.

of the shipment. Think it really doesn't matter? Let's suppose a trucker delivers a shipment of metals worth $1 million. If the funds are transferred from the buyer to the seller 24 days after delivery, the buyer could earn over $4,600 on the float (assuming they could figure out how to earn 7 percent on the seller's money during that 24-day period).

That is $4,600 of additional profit the seller could have *per shipment*. A large metals company will make thousands of shipments per year, meaning millions of dollars in additional profit just by obtaining their funds immediately upon delivery.

What is required to make immediate remunerations happen? The shipper has to be able to provide immediate proof of delivery, and the buyer has to immediately acknowledge receipt. With these two inputs in place, the bank or other paying agent has no barriers to executing the payment. This delivery process is roughly the same as the process followed by car rental agencies. Upon returning their car, some information is obtained by an employee wirelessly connected to their enterprise. This enterprise on the other side of the process can generate a bill and obtain funds from your credit card immediately. The car rental return process has been made essentially paperless, with the exception of generating a receipt. The only difference between that process and the shipper is the magnitude of funds typically involved.

Summary

This chapter gave an overview of the data requirements generated by the business processed outlined in Chapter 8. We took those processes, defined the interactions between them, and applied the company's information principles to create a data architecture. We focused on three areas of the application capable of providing the most value from wireless enablement: source data collection, real-time monitoring and event-driven feedback, and dynamic supply chain optimization.

Not only did we look at the information requirements of the system as a whole, but we also considered the principles guiding these three parts of the application.

One of our objectives was to highlight some areas in which wireless enablement could provide the highest return on investment and describe the associated data requirements. We showed how efficiency could be gained, obstacles more easily overcome, errors more easily corrected or avoided, and access facilitated between the mobile employee and the enterprise.

We are one step closer to completing the architecture work necessary to create a scalable, well-planned software solution that meets all of our business requirements. The data architecture we developed will serve as an input to the information systems architecture to be described in Chapter 10. The information systems architecture is essentially the application layer. Since we have a clear definition of the data needs, fueled by the clear definition of the business needs, choosing the applications to meet our requirements becomes simpler and less prone to error. This reduction in risk is one of the key characteristics of our architecture framework. We can feel more secure in our choices as we create the information systems architecture, which we will do in Chapter 10.

Information Systems Architecture of a Wireless Exchange

Introduction

The intent of this chapter is to pick up where Chapter 9 left off by digging into the details of the key components that make up the wireless information systems architecture. This chapter continues with the discussion of the MetX Exchange architecture by answering the question, *How is the application built?* Remember the three major dimensions in a solid architectural design? Let's do a quick refresher:

Conceptual. Answers the question of what the overall system is supposed to accomplish. Frankly, if you don't know what to design, how are you going to design it? Chapter 8 covers the conceptual business architecture of the MetX Exchange.

Logical. Answers the question of how the system is built. As architects, you'll want a design that reflects the conceptual business requirements as completely as possible.

Physical. Essentially, the physical architecture is the implementation of your design. In other words, the physical dimension will tell you with what you are going to build your architecture.

You may have already guessed that this chapter will encompass an understanding of the logical components of the information systems architecture.

Later in this chapter, we will take a closer look at the logical architecture and how all of the components fit together. In the MetX Exchange information system, all of these architecture components are critical to understanding the design of the enterprise application. As was described when we determined the conceptual business architecture, the wireless exchange provides three primary wireless services:

1. Source data collection.

2. Real-time monitoring and event-driven feedback.

3. Dynamic optimization.

The first part of our discussion will focus on the high-level information systems architecture in a prewireless state. In other words, we want to determine the key components of the exchange as they would appear in a non-wireless environment. This is also called the *current state architecture.*

On the other hand, the future state architecture defines the exchange utilizing the wireless services defined in the conceptual business and data architectures. We will cover the future state design of the exchange, and specifically the architecture relating to the aforementioned three wireless services, in the second half of this chapter.

Chapter 11 will complete the fourth architecture required by our framework, the technical infrastructure architecture. It will spell out the same wireless services as the other architectures, but from the standpoint of how they will actually be delivered. Right now, let's look at the current state of the MetX Exchange from an application perspective.

Current State Architecture

As this section title suggests, the current state design of the MetX Exchange describes each of the information systems components in greater detail. Again, the current state architecture covers the logical views in a prewireless state. The objective here is to understand how the components and objects integrate *prior* to a wireless implementation. This is important, as a good number of wireless enterprise solutions start from some sort of existing Web-based application, enterprise resource planning (ERP) system, or a combination of other existing technology infrastructure. Perhaps one way to begin this section is to illustrate the high-level current state logical diagram and to briefly explain the components. Figure 10.1 illustrates the logical architecture view of the MetX Exchange.

As you can see, there are quite a few complexities that need to be explained in our logical picture. Let's go through some of the major compo-

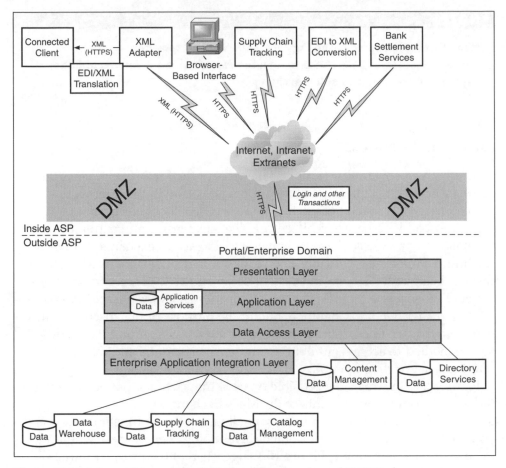

Figure 10.1 Information systems logical architecture view of MetX Exchange.

nents of this diagram, which should give you a good understanding of how the objects integrate with the overall system.

Interfaces to the MetX Exchange

Every enterprisewide system, large or small, must provide its users (or other applications) with access in one way or another. Because access to a system's functions or back-end data is what renders the application useful in the first place, all system designs must contain access points, commonly known as *interfaces*. In Figure 10.1, the interfaces of the MetX Exchange are clearly labeled on the upper portion of the illustration. If you recall the various layers of the logical framework, interfaces exist within the presentation layer. The exchange shows two primary subcategories of interfaces: front-end interfaces and back-end interfaces. Let's discuss each type separately.

> **NOTE**
>
> It may seem that some of the logical architecture actually borders on the physical architecture. We chose to be more explicit in the diagrams to better explain the complexities of the exchange.

Front-End Interfaces

Probably the most familiar front-end interface of the MetX Exchange is the browser-based client. Because of the exponential growth the Internet has experienced in the last several years, the browser-based client has been integrated into almost all Web-based architectures. The inexpensive and extremely intuitive nature of a browser-based client as a tool to access exchange data over the Internet makes it an obvious choice for our logical design. In addition to the tool being cost effective and easy to use, it allows for secure transactions via HTTPS (secure Hypertext Transfer Protocol). The topic of security is covered in greater detail in Chapter 6, but for the purpose of this discussion it is sufficient that we understand that the MetX Exchange community needs a secure method to access confidential intra-enterprise data, and that all the interfaces in our diagram utilize HTTPS for secure transactions.

Figure 10.2 illustrates the possible types of transactions that occur between the various logical components of the MetX Exchange.

The application transaction list essentially mirrors our functionality list. If you are thinking in an object-oriented methodology sense, these would be the use cases. For example, Figure 10.1 suggests that the browser initiates *login* and other related transactions. Returning to Figure 10.2, our transaction list indicates a small box equivalent to the *login* functionality of the system. Subsequently, the next box below it is equivalent to *authentication*, and so on. This book will not go into great detail on each of the use cases, but will refer to them and map them to the necessary components of the system.

The browser front end also provides access to the logistics/supply chain components of the exchange. Specifically, the supply chain tracking (SCT) Web-based front-end system gives supply chain managers the ability to link with carriers, such as rail, shipping, and trucking. The main function of SCT is to have the ability to track shipments and orders.

> **NOTE**
>
> The supply chain tracking system actually uses the same Web-enabled front-end tool. In effect, you are simply accessing a different area of the same portal.

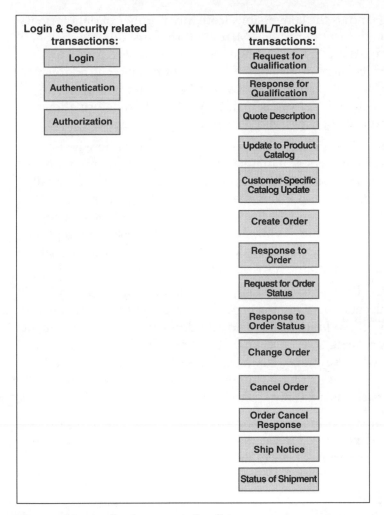

Figure 10.2   Application transaction list.

Client and Back-End Interfaces

The reference to client and back-end interfaces can be somewhat misleading. When we talk about a back-end interface as a type of interface, we are talking about a backdoor mechanism of entering the system. Traditionally, *back-end* systems refer to data servers, legacy systems, and other applications designed to process transactions of the back office. These areas will be covered in the enterprise applications integration (EAI) and enterprise resource planning (ERP) section of the current state architecture. The two primary back-end interfaces are the XML adapter and the connected client. The following is a description of the components that make up these interfaces:

XML adapter.   Handles the conversions from any number of data formats into MetXXML, the MetX-specific XML data format used by its back-end processes. For example, it manages the XML processing required to translate electronic data interchange (EDI) function calls. The adapter also handles back-end clients, such as a MetX customer who wants to access exchange data. It performs XML-to-MetXXML transformations using XSL-T to ensure a uniform data format throughout the system.

Connected client.   A connected client is any client (a *client* here is simply the organization, not a system) that has an existing relationship with the MetX Exchange to access data from the ERP back end and to potentially access business information systems such as the ERP business warehouse. For instance, if Acme Metals is a member of the MetX Exchange community and would like to view accounting/purchase order reports on a regular basis, the company would deploy a connected client that would communicate with the XML adapter.

Bank settlement.   The last back-end supply chain and logistics component is the bank settlement interface. As the component title suggests, someone must get paid when all is said and done. It covers the payment process between carriers and exchange members. The bank settlement tool integrates with the supply chain tracking system to automatically perform financial transactions as orders are placed and filled. Transactions are processed via a flat file.

These interfaces are the primary mechanism for exchange members to access EAI and ERP components, such as the ERP business connector. We will talk more about these interfaces later in this chapter.

Web Components

Of course, no exchange architecture would be complete without Web components or Web architecture. Figure 10.1 shows that the primary systems necessary for an Internet transaction reside in and around the demilitarized zone (DMZ).

The DMZ defines an environment that is more secure than the Internet and that is located within the domain of an application service provider (ASP) or a corporation. Generally, the DMZ holds servers that provide functionality,

WHAT IS A DMZ?

The DMZ, the demilitarized zone, is the area within the Web infrastructure that protects the internal systems from the hackers and crackers of the world. It is specifically designed to allow only the users that can authenticate with the system to access data behind the firewalls.

but not data. You'll also note that the line separating our front-end interfaces from our Web servers and ultimately the back end is labeled "inside ASP" and "outside ASP." Obviously, an ASP model was used in this architecture, but that is certainly not a requirement. The enterprise machines in the DMZ are connected to the Internet, so they are more likely to be attacked by hackers. The whole concept of the DMZ is to limit the damage that unwanted access can cause—hackers on those machines are blocked from sensitive data, which is hidden behind enterprise firewalls.

The systems that we want to discuss include the following.

Web server.   You guessed it—we can't have a Web platform without the traditional Web server. The Web server resides within the DMZ and acts as one of the first points of contact, sending and receiving HTTPS transactions and communicating messages to all the other subsystems in our design. The Web server obviously is a critical component of the architecture.

Application layer.   The application layer simply manages and houses the application and business logic of the Web subsystem. It is also the main work flow management, commerce, profiling, and personalization mechanism for users accessing the exchange. The application layer is obviously a very common and necessary piece in eCommerce design.

Content management.   Manages how content is entered into and presented by the system. To accomplish this, we will need a front-end interface.

LDAP server.   A Lightweight Directory Access Protocol (LDAP) server is used to manage user rights in the system. Thus, one of the primary functions of the LDAP server is user authentication and access control rights to applications and data.

Logistics/Supply Chain Components

The MetX Exchange would be incomplete without the logistical and supply chain functionality that manages purchase orders and maps customer demand to warehouse supply. The three main interfaces of the exchange that relate to logistics and supply chain processes are as follows:

Collaboration engine.   This very important component of the supply chain subsystem interfaces with the Web services layer, as Figure 10.1 indicates. The functionality that the collaboration engine provides entails supply chain/logistical demand planning, forecasting, and supply-level reporting.

Transportation, arrangement, and optimization (TAO) services.   The next piece in the supply chain/logistics architecture is called TAO services. This critical component is the heart of the supply chain and logistics architecture. It acts as the central communication mechanism that reaches the other supply chain/logistics and EAI/ERP systems. The

> **NOTE**
>
> For the most part logical diagrams stay away from naming specific products or solutions, unless they are absolutely critical to the architecture. As you'll see, the term *ERP* is used fairly extensively in our design. The reason we did not name this package is that almost any ERP system could fit in this situation. Some of the leading package ERP vendors out there are SAP, PeopleSoft, Oracle, and BAAN. From an architect's perspective, the *ERP system* is the general term we'll use.

TAO manages the notification surrounding all the supply chain transactions and sends them to their proper destination.

Shopping/Catalog Management Components

Another critical component of the MetX Exchange is related to the system's main catalog management interfaces. *Catalog management* refers to maintaining a directory or register of all items the MetX Exchange is selling to its customers. For example, when a customer accesses the site, a catalog management function presents the customer with a list of all available inventory. Figure 10.1 refers to the catalog management as a *layer*. The two key areas of interest are the following:

Catalog management.   The primary manager of inventory data and information for the exchange.

Application server.   Manages application-layer logic and communicates directly with the EAI/ERP systems. Some of the functionality needed here is to synchronize with the ERP catalog.

Enterprise Application Integration (EAI) Components

Enterprise application integration (EAI) is a key process of most any enterprise application. EAI is defined as the unrestricted sharing of data and business processes throughout the networked applications or data sources within the enterprise. It's very rare to design a system architecture without considering how other IT components of the organization will fit into the mix. Most

> **NOTE**
>
> In the exchange architecture, you'll note two application layers. The application layer here is leveraged specifically for the catalog management components.

NOTE ON ERP/EAI

It's important here to recognize the distinction between ERP systems and EAI systems. Figure 10.1 refers to the EAI and ERP components in the same layer. However, the two are very different and separate systems. The ERP components are not directly labeled because our exchange leverages those systems in the EAI mechanisms.

large enterprises have invested a great deal of time and money in existing systems that were originally designed to run on a stand-alone basis. Some of these systems—say, legacy accounting or payroll systems—are very large and can contain terabytes of data, while others are small, such as a Web server running very specific Java applets that happen to be critical for day-to-day business functions.

Regardless of the complexity of these existing systems, the job of integrating them into the new enterprise architecture is an absolute must. The MetX Exchange, of course, is no exception. Referring to Figure 10.1, notice that the main EAI components in the exchange architecture are referred to in the enterprise application integration layer. The following list illustrates the EAI components of the exchange.

Data warehouse.   Repository that manages reporting for the connected clients.

Web reporting.   An ERP component that performs the necessary work to translate application function calls and display formats between the ERP and Web environments.

Middleware services.   Essentially acts as the MetXML and transaction verification engine. Receives inbound/outbound messages and releases them to the EAI/ERP components.

Web translation mechanism.   Not a necessary component, as a general rule, but required in the MetX architecture to convert a client/server design to more of a Web-based design.

We have concluded the overview of the information systems (IS) architecture of the current MetX Exchange. Now for the fun part: a description of the future state of wireless IS architecture.

Future State Architecture

By now you should have a good, high-level understanding of the logical architecture of the MetX Exchange. The next step is to provide an overview of the future state design of MetX by illustrating three common wireless services within the general exchange architecture. We will continue with demonstrating how each of these wireless services is designed at a more granular level. Recall that Figure 10.1 illustrates the current state architecture, assuming the wireless pieces of the application are not yet designed. On the flip side, Figure 10.3 describes that same MetX Exchange, but this time including the wireless dimension.

As you can see, the architecture of the wireless solution at a high level

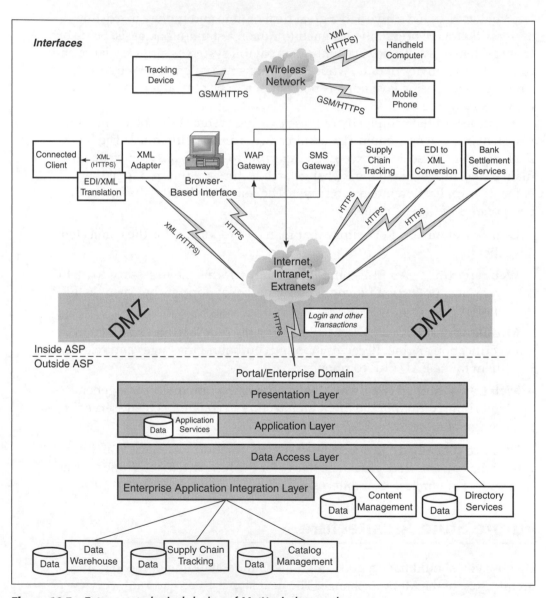

Figure 10.3   Future state logical design of MetX wireless exchange.

really is not that different from its tethered version. However, there are some critical changes, including the following.

- We have added several access devices. You can see that customers can now access MetX through their mobile phones and handheld computers in addition to Web browsers.

- A Wireless Access Protocol (WAP) gateway is needed to support WAP access to the services provided by MetX. The WAP gateway is not operated by MetX, but by an external telecommunications provider. This will become more important when we describe the technical infrastructure components.

- A Short Message Service (SMS) gateway is needed to send alerts to customers, assuming that most alerts are sent to mobile phones.

To illustrate the major differences of the wireless solution even further, we have to look a little deeper into the solution and its components. Let's start with a simple picture. Figure 10.4 illustrates a typical application layering.

Before describing Figure 10.4, please note that this picture does not focus on the following:

- *System services.*   This includes services such as session handling, how to provide access to data, and application security. System services are not

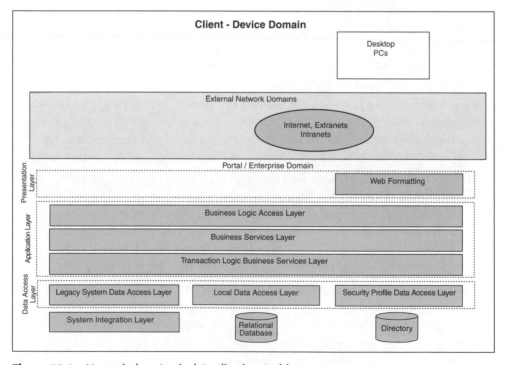

Figure 10.4   Non-wireless Logical Application Architecture.

included in the discussion, as they are not unique to wireless and since these can be purchased as part of a product such as an application server.

- *Commerce products.*   This includes electronic wallets, auction components, and so on. Commerce products are not included, as this functionality can also be purchased.

- *Personalization functionality.*   This includes how to customize the front end. Personalization functionality is also not included, as it can be purchased.

It is advisable in all systems implementation and integration efforts to evaluate which functionality should be bought and which should be built. The answer differs from case to case.

Figure 10.4 shows several different application components:

Formatting layer.   In this nonwireless application environment, the formatting layer is really a Web formatting layer, as users can only access the system using a Web browser. This layer is responsible for creating HTML pages and handling requests from Web browsers. It has no knowledge of business logic, but communicates with the business logic access layer to execute the appropriate business logic.

Business logic access layer.   This layer is responsible for knowing where the business logic is located. For instance, if a Web browser requests information about a particular financial transaction, the Web formatting layer would send the request to get information about the transaction to the business logic access layer. The business logic access layer then knows which business components must be called to run the appropriate checks and calculations and to get the data. It's important to note that this layer has no business logic in itself.

Business services layer.   Contains all business rules and logic to be executed in the application. The only exception to this rule is certain error checks that can be performed on the front end of the application. An example of this is checking whether a postal code has the correct format.

Transaction logic business services layer.   This layer is responsible for getting the data that the business services layer requires. For instance, if we need information about a transaction in the business services layer, the business services layer would send a request to the transaction logic business services layer. This layer would then know where the data can be found and would forward the request to the appropriate place.

Data access layer.   This layer is responsible for retrieving the data using specific data access methods for the mechanism used to store the data. In this case, it is broken down into legacy system, local data, and directory access.

System integration layer.   Provides middleware-based access to legacy systems when required.

Figure 10.5 shows this picture in the context of wireless services.

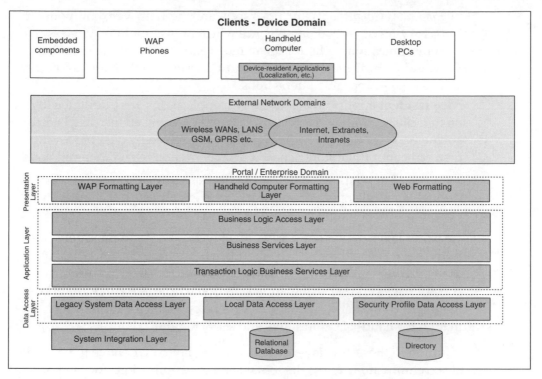

Figure 10.5   Wireless logical application architecture.

Several pieces have been added here:

New channels.   You can see that we now have channels for WAP phones and handheld computers. Here is a description of all three channels in the new architecture:

- *WAP phones.* This interface will be used primary on mobile phones and can be activated only if the user is online—no offline capability is implemented. In addition, this interface can display only text and rudimentary pictures. It is not capable of multimedia applications.

- *Desktop PCs.* Probably the richest interface today, the PC-resident Web browser is capable of showing text, multimedia, and so on.

- *Handheld computers.* At the writing of this book, Palm Pilots and PocketPC devices such as the Compaq iPAQ are good examples of this form factor. Handheld computers are capable of running online and offline applications as well as displaying text and multimedia. Aside from the size differential, the main difference between a desktop PC and a handheld computer is that the handheld device doesn't have as much processing power and memory. Furthermore, the input mechanism is generally not a keyboard. As a result, applications for handhelds are generally more lightweight and of lesser complexity.

■ *Embedded Components*. Includes software and hardware componentry of varying degrees of intelligence which allows mobile, wireless, communication without human intervention. This class of device would include most remote diagnostic services, telematics and other machine-to-machine interactions.

Device-resident applications.   Most Web applications were designed to run on thin clients. With wireless, however, we are seeing an increased need and demand for device-resident applications, driven by data processing requirements, offline storage needs, and connection reliability limitations.

New types of networks.   Wireless applications will access the system through a mobile operators network, whereas access to the system was provided only through the Internet before.

Formatting layer.   Two new formatting layers, WAP and handheld computer, have been added to accommodate new channels. The WAP formatting layer is responsible for creating Wireless Markup Language (WML) pages and servicing application calls from WAP devices. The handheld computer formatting layer is responsible for formatting information for handheld computers and servicing their requests.

In addition to these new layers, some existing layers will have to be modified to accommodate wireless devices when new applications focused on deploying wireless services are created. If only existing services are made accessible via wireless devices, you may not have to change much here, with the exception of expanding the user profiling to accommodate wireless device–based information and preferences.

You can see that there are wide-ranging effects when adding wireless services to an existing system. However, the real work is adding new channels: creating new network connections, modifying presentation layers, and expanding the user profile. A well-designed system will allow you to add these services without forcing you to change a large percentage of your existing application, data, or integration layers. That is, unless you are introducing entirely new functionality that doesn't exist in the wired version of your system.

The rest of this chapter will focus on the three application areas that were introduced in Chapter 8 and how they can be implemented in the context of this architecture. More specifically, the following sections will discuss which particular components must be introduced and where they need to be introduced. Remember the three application areas:

1. *Source data collection*.   Collecting data from many sources and feeding this data back into the system.

2. *Real-time monitoring and event-driven feedback*.   The ability to receive and react to critical real-time events that may require immediate action. Also, receiving source data to assist in monitoring the supply chain and thus the ability to act on business needs.

3. *Dynamic optimization.* The integration point between all three of the wireless services. Essentially, dynamic optimization refers to the ability to streamline a business process (the supply chain in our example) through real-time information attained from source data collection and event monitoring and feedback.

To a certain extent, each of these wireless services is integrated in such a way that one application may feed information and data into another application. In other words, dynamic optimization of the supply chain requires that some data be collected from the system (source data collection) and some event triggered (real-time monitoring and event-driven feedback) to ultimately react to a situation. For example, you are a manager of a certain component within the supply chain and are responsible for making sure a product is shipped to the proper location. Suddenly, you are notified of a break in the supply chain through a hot alert sent to your cell phone. You immediately check where the break in the process occurred by accessing source data. Finally, you react to the event by initiating a specific action through your wireless device.

Let's go through each of the wireless service areas to give you a more detailed picture of how these components work and collaborate with each other.

Source Data Collection

One of the main features of the MetX Exchange is source data collection, referring to a user being able to collect data from many different sources and feed this data back into the system. Once in the system, the data is distributed to all the components that require the updated information. Table 10.1 shows the different areas from which data is collected.

The data is collected using the four device classes listed previously. For example, the handheld computer may be equipped with a bar code reader that scans shipment information at various stages within the shipping

Table 10.1 Data Sources for MetX

OPERATIONS	INVENTORY MANAGEMENT	TRANSPORTATION
Order events	Inventory levels	Rail car location
Shipment events	Order events	ETA
In-transit inventory	Alarms and alerts	Product in car
Itinerary exceptions	Inventory events	Load status
Location and ETA	Location	
Planned orders	Events	

process. No matter whether the information is read when the device is online or offline, the handheld device sends this information to the back-end system at the next synchronization event. In addition, each of the data sources mentioned in Table 10.1 contributes to dynamically optimizing the supply chain process. For example, load status and inventory events may be used to redirect a planned shipment because of a possible break in the supply chain, such as a fire in a warehouse.

Let's look at this from a systems perspective.

Figure 10.6 shows our picture including some of the components required to accomplish the source data collection objectives. Several components have been added here:

Source data collection.   Source data collection consists of several application components. For instance, you will run one specific application to record location information while the product is being shipped. This application is completely different from the one that receives an alarm or alert. Specifically, the components include location tracking, order status, inventory status, and alarms and alerts.

Offline application.   The handheld computer now supports both online and offline applications. This is necessary because much of the data we

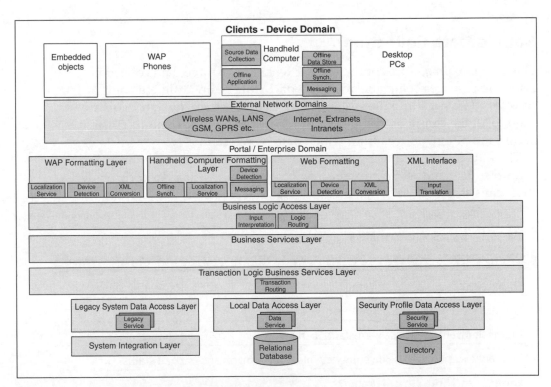

Figure 10.6   Application components needed to provide source data collection.

are concerned with is collected offline by operations personnel, warehouse staff, and other personnel. The offline application component performs two functions: First, it determines whether the device is operating online, and, second, it allows the user to work in both modes.

Offline data storage.   This component is responsible for storing data when the device is offline.

Offline synchronization.   After the user operates offline, the data stored in the offline data storage must be synchronized with the back-end application. The offline synchronization component is responsible for determining what data has to be synchronized and for performing the synchronization. Please note that this component is located on both the front- and the back-end systems.

Messaging.   This component is responsible for communication between the device and the back-end system. Please note that there are messaging components on both the front- and the back-end systems.

Localization service.   Users of the system are going to be located in several geographies or simply prefer to interact with the system in one language rather than another. For example, a warehouse worker in California is shipping some metals to a warehouse in Mexico City. Both workers are running the same application, yet the American employee wants the application to be in English while the Mexican employee wants it to be in Spanish. One day, the Mexican employee decides that his preference is now Portuguese because he wants to practice his Portuguese language skills in anticipation of a shipment he must arrange for a new customer in Brazil. The application allows him to state this preference and to run the application in the language of his choice. The localization component automatically detects the physical location of the user and determines the standard settings. It then compares the standard settings to any preferences the user has specified and makes an adjustment if necessary. Please note that localization components reside both on the device and on the back end.

Device detection.   The device detection component located on the back end is responsible for identifying the device that is requesting information. For example, if you are accessing the system via a Palm Pilot, you would be routed to the handheld computer presentation layer, where the device detection component would realize that your Palm Pilot is of make X and has Y characteristics.

Input translation.   All devices can communicate with the back-end system using XML-based messages. This component is responsible for reformatting the requests so that the rest of the application can understand them. In many cases, this may be a different XML format.

XML conversion. This component is responsible for translating the received data into XML format for transfer to and between layers in the back end.

Input interpretation. After extracting the input, we have to figure out what to do with it. Given that we are running several applications that are used to collect data, we need several corresponding application components to determine how to use the information. For example, if our system includes a bar code reader, we can determine a shipment's location by scanning the code at various physical stages throughout the process. Once scanned, the data is sent to the back-end system, either in real time or during the synchronization process. When it gets to the input interpretation component, the information is routed to the correct business services layer components to run the appropriate business logic on the component. From here, the data is routed to the correct storage mechanism, if required. For example, the data might flow to the ERP system to update the order status, or it may flow to the supply chain system to update inventory levels. This example illustrates that the input interpretation component consists of four mechanisms: location tracking information routing, order status information routing, inventory status information routing, and alarms and alerts information routing.

Transaction routing. This component is responsible for knowing what systems require which data, and how the data is shared. As before, we have one component for each of the applications. In this case, they are location tracking information sharing, order status information sharing, inventory status information sharing, and alarms and alerts information sharing.

Legacy/data/security service. These components load the collected data into the local data storage. The legacy service loads data into legacy systems, the data service loads data into local storage, and the security service loads data into a directory. Again we have components for each of the applications in these categories: location tracking information load, order status information load, inventory status information load, and alarms and alerts information load.

Now we are ready to explore how these components communicate. Figure 10.7 illustrates how the components are connected.

As you can see, the device-resident components communicate with the appropriate formatting layer components. The reason for this is that the devices must have the appropriate formatting layer create the front-end component. The devices make a request to the back-end system and are routed to the appropriate formatting layer.

As noted before, the formatting layers contain a localization component to ensure that the presentation of information is appropriate. The device detection component identifies the device that is asking for information. Both com-

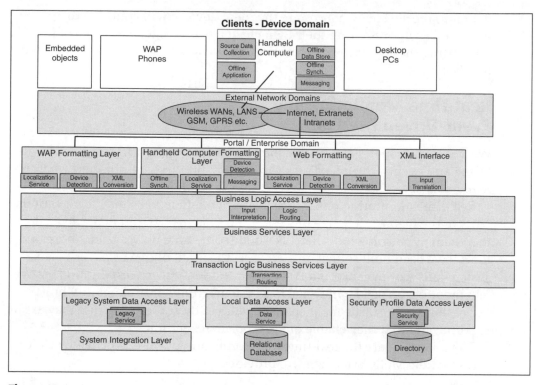

Figure 10.7   How components are connected.

ponents must determine the appropriate localization and personalization parameters by calling the appropriate back-end logic.

The formatting layer is in charge of sending individual pages or an application back to the device, where the user now can access them. When data is collected, it is sent from the user's device to the XML interface over an HTTPS connection. This happens only if an application has been downloaded to the device. If no application was downloaded to the device, the data is first sent back to the presentation layer, where it is translated into XML format and then forwarded to the input translation component. The input translation component verifies the result and sends the information to the application layer, where the input interpretation component for that application decides what to do with it. Finally, the information is forwarded to the legacy/data/security components for the particular application.

Please note that we have not gone into detail about how the application layers connect and how the integrated systems import this data. We will do so in the dynamic optimization section. At this point, we are only concerned with those components that are interesting from a wireless perspective. So, what is important here is the means by which we are able to enter data into the system. Chances are that new devices will continue to appear on the market, presenting their own challenges. The general concept for integrating these devices, however, will be very similar to what is outlined in this chapter.

Real-Time Monitoring and Event-Driven Feedback

Wireless technologies also enable the MetX Exchange to add real-time monitoring and event driven feedback capabilities. But first, let's discuss what this really means.

Real-time monitoring refers to using technology to keep an eye on various parts of the metals exchange processes and systems. We can make an analogy to system monitoring tools here. For quite some time now, system monitoring tools have been built so that an alert is sent to a cell phone or pager when a system fails. For instance, if a Web server suddenly stops responding, the system monitor can send a message to a system administrator to notify him or her of the problem. The main characteristics of these alerts are that they are in real time and that they enable quick problem resolution.

The following are the real-time monitoring and event-driven feedback features that the MetX Exchange will provide:

Location services. A tracking device can be attached to a shipment, for example, to find out where it is at all times.

Event-driven alerts. These alerts are launched whenever something out of the ordinary happens. An alert can be defined for several different purposes, but the general functionality would be to make sure the customer and the operations staff know when an order will be delayed. More alerts with more detail have to be sent to the operations staff, as they will have to address the problem as quickly as possible.

Employee-driven alerts. Authorized employees at the MetX Exchange should be able to send alerts to users or user groups. For instance, if a customer service agent notices that MetX Exchange just received a large amount of liquid nitrogen, she may send an alert to her large customers to inquire if they would be interested in the inventory.

Customer-driven alerts. Any customer should be able to set up alerts in the system if certain products they are interested in become available or go on sale. To continue with the liquid nitrogen example, a customer may proactively sign up with the supplier to receive such alerts whenever a new shipment comes in.

So what does this mean from a technical perspective? Let's refer to Figure 10.3. Three components in this diagram refer directly to the real-time monitoring and event-driven feedback:

- An SMS gateway is needed to send alerts to the customers. This assumes that most alerts are sent to devices capable of displaying SMS messages, such as mobile phones.

- GSM-based tracking devices are used to provide location services for shipments.

- A Simple Mail Transfer Protocol (SMTP) gateway allows the sending of email alerts to the customer.

Handheld computers are also important from a real-time monitoring perspective, since the MetX Exchange employees will capture data with these.

Figure 10.8 shows the application components that are needed to provide these services. This figure is an extension of Figure 10.6 with the following additional components:

Device tracking.   An application component has been added to the handheld device to provide location tracking services. You can buy off-the-shelf applications to provide some of this functionality.

Location services.   This component is responsible for running the location service on the back end. This is most likely an integrated off-the-shelf application.

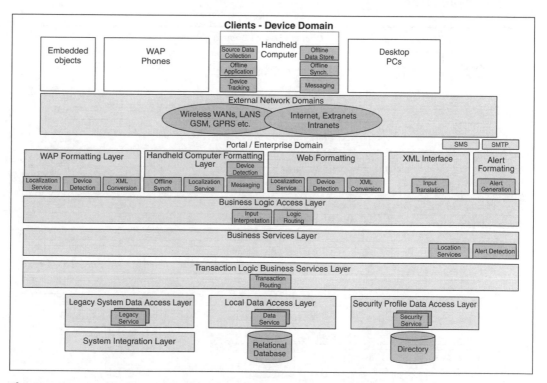

Figure 10.8   Application components needed to provide real-time monitoring and event-driven feedback.

Alert detection. This component is responsible for running the business logic needed to determine whether an alert should be sent. This component should be broken down into all the different sourcing application components. The components are location tracking, order status, inventory status, and alarms and alerts.

Alert generation. This component takes a request for an alert, formulates it for the alert mechanism being used, and sends it on its way. In this case, we only have SMS and email-based alerts, which means that we need one component to formulate the SMS alerts and communicate with the SMS server and one component to compose the email and communicate with the email server.

Please note that all of the data sourcing applications are still exactly the same as before. We are using the data to determine whether alerts should be sent. Figure 10.9 shows how these components communicate.

The data source collection communication is the same as before, with the exception that the alert detection component is placed in the application layer and that all inbound data is analyzed by it. If an alert needs to be sent, the alert detection component sends a message to the alert generation component that then generates the alert in SMS or SMTP format, depending on the user preference.

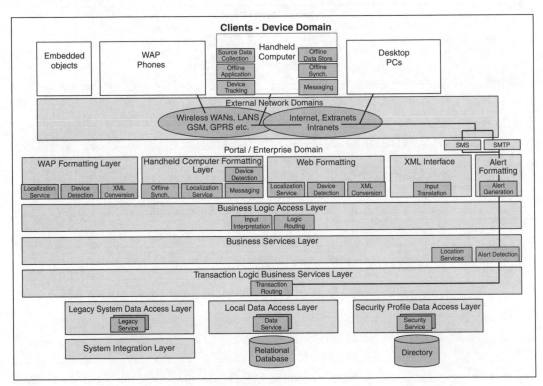

Figure 10.9 How components are connected.

Though not shown in Figure 10.9, the user also interacts with the alert management component to set up alerts. This can be accomplished from all three interfaces. There are no offline components. The information entered in this component is used by the alert detection component to determine whether an alert should be sent.

Dynamic Optimization

Since we now have a lot of data provided by the source data collection components, we are in a position to feed this information back into some of the other system components. We will take only a very high-level look at this since it is not wireless. However, this section does show how to extend your wireless services further into the enterprise.

Please refer to Figure 10.8, which shows all the different components that are responsible for executing the application functionality. One aspect we didn't spend much time on in the discussion of the functionality is data access. As you can see in Figure 10.8, we have three different data access layers. The one we are really interested in here is the legacy system data access layer. This is where much of the integration to other systems, such as ERP and supply chain systems, will take place.

The integration to these systems varies from system to system and we cannot say here how you should do it. However, in some cases you will need message-oriented middleware or some other middleware to run the transaction. This will be the case if the updates are in real time. If they don't have to be in real time, a batch data transfer process may suffice.

The technology decisions here will depend on the constraints on the products you are integrating. You will need to evaluate this integration when looking at the architecture of your solution. The solution can be tricky.

Summary

This chapter illustrated how to take an existing system and modify some existing services to be available in a wireless environment. The important lessons to be taken away from this chapter include the following:

- It is not necessary to replace an entire system to make some of its services wireless. The system was designed and built in such a way that channels and components could be added in a relatively standard way.

- The effort to add the wireless services was minimized, as we were able to add components and channels to the existing architecture.

- The biggest challenge was to integrate the new wireless services into the existing legacy applications. A good example is the ERP integration for data sourcing. Much of the data was fed into the system through the front end and several ERP components that were already in place. However, we did have to go through several systems to accomplish this.

To wrap up, you saw that it is possible and cost effective to use as many of your legacy systems as required to provide wireless services. Leveraging existing systems can eliminate the need for large process changes and complex deployments of entirely new technology infrastructure.

We will explore those technology infrastructure decisions in Chapter 11, where we create our fourth and final architecture for the MetX Exchange.

Technical Infrastructure Architecture

Introduction

In the previous chapters, we reviewed the logical business processes, the data requirements those processes generated, and systems information logic that is required to build our wireless visibility application. We will now look at the technical infrastructure that is needed to support the application and the MetX Exchange in general. Chapter 10 talked about how the wireless application brought additional functionality to the exchange while leveraging many of the traditional business functions and information systems already in place. The same concept holds true for the technical infrastructure. It is necessary to consider the entire physical infrastructure of the exchange to properly design its individual mobile components.

First, this chapter will take an overview snapshot of the entire system as it exists today. It will also look at some of the design and data gathering requirements that feed the design principles. The chapter will then explore the physical infrastructure design itself in terms of systems and data center architecture, network architecture, and security systems required to maintain the exchange.

Next, we will see the effect a wireless channel has on the technical infrastructure of the MetX Exchange.

Existing MetX Exchange Technology Infrastructure Overview

Figure 11.1 illustrates a high-level overview of the as-is technology infrastructure.

From this picture, we can deduce the following:

- Actors have fixed locations, or at least a fixed system of entry points. Actors can be any system participants, including users, service providers, data gatherers, transmitters, and receivers of all types that are viewed as separate from the users.

- System infrastructure consists of various presentation and systems administration servers, application servers, data servers, and storage systems.

- Network connections entail both local and regional as well as wide area networks that encompass a global network of actors.

The success of the exchange relies heavily on utilizing and exploiting the uninterrupted, protected, and managed flow of enterprise information across

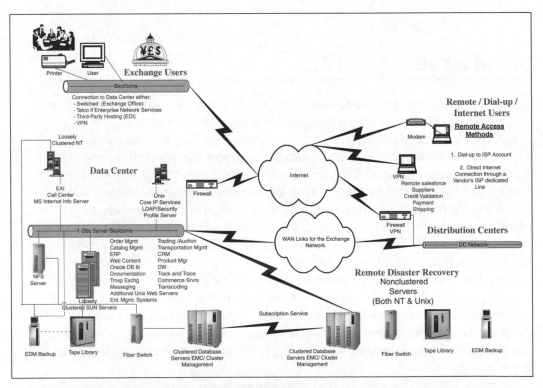

Figure 11.1 As-is MetX Exchange technical infrastructure.

its network of customers, employees, partners, and suppliers. Fast and reliable transparent access to this information from any source at any location leads to faster decision making, improved collaboration, and enhanced efficiency.

The physical infrastructure design in this document consists of:

- A three-tier architecture from Oracle
- EMC enterprise storage and software
- Cisco networking
- Sun Microsystems Solaris and Data General NT platforms
- Other third-party host hardware and software

The plan details a highly available data center platform for the exchange. It is focused on high availability and data protection, with linear scalability and consistent performance delivered in a flexible, secure infrastructure. It also develops agility for the exchange's networks by provisioning for higher-level services to manage the data-flow of these applications in real demand time.

The high-level view shows the overall physical infrastructure that supports the exchange. It is divided into five main areas:

1. Suppliers, buyers, and other third-party exchange users.
2. Remote dial-up and Internet users.
3. Distribution centers.
4. A data center.
5. A remote disaster/business recovery data center.

The design portion of this chapter focuses on providing a dynamic business exchange with a highly available data center and network platform. It is based on the concept of a single data center site that is reliable and available with high performance. It expects input via a variety of wireless service providers and also provides for wireless application service providers (ASPs). The design includes redundancy of hardware and software. In addition, this solution provides nondisruptive backup of the database using business continuance volumes (BCVs), server-based I/O channel load balancing, and failover.

Including redundant components throughout the entire stack (network to storage) is key to achieving constant availability. The design includes redundancy in all hardware and software products listed in the data center and network architecture sections. Building redundancy into the backup and recovery process further ensures high availability.

Design Requirements

When we begin designing the technical architecture, there are several initial infrastructure, integration, and testing questions we need to ask. For example, does the software meet our information system requirements? As we gather the data to answer this query, it also helps us to assess the technical requirements in terms of sizing the infrastructure to meet specific application hardware, connectivity, and throughput demands. This will enable proper sizing of the system design to address the current requirements of the system and provide ample growth over a predetermined period of time with minimal cost to the users. The phenomenal growth rate of information technology (IT) demands that we plan well ahead for infrastructure development and perform a reality check at least every six months. This exercise will reduce the possibility of an IT plan becoming rapidly obsolete due to industry developments. The methodology for information gathering and project definition can be applied to many types of IT-related undertakings, as they all will have similar concepts but varying degrees of complexity and client requirements.

When we gathered the technical requirements for our current state of the MetX Exchange, there were hundreds, if not thousands, of technical decisions to be made and details to work out. Following is a long, but by no means exhaustive, list of the categories of technical requirements along with the factors that were considered prior to implementation.

Hardware and Software Requirements

PHYSICAL SERVERS (PLATFORMS)

- Supported tiers (database server, application servers, Web servers, etc.)
- Server load-balancing strategy
- Memory utilization model
- Server load-balancing options supported
- Availability of a transactional capacity planning model/worksheet

OPERATING SYSTEMS

- Supported operating systems and versions
- Supported clustering/failover software

RDBMS

- Supported remote database management system (RDBMS)
- High-level data model
- Failover support (e.g., Oracle Parallel Server)

WEB SERVERS SUPPORTED

- Apache
- BEA
- Microsoft IIS

- Netscape
- Websphere

SUPPORTED BROWSERS

- Microsoft Internet Explorer
- Netscape Navigator

ENTERPRISE SYSTEMS MANAGEMENT (ESM)

- Supported (ESM; e.g., HP OpenView, BMC Patrol, CA-UniCenter, Tivoli)
- Standard message information broker objects provided

STORAGE SYSTEM

- Backup and recovery packages supported
- Potential restrictions on third-party disk subsystems (i.e., EMC)
- Preferred disk I/O strategies

SECURITY OPTIONS

- Lightweight Data Access Protocol (LDAP) support
- Active Directory support
- Support of other single-login security systems
- Support for a security profile management system

INTEGRATION

- Integration standards to be supported by the application: XML, EDI, HTML, CORBA, DCOM, EJB, SOAP, etc.
- Does the application come with a repository of standard integration objects?
- Does the application provide internal tools for modifying, extending, or creating standard integration objects?
- Does the application vendor have a formal partnership with one or more third-party integration/middleware providers?
- What information on initial data loads can you provide today?

SPECIFICS FOR THE DATA WAREHOUSE

- Is the data warehouse provided with a standard extraction, transformation, and loading (ETL) tool?
- Supported third-party ETL tools
- Links to enterprise resource planning (ERP) information systems and data structures

- Standard business intelligence schemas
- Tools for defining custom schemas
- Supported online analytical processing (OLAP) tools (including third party) and related presentation strategy (Web-based, PC-based, server-based)
- Supported data mining tools

TESTING ENVIRONMENT

- Recommended instance strategy (development, quality assurance/quality control [QA/QC], production)
- Internal release management and version control tools
- Support for automated testing tool environment (i.e., Mercury)
- Provision of an internal facility for documenting business processes and transactions

SYSTEM ADMINISTRATION REQUIREMENTS

- Enterprise systems management
- System administration resource requirements
- Functional and technical skills required to administer the system

Type of Access Needed

NETWORK SCENARIOS

- Local area network (LAN) access
- LAN access via wide area network (WAN)
- LAN access via Internet
- Dial-up access
- Disconnected user
- Wireless
- Virtual private network (VPN)

NETWORK PROTOCOLS

- 802.3
- 802.11b
- 802.15
- Asynchronous Transfer Mode (ATM)
- Frame Relay
- Bluetooth

- TDMA
- CDMA
- GSM
- GPRS
- iMode
- EDGE
- CDMA 2000
- WCDMA

BANDWIDTH REQUIREMENTS OR CAPABILITIES

- 9.2 to 14.4 Kbps
- 57.6 Kbps
- 384 Kbps
- 1.5 Mbps
- 45 Mbps
- 10–100 Mbps
- 1000+ Mbps

CLIENT'S BROWSER TYPE

- WAP
- WebClipping
- UP.Browser
- iMode
- Netscape
- Microsoft

CLIENT OPERATING SYSTEM

- PalmOS
- Symbian
- Windows
- Unix
- Apple
- EPOC
- Windows CE
- UP
- FLEX

APPLICATION CHARACTERISTICS

- Single CPU or SMP
- Bandwidth needs (request and return)
- Any application quirks (such as an app that makes a DNS call for each dropdown box clicked)
- Browser-enabled capabilities

SERVER REQUIREMENTS (NOT MINIMUM)

- DASD needs
- DASD per project, per product, per whatever
- DASD per level of detail for each of the above
- DASD per resource, etc.
- Concurrency expectations
- Types and size of stored procedures
- Number of SQL transactions per second per request
- Number of cache reads per transaction
- Number of writes per transaction
- Memory needed per user accessing the system
- Memory needed by application
- Memory needed by application process, etc.

SECURITY NEEDS

- NT Domain/Novell NDS requirements
- Internal security needs
- Direction on directory services/LDAP
- Profile management

INTEGRATION

- ODBC capability
- Direction on OLEDB, SQL/Net, etc.
- APIs
- Capability to load multiple versions on same CPU (same or different levels)
- Initial loads, conversions, white papers, customers, etc.

What we are trying to do here is get a thorough understanding of the capabilities and requirements of various software and hardware components that will make up the exchange. We also need to be clear on how they will be integrated—or not—because of compatibility links with the overall system design. Here are some more examples of what to look for when sizing and designing the system:

- The communications requirements of the application, such as:
  - Internet versus intranet versus extranet system access
  - VPN to allow private transactions and data transfers over public networks
  - DNS/DHCP providing logical name lookup capability for finding other parties in the system
  - Firewall/gateways/proxies to restrict network access to only those intended to participate
- Vendor-specified capacity requirements and justifications
- The operating system for a particular function
- Software components required for the application and function
- The number of users who will access the function, at what times, and from where
- Transaction details and bulk data transfer information
- Projected disk space requirements
- Database(s) to be used (DB/2, Informix, Oracle, SQL Server, none, etc.)

What are the backup requirements? Backup requirements are frequently overlooked and are even more often underscaled for the business requirements. Backups are key in establishing the time to restore after disaster and the restore level achievable in the event that a segment of data needs to be restored from a previous state. For example, suppose it is agreed upon that backups should be daily. This decision will most likely be made by doing a cost-based analysis, though it needs to be understood that there is the possibility of losing 23.999 hours worth of transactional data with this model. If this loss is acceptable, then the daily backup strategy should be used.

Availability Requirements

- 7×24×365 (never down, 99.999 percent available)—extremely expensive
- 7×24×365 (highly available, 99.99 percent)—standard requirement
- 5×12×360 (business day)

Requirements for a Development Test and Training Environment

Another one of the typically most neglected areas, the test environment needs to mimic the production environment as nearly as possible. How often have systems performed perfectly in testing, only to have the exact same transactions blow up in production? Differences in hardware can cause lockups to occur in one environment and not the other. Testing the scaling limits of the system is difficult without mirrored environments, as well. Though it is usually not reasonable for an organization to buy multi-million-dollar test servers, having the environments as closely simulated as possible can save an IT department tens of millions of dollars.

Once we have a complete picture of what is required from a computing data center perspective, we also need to compile the appropriate network information that will support all exchange LAN/WAN environments. The following network requirement questions help to provide us with examples of how this might be approached for each site. It is assumed that some network environment already exists for the organization and/or ASP host.

Network Requirements

There are many different characteristics of the network needing elaboration for our exchange to work. Understanding the data flows, the user types and volumes, and the times at which network traffic peaks are but three such characteristics. We will show how the network requirements were understood when we put together our existing wired network version of the exchange. For instance, we had to answer the following types of questions:

Number of endpoints

CURRENT
10BaseT:
100BaseT:
1000BaseT:
ATM:
Other: specify types and number of end points

FUTURE (2 TO 5 YEARS)
10BaseT:
100BaseT:
1000BaseT:
ATM:
Other: specify types and number of end points

Number of Clients/Servers

CURRENT
Clients: Unix Clients: Specify numbers
PCs: Specify numbers
Others: Specify types and numbers
Servers: Specify types and numbers

FUTURE (2 TO 5 YEARS)
Clients: Unix Clients: Specify numbers
PCs: Specify numbers
Others: Specify types and numbers
Servers: Specify types and numbers

Service-Level Requirements. Special requirements on network availability, fault tolerance/redundancy, management, operation support, QoS, Class of Services, Types of Services, content flow, caching, voice and video.

Other Requirements. Other requirements not specified above.

Once we understood these physical network requirements, we were able to look at other aspects, such as the applications using the physical networks and the patterns of use those applications exhibit.

Application Profiles

We need to know specifically how the various applications that make up the MetX Exchange will impact the network. Such an analysis will be helpful to produce a baseline for system loads during different periods, especially peak periods, of the exchange's operation cycle. Monitors can be set up across the network that will graph utilization of the servers, other key components, and the network as a whole. The same study can be performed after systems integration to determine the positive or negative impact to the systems. This process will also produce a qualitative means of demonstrating the level of infrastructure improvement achieved.

Network Protocols. Enter types of protocols running, current and future, such as TCP/IP, IPX, SPX, and NetBIOS, and how each impacts the network.

Applications. List applications, current and future, and how each impacts the network. For instance, data backup, WWW, videoconferencing, and voice. Point out applications that are primarily used outside of primetime.

Enter any special application requirements on the network, current and future, which require attention

Network Profile (Current Network)

At this point, we provide a detailed network profile of the existing network.

EQUIPMENT TYPE	DESCRIPTION
Routers	Enter the number and type of router, and number of router ports in use.
Hubs/concentrators	Enter types, makes, and number of hubs and concentrators in use. For instance, 10BaseT, 100 BaseT C/FDDI, etc.
Switches	Enter type, makes, and number of switches in use. For instance Layer 2, Layer 3 switches, 10 BaseT, 100 BaseT, 1000 BaseT, FDDI, ATM.
Wiring	Enter type, makes, and number of switches in use. For instance Layer 2, Layer 3 switches, 10 BaseT, 100 BaseT, 1000 BaseT, FDDI, ATM.
Wiring closets	Number of wiring closets:
Network connections	Enter the number of nodes with the following connection types:
10BaseT:	
100BaseT:	
1000BaseT:	
ATM:	
C/FDDI:	
Other: Specify type and number	

ADDRESSING INFORMATION	DESCRIPTION
IP Addresses	Enter class B and C addresses allocated for the current network
Netmasks	Enter network masks used

The People Aspect

The same level of questioning that goes into the physical architecture must also be applied to the users of the infrastructure to allow the system integrators to determine and minimize the risk to users and the systems they depend on to complete their daily work. Failing to consider the people aspect can have a serious impact on the business in the form of lost revenue or diminished employee satisfaction. Testing staff views of the system before and after the systems integration will be a key marker in determining the level of success of any infrastructure project. Success can be measured by tallying the quantity, quality, and severity of calls to the help desk or problem-reporting facility inside the organization.

It is critical to consider the solution's impact on people and their systems so that proper planning can be applied with ample recourse time to address potentially misbehaving solutions that do not want to play ball with your migration or integration. Risk mitigation has to start with the consideration of the users. If your users are well taken care of, you can rest assured that the risks to physical systems are also diminished.

The diversity of applications and their use from corporation to corporation has made it very difficult to employ any single method of soliciting information from users in an effort to develop a good project plan. It is better to understand the applications that are key to your organization and then to develop a line of questioning that will glean the required information that matches each specific situation. Tailoring your inquiries will constitute a study in itself and consist of many meetings with the leaders of the corporation's user groups. One of the side benefits of this approach is an increased acceptance of the changes systems integrators are implementing. If the people who will be affected are included in the planning phase of the project, they are more likely to buy into the concept, knowing that their concerns are being considered.

Problem Summary

Here we provide detailed information on known problems and shortcomings that exist with the current network, especially those that must be solved. They are ranked as Severity 1 (system cannot function properly until the problem is fixed), Severity 2 (system functionality is diminished, but there is either a workaround or noncore functionality is lost), Severity 3 (a defect is present, but it does not prevent system usage in any meaningful way), and Severity 4 (very minor usability issues).

Obviously, the problem-solving effort focuses mainly on Severity 1 and Severity 2 fixes, and, when everything is quiet, Severity 3 and Severity 4 problems are explored. As a rule, all Severity 1 and Severity 2 problems have to be closed satisfactorily for a new version of the code to be considered ready for release.

Scope

The scope of infrastructure projects may be determined by the summary of problems experienced in the environment. The problem summary document serves the purpose of providing the status of the IT infrastructure, identifying problems, and recommending the upgrade path. Clearly defining scope will not only assist in prioritizing projects, but also allow for better risk management during the implementation of projects. The recommendations contained in this document should outline practices that will allow for growth in the infrastructure, using the most current principles and equipment. The problem summary is not meant to be a final analysis or recommendation, but a starting point for discussing and planning a more functional, efficient, and secure infrastructure. The document is meant to be a living document that may change with the environment to keep up with the most current technology developments.

Network Management

Here we provide information on what tools are or will be used to perform network management. Escalation protocols should be developed and in use at the site in order to manage the communication of changes, timelines, and affected parties of pending changes. Proper management will improve overall system integrity and customer satisfaction, as users will be informed and aware of what is happening in their environment.

When working with network equipment, change management practices are essential because of the huge impact even just a single wrong word in a configuration line item can cause. Changes to network equipment should be documented at every step of the way to avoid serious denial of service that can occur even from only minor changes. If at all possible, very time-intensive changes may warrant having users on site that are capable of stress testing systems that relate to their areas of expertise. Their tests performed during network modifications will reassure the system integrators that the system will function as desired after the changes are completed and that the company can pick up its operation without delays.

Tools Used to Monitor the Current Network

TOOLS	DESCRIPTION
Network management	e.g., HP Openview, SunNet Manager, CiscoWorks, etc. Briefly describe how they are used.
Data collection tools	Netscout, NetMetrix, RMON software, etc. Include number and how they are used.
Troubleshooting tools	Sniffers, Fluke, etc. Include number and how they are used.
Others	

Security Requirements

Finally we need to know what types of security requirements exist. For example, all applications have a TCP/IP port number associated with them in order to traverse the network. By default, the firewall will *deny* all traffic. List *all* applications (or protocols) that need to pass through the firewall for this project. Also after each application, list the TCP/IP port number associated with the application or protocol and whether it is TCP, UDP, or ICMP.

APPLICATION NAME	PORT NUMBER	TCP/UDP
EXAMPLE: HTTP	*80*	*TCP*
HTTP	80	TCP
SSL	443	TCP
Lotus Notes	1352	TCP, UDP
Planview	80, 1269	TCP
Legato	7937, 7938	TCP
Lucent Message Manager	111	TCP & UDP
DNS	53	TCP

In order to allow the external network to access the exchange network, we must know where the users or systems are coming from. Therefore, a list of Internet Protocol (IP) addresses must be mapped to the external (remote) vendor/contractor's network or hosts that need to communicate with the exchange. A comprehensive list of *all* external IP addresses is needed because these are the only addresses that will be allowed through the exchange firewall. A list might look like this:

NAME OF HOST OR NETWORK	IP ADDRESS	SUBNET MASK
EXAMPLE: vendor network	*200.100.50.0*	*255.255.255.0*
NITRO	199.177.41.0	255.255.255.0
NITRO	199.177.42.0	255.255.255.0
NITRO	199.177.43.0	255.255.255.0
NITRO	199.178.41.0	255.255.255.0
NITRO	199.178.42.0	255.255.255.0
NITRO	199.178.43.0	255.255.255.0

In order to allow the external network to access the exchange network, we must know what areas of the exchange network are affected. Therefore, a listing of the IP addresses for servers, workstations, and networks that the external vendor/customer will need to access must be provided as well. Here is an example:

NAME OF EXCHANGE HOSTS OR NETWORKS	IP ADDRESS	SUBNET MASK
EXAMPLE: Exchange-Web server	*20.200.100.50*	*255.255.255.255*
All exchange Web sites via port 80	All	
(Legato Backup) Exsrvp01	20.97.96.21	255.255.255.0
(Legato Backup) Exsrvd01	20.97.96.22	255.255.255.0
(Legato Backup) EXCsrvp01	20.162.240.21	255.255.255.0

To properly configure security rules, we must know the data flow (who originates the traffic). In the following example, the *Source* is the side that initiates the connection, and the *Destination* is the side that receives the information and replies.

SOURCE (Network or Host from Questions Above)	DESTINATION (Network or Host from Questions Above)	APPLICATION (Questions Above)
EXAMPLE: vendor-network	*Exchange Web server*	*HTTP*
Trusted site-Exchange workstations	NITRO Lotus Notes	1352
Trusted site-Exchange workstations	NITRO Mobile Device	80, 1269
Trusted site-Exchange workstations	NITRO Message Manager	1324
NITRO/Exchange File/Print server	NITRO Legato	7937, 7938
NITRO mobile device	Exchange Web sites, Exchange DNS	80, 53
RO Legato server	Exchange file/print servers	7937, 7938
Trusted site-Exchange workstations	NITRO Web sites	80, 443

We also need to know the file size and frequency of a transmission.

File size	(e.g., 10 MB) LAN
Frequency of transmission	(e.g., daily) 24×7

Does the business case for the project require high availability? If high availability is needed, we would use two firewalls, located in the same building, for automatic failover purposes.

What type of connectivity is being used between the exchange and the external customer/vendor?

- Point-to-point (T1, 56 Kb, etc.)
- ISDN
- Frame relay
- Internet

Does the external customer/vendor currently have a firewall?

x	Yes	Brand and type	Cisco Pix 520
	No		

Does this project require a firewall-to-firewall VPN? If so, the current information about the firewall is needed.

Firewall software type (Raptor, Checkpoint)	Checkpoint
Firewall version	4.1
Service release	SP 1
U.S./global	U.S.

Also needed is the preferred firewall-to-firewall tunnel type.

- IPSec-Default
- ISAKMP
- FWZ-(Checkpoint only)

We need to specify the preferred level of encryption to be used for the firewall-to-firewall VPN connectivity.

- 40 bit
- 56 bit (DES)
- 128 bit (DE3)

And we note the preference of the supplier regarding a shared secret (if used).

- Exchange security
- Customer/supplier
- Buyer
- Seller
- No preference

We hope this paints a good picture of the detailed analysis that must be undertaken in order to ensure the success of an exchange, characterized by heavy reliance on utilizing and exploiting the uninterrupted, protected, and managed flow of enterprise information across its Web and mobile-based network of customers, employees, partners, and suppliers. It would be impossible to create a well-integrated environment that functions across many different silos in an ebusiness enterprise without thoroughly understanding all these factors. Now we need to see what changes must be made to enable the MetX Exchange to leverage a wireless channel.

Wireless MetX Exchange Technology Infrastructure Overview

Figure 11.2 depicts what would need to be added to the infrastructure to allow the wireless channel to exist.

If you recall, we used the high-level view in Figure 11.1 to show the overall physical infrastructure that supports the exchange. It was divided into five main areas:

1. Suppliers, buyers, and other third-party exchange users.

2. Remote dial-up and Internet users.

3. Distribution centers.

4. A data center.

5. A remote disaster/business recovery data center.

Our new wireless-enabled infrastructure adds a sixth area: wireless system components and exchange users. This area will interact with the other five areas exactly as they interact with one another in the current state architecture; they simply originate from a wireless piece of technology. We will now look at how those other parts interact with one another, and with their wireless partners. Let's take a closer look at the data center

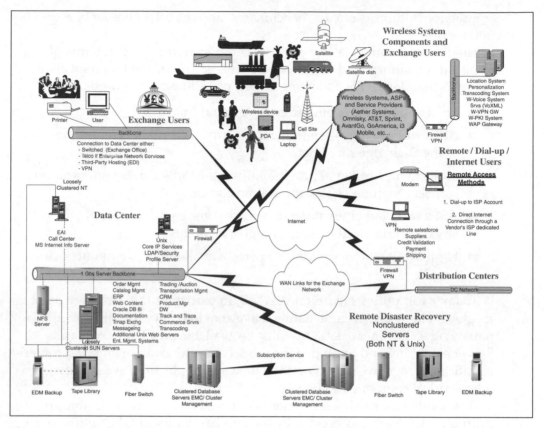

Figure 11.2   Wireless MetX Exchange technical infrastructure.

and network components supporting the exchange and our visibility application.

Assumptions

First, there are a few assumptions that we need to consider. One assumption is that the systems represented in the data center view exist as one site that is connected to a disaster backup recovery location. In this case, we do not perform geographic load balancing (mirroring) across multiple data centers. However, even in this scenario we might outsource certain portions, such as the wireless components, to an ASP. Figure 11.2 illustrates this case.

The next assumption asserts that certain basic requirements have been met in regard to the availability of adequate bandwidth and system sizing, based upon some of the information previously discussed. Included here would be transaction volume estimates, benchmarks, and the total number of requests and reports.

Finally, we assume that our physical component mapping in terms of application platform and network support is accurate, and that high availability is mandated. The following is a list of additional assumptions:

- The environment is built from the ground up.

- ASPs will use their own Internet security and will not connect directly to the Exchange's intranet.

- Links to data center(s) provide sufficient bandwidth to support local contact center and CRM application clients.

- ASPs will offer alternatives or replace some components but will, in any case, address the issues of high availability.

- High availability means 99.999 percent availability (= 5 minutes' downtime?).

Today's computing environments, and, in particular this exchange due to its global and transportation nature, have zero tolerance for unplanned system outages. In general, systems must be available on a 7×24×365 basis. Server clustering and redundant network computer components and routers significantly improve mission-critical data and application availability, even when critical components fail.

Availability refers to the computer systems' ability to deliver usable applications to the users, represented by the number of nines in the uptime percentage. For example, a high availability requirement of 99.9 percent translates into an unplanned application downtime of approximately 9 hours per year using the following calculation:

365 days × 24 hours = 8,760 total hours

8,760 hours × 99.9 percent = 8,751 hours

8,760 − 8,751 = 9 hours unplanned downtime per year

The following table shows the annual downtime in a progression of additional nines.

UPTIME PERCENTAGE	NUMBER OF NINES	DOWNTIME
99 percent	2	3.5 days per year
99.9 percent	3	9 hours per year
99.99 percent	4	1.0 hours per year
99.999 percent	5	5 minutes per year

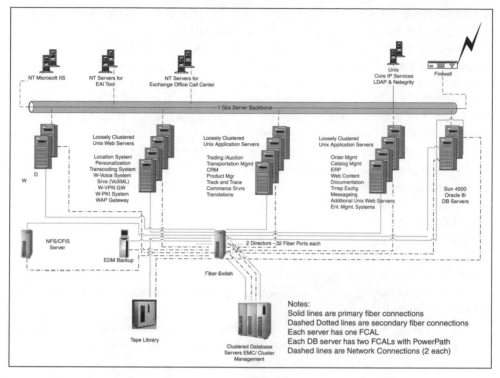

Figure 11.3 MetX Exchange production data center.

Bearing this in mind, let's look at the various systems of our data center, as shown in Figure 11.3. Here we see the major exchange's information systems application servers as well as those that would or could be used by the exchange.

Data Center Architecture

Figure 11.3 depicts the data center sitting on a trunked gigabit Etherchannel backbone, extending out into the LAN and WAN environments. It also shows the storage area network (SAN) segmentation for the Web, application, database servers, and network storage arrays. The SAN is provisioned with dual redundant systems in the EMC Connectrix switch. Although we could use other vendors, especially if cost is a major consideration, the EMC-Cisco-Oracle combination offers a high level of sophistication, flexibility, and scale. Symmetrix is the industry leader in open storage solutions, found in most successful ebusiness data centers. Symmetrix is deployed in this design primarily for external database and Oracle software storage and business continuance volume (BCV) devices used for backup and recovery, as well as a potential environment for development and testing.

The SAN, EMC Network Storage, and Connectrix Switch are managed through the EMC ControlCenter. The ControlCenter provides centralized information management for the Symmetrix Enterprise Storage and the Connectrix Enterprise Storage Network systems. These comprise the monitoring, configuration, control, tuning, and planning capabilities of the information infrastructure, while keeping track of changing requirements. Status, configuration, and performance data, as well as key control features such as BCV control, are also available to system administrators through application program interfaces (APIs).

In this environment, EMC PowerPath software provides multipath access to disks and dynamic load balancing for enhanced performance. PowerPath can leverage the alternate path to increase performance by routing I/O requests on the fastest access path and avoiding delays that are a result of contention. It also protects against failure of a server I/O channel by failing over to a secondary channel and providing load balancing of the server I/O between channels.

The exchange's mission-critical ebusiness operations make it important to have point-in-time copies of the production database in order to off-load certain tasks. For example, the copy can be used to:

- Extract data for a data warehouse.
- Run reports.
- Test upgrades.
- Run database consistency checks.

The proposed recommendation requires both a physical and a logical backup of the database. EMC TimeFinder software uses the backup host to move data to tape. Once that process is completed, the entire contents of the database can be exported and copied to tape. The backup host off-loads and isolates tasks from the production database host, resulting in greater system performance and higher availability of the data. Using this approach for maintenance tasks also enables greater frequency of backups and integrity checks to be run with less impact to the production database's availability.

The EMC TimeFinder also allows system and storage administrators to create, in background mode, independently addressable BCVs for information storage. BCVs are point-in-time mirror images of active production volumes and can be used to run simultaneous tasks, without disruption to the production environment. This parallel processing capability offers workload compression and allows increased availability of the application on the resilient platform. Once BCVs have been created, they can be split from their production mirrored volumes and used for backup. When the backup is completed, the BCV may again be mirrored to the previously paired production device.

To also support the high availability of the database, an Oracle Parallel Server (OPS) interfaces with a platform-specific software component known as the Cluster Manager (CM). The Cluster Manager monitors the status of various resources in a cluster, including nodes, interconnected hardware and software, shared disks, and Oracle instances. The Cluster Manager provides the ability to fail over host processes, other than OPS, to the secondary host. One of the OPS key components is the Distributed Lock Manager (DLM). The DLM interacts with the Cluster Manager to track cluster node status and to keep the database aware of which nodes form the active clusters.

The Sun 4500s are positioned as the primary and backup data servers. The secondary database host is configured in a cluster and shares Symmetrix disk volumes with the primary database host. The secondary host runs an Oracle Parallel Server instance to provide node failover. The secondary database host running an OPS instance will be idle most of the time in anticipation of a primary failure. This node may be utilized for backups from Symmetrix BCVs or other non-Oracle tasks. The other set of 4500s are used for data warehousing.

Oracle8i is the selected database because it was specifically designed for Internet application development and deployment. The Oracle Application Server (OAS) provides an open, standards-based architecture that is ideal for developing and deploying real-world business and commerce applications. By moving application logic to the application servers and deploying network clients, the exchange can realize substantial savings through reduced complexity, better manageability, and simplified development and deployment.

The Oracle Parallel Server is used to protect against downtime in the event of a loss node. An OPS provides a node failover solution in case the primary host fails. The hot secondary database instance running on the secondary host can take over for the primary instance without lengthy start-up or data recovery. If an application uses a preconnected session with transparent application failover (TAF), almost no delay occurs while reconnecting to the second server.

Located apart from the Symmetrix Network Storage systems and Connectrix Switch is the EDM Backup Manager and tape library. The EMC Data Manager is a centralized, high-performance backup and restore system. The system combines software, hardware, and support services to provide solutions that increase productivity and facilitate business continuity. The EDM server is a multiprocessor system that handles the transfer of data to an automated tape. A direct SCSI or Fibre Channel connection is used for backups to help off-load the SAN.

The file server is designed to be a high-performance, high-availability, high-capacity, and scalable server for enterprise file storage. The system achieves this through a cluster of discrete nodes within a common enclosure cabinet and with a common management environment. It is positioned here as an Network File System/Common Internet File System (NFS/CIFS) server.

Each set of application servers in the data center consists of four loosely clustered Sun 420R servers. Server load balancing is addressed through the Cisco/Arrowpoint local director. Servers are clustered in fours in order to provide 75 percent capacity in the event of a single server failure. This arrangement is also deployed for a set of four Sun 220R Web servers. NT application and Web servers are also deployed in a similar fashion. Certain NT applications require their own Virtual Local Area Network (VLAN) in order to isolate their bandwidth-intensive traffic and any non-compliant protocols. Fiber connects each server to the distribution layer network via gigabit Ethernet as well as through the Connectrix SAN via Fibre Channel Arbitrated Loop (FCAL). Database servers get two FCALs with PowerPath loaded on them. Each application and Web server has one FCAL.

Also shown in Figure 11.3 is the secondary firewall, a VPN service supporting wireless access, a WAP gateway, and an integrated access device supporting Voice over IP (could be ATM if needed). The wireless application servers include:

- Location systems
- Personalization
- Transcoding systems
- W-Voice system servers (VoXML)
- W-PKI systems

Because of the global nature of this exchange, the use and control of single mobile devices are beyond its capability and are not desirable since it would discourage use. It is safe to say that a plethora of wireless devices and network protocols will be used. Therefore, an extremely robust and exhaustive transcoding system will be an absolute in the systems design.

The ability to locate items and track shipments will incorporate a variety of devices such as bar codes, RFID, GPS, and GIS. Again, a comprehensive, robust system will be deployed and comprise several integrated tracking systems. This system will also tie into the personalization parts of the exchange or, rather, user profiles and business rules, knowledge, and event managers, as outlined in Chapter 3.

Speech recognition and voice-to-text is required from a human factors point of view, as well as the multiple languages used by the global exchange. Any system that does not support at least 15 types of languages would simply not qualify.

Finally, security must be implemented at a variety of levels, including a wireless VPN solution implemented across a number of localized and global carriers. A wireless private-key infrastructure with the use of digital certificates is mandated.

Network Architecture

The next component in the technical architecture is the network. This section will cover the various network models that were used to enable connectivity between the enterprise systems.

Topology

Figure 11.4 illustrates the overall network infrastructure architecture that supports the environment. The basic overall topology can be divided into three sections:

1. A demilitarized zone (DMZ) with external and internal firewalls, Web portal front end, and various FTP, mail, and security servers/devices.

2. An enterprisewide network encompassing LAN/WAN connectivity.

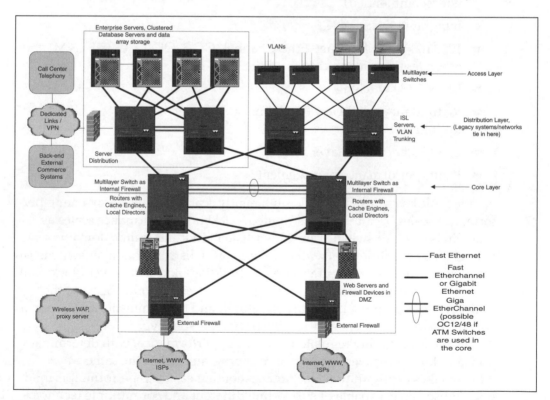

Figure 11.4   MetX Exchange network architecture.

3. A server farm back end hosting an integrated ebusiness suite of database-driven applications that provide enterprisewide connectivity to customers, suppliers, and employees. This section also includes a DMZ.

Multilayer Model

The model used in the topology is aimed at providing ubiquitous, end-to-end network services. It follows a 20/80 rule, where 20 percent of traffic remains within the local domain and the other 80 percent traverses the entire diameter of the network. Layer 2/3 switching is incorporated with VLAN trunking in the multilayer hierarchical relationship of an access, distribution, and core design. The distribution and core layers, when implemented across the enterprise, hold the network diameter to a maximum of two router hops. Ideally, these hops are implemented in hardware via layer 3 switches, occurring at near wire speed, thus reducing latency to within microseconds. Typical technologies found in this model include:

- Contiguous address schemes with 24- or 23-bit subnet masks (wider masks of 22 and 21 bits, i.e., 1024 and 2048 endpoints per domain, are not recommended)
- Integrated DNS/DHCP
- 100/1000 Base Ethernet, EtherChannel (OC-3, OC-12, OC-48 ATM) and other, less dominant, carrier protocols
- Layer 2/3 switching
- Redundancy/resiliency (including such protocols as Cisco's HRSP and uplinkFast)
- Load balancing (at layer 2/3)
- Fault/performance management

This multilayer design model is inherently scalable. Layer 3 switching performance scales because it is distributed. Backbone performance scales as more links or more switches are added. The individual switch domains scale to over 1,000 client devices with two Cisco 6500 distribution-layer switches in a typical redundant configuration. More building blocks or server blocks can be added to the campus without changing the design model. Because the multilayer design model is highly structured and deterministic, it is also scalable from a management and administration perspective.

This model is highly regulated in terms of performance, path determination, and failure recovery. Each device is programmed in the same way, which makes configuration and troubleshooting easier to perform. It is modular and scalable, capable of supporting different and/or multiple technologies such as Token Ring, FDDI, Ethernet, and ATM at the core. It is also the distributed nature of the multilayer model, when enlightened, that enables higher-level service.

This provisioning addresses the need for intelligent services that can scale via Cisco's ContentFlow TM architecture. The ContentFlow architecture consists of hardware (e.g., a local director and cache engines) and software components resident in all devices in the network, providing scalable, reliable network designs. The ContentFlow architecture has two goals: to distribute scalable and highly available services across the platforms, including switches, routers, and appliances, and to enable more informed networking decisions based on dynamic content status.

Server application feedback is delivered through technologies such as the Dynamic Feedback Protocol (DFP) and the Web Cache Control Protocol (WCCP). These feedback technologies enable the network to make informed content-based routing decisions based on server status and application availability. Flow classification is provided by technologies including authentication, authorization, and accounting (AAA) mechanisms and QoS assignments based on IP/type of service (ToS) and Network-Based Application Recognition (NBAR).

Security

The final component within the MetX Exchange technical infrastructure is security. Let's review some of the ways network security was designed along with the critical layers of a secure wireless network.

Multilayered Approach

The most highly sensitive and valuable data within the infrastructure is housed on back-end database and application servers. Information including customer account histories and profiles, product inventories, and financial transaction details must be secured from unfriendly activity at all costs. The security architecture must take a layered approach. No single layer will, in and of itself, prevent intrusion from unfriendly sources. The Multilayer Security Model includes:

- Password authentication of all customers and participants
- Secure Sockets Layer (SSL) encryption of member connecting pages for credit, data, and personal information
- Encryption of sensitive payment transactions
- Firewall/router access controls (new protocols for wireless protocols must be supported—see Figure 11.4)
- OS hardening (system servers as well as clients both fixed and mobile)
- Malicious content scanning (antivirus)
- Security monitoring

- Limited data storage
- Intrusion detection
- Application sessions and access controls, application password management
- Physical and personnel security controls
- LDAP
- Password authentication of all back-end eCommerce connections; SSL encryption of all credit transactions
- Use of digital signatures, certificates

A site can be secured through the implementation of multiple layers of security. However, the approach we take for this application does not address all aspects of security architecture such as the following:

- Customized security policies, standards, and procedures
- Organizational and operational issues relating to the deployment, implementation, and maintenance
- Operational technical and administrative specification

Layer 1: Routers

Routers offer a first layer of defense to prevent intrusion through domain segmentation and because of their ability to act as gateways, as illustrated in Figure 11.5. One key area of security is found in the area of VLAN implementation. A good example of this is Cisco's private VLAN feature. This feature, available on the Catalyst 6000 Series and Catalyst 3500 Series switches, is a layer 2 feature targeted toward providing port-based security between adjacent ports within a VLAN. It is a feature in which access ports are allowed to communicate only with certain designated router ports. Private VLANs and normal VLANs can exist simultaneously in the same chassis. The security implementation with a private VLAN is conducted at the hardware layer and does not allow any frame to pass between adjacent access ports.

In the designed network infrastructure, front-end Web servers are connected to a multilayer switch. These Web servers gain connectivity to the Internet through the switch to accept connections from various stakeholders. The Web servers must connect to two services, namely the Internet and back-end application and database servers. The private VLAN connects the Web servers in a dual-homing arrangement, whereby each network interface card (NIC) accesses each of these services. On the front end, Web servers only need to connect to the Internet through the default gateways. The private VLAN feature allows for connectivity from these servers to the default gateway routers while preventing any communication between adjacent servers.

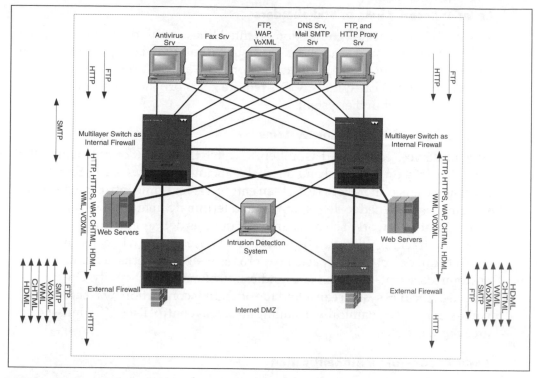

Figure 11.5 MetX network security.

This ensures that even if one Web server is compromised, access cannot be gained to other back-end application and database servers.

Layer 2: Firewalls

To address security concern, state-aware firewall services are also used to secure connections from front-end Web servers to back-end database and application servers. The firewalls provide stateful inspection on all connections and allow only the Web servers and WAP gateways themselves to access these servers on authorized UDP or TCP ports. The firewalls are typically used to allow only certain protocols to traverse them. These typically include HTTP, HTTPS, SMPT, VoXML, CHTML, HDML, WML, and FTP. They are supported via their appropriate proxy servers, which function from a security viewpoint in the same manner as the Web servers. In addition, as they are used in pairs, the firewall has the ability to perform stateful failover should a single firewall fail.

Layer 3: Additional Firewalls

Firewalls are implemented in a two-tier fashion, where one set acts as an exterior line of defense and the other as a second interior level gateway— each on either side of the DMZ.

Layer 4: Secure Links to Stakeholders

Additionally, dedicated connection-oriented lines are used between the external suppliers, partners, and payment systems, and they help prevent man-in-the-middle attacks. Virtual private networks are another, cost-effective, alternative to dedicated links. VPNs require x.509 private-key infrastructure to be present within the system.

Layer 5: Intrusion Detection Systems

Various devices can be used as sniffers to watch the network traffic within the DMZ for detecting external and internal malicious activity. Cisco's Secure Intrusion Detection System (IDS) is an enterprise real-time intrusion detection system. It is designed to detect, report, and terminate unauthorized activity throughout a network. The most granular options define signatures and trends for tracking within an IP network and raise alarms when discovered. The IDS can detect default packet patterns as well as patterns generated from customized scripts that are a clear indication of malicious activity. Once detected, the IDS system can avoid the offending connection by resetting the connection and dynamically installing an access control list (ACL) into the area border router.

Layer 6: Two-Token Authentication

Token authentication schemes depend upon authentication of a token associated with a given individual or device before granting access to services on a defined server. Security Dynamics ACE/Server and the SecureID token card are examples. Technologies such as biometrics fingerprint identification represent a further enhancement of this theme. Token authorization is typically used in the area of remote access and becomes very critical in the use of mobile devices.

Layer 7: Encryption

A type of security used extensively throughout applications and environments entails encryption algorithms, which convert messages to ciphertext. Examples of program-based algorithms include PGP and S/MIME. Protocol-based algorithms include SSL, S-HTTP, DNSSEC, IPSec, and Kerberos.

Layer 8: Digital Certificates

Digital certificates offer proof of origin and receipt. Thus, an originator cannot deny having sent the data, and a recipient cannot deny having received the data. Digital certificates often use a third party for verification. Types of certificates, usually x.509 v3-based or PKI, include authority certificates, server certificates, personal certificates, and software publisher certificates.

Layer 9: Profile Management

Using multiple levels of identification, authentication, and authorization (I&A) is a key element in the designed security architecture. The Lightweight Directory Access Protocol (LDAP) is an information model and protocol for a common A&I lookup method. Profile management is integrated across the enterprise via an LDAP server accessed via Netegrity's SiteMinder Portal Management Solution. SiteMinder provides a wide suite of entitlement and authentication management, with a single sign-on, across platforms, applications, and Internet domains. The last also provisions affiliate services, linking profile management across the exchange of suppliers and buyers.

Layer 10: OS and Platform Hardening

Operating systems are notorious for being vulnerable to security breaches. This is even more true for the wide range of handheld devices. Efforts must continuously and consistently be made to deploy known patches that fix these liabilities across all software and application versions. In addition, root login should be disabled, and hosts and password files should be hidden and not contained in the typical default locations. Default passwords should be removed or disabled in all system components.

Policies and Procedures

Finally, no infrastructure will ever be effective without guidelines for policies and procedures related to information systems functionality at the exchange, the work group, and the stand-alone levels. Although policies and procedures are not intended to replace instructions by various vendors regarding the operation of their software or hardware, they are absolutely critical to the overall health and performance of the technical infrastructure and associated information systems. The subject has already been covered in Chapter 6, but warrants mentioning at least at the high level to again stress its importance to the maintenance of the overall architecture.

Porting Applications to Wireless Channels

Porting an application to a new operating system is never trivial. Even with the universal nature of languages such as Java, the virtual machines running these applications have slight nuances between implementations, requiring the attention of both testers and developers.

For other languages, the task is even more difficult. Suppose we have an ERP system distributed client interface written in C. Even if very well written, a great deal of work will have to be performed to get the same level of functionality to work with a Palm device as with a desktop machine. Porting the application between the PalmOS and Windows CE also requires a great deal of skill, though some Visual Basic (VB)-like tools exist to try to facilitate this process.

This porting problem can be avoided for the most part if a custom client/server application is discarded in favor of a Web browser model. But even this concession will not solve all of the developers' problems, because a wireless device like a WAP phone does not have the capability to process HTML like the Palm and Windows CE devices can. This leaves two choices: Either leave yourself at the mercy of whatever HTML-WML translator the WAP gateway possesses, or take on the overhead of writing the translator in the application itself. The translator becomes more difficult by the fact that the optimal format of the WML depends on the client hardware (different phone manufacturers are geared toward different levels of WML, and some can't handle it at all).

All of this leads to headaches for bringing the wireless channel into the MetX Exchange. The benefits of the channel are so great, however, that the pain is easily outweighed by the reward. In the future, as wireless clients become more robust and standards in their functionality become more widely adopted, the porting issues will decrease. As this pain diminishes, so will the barriers to acceptance, and wireless will become as integral a part of the exchange solution as email and voice mail are today.

Summary

This chapter first explored some of the complexities and requirements surrounding up-front data gathering to determine our application's principle design. The chapter continued with a look at the physical infrastructure blueprint in terms of systems and data center architecture, network architecture, security, and procedural systems required to run and maintain the MetX Exchange. Specifically, we discussed the following:

- Fixed and mobile actors, that is, users, service providers, data gatherers, and transmitters and receivers of all types of information

- System infrastructure, consisting of presentation servers, system administration servers, application servers, data servers, and storage systems

- Local, regional, and wide area network connections

Most of this work could not have been performed without first understanding the specific business requirements that drove the need for the wireless application. Understanding the conceptual business architecture of the exchange, we had to look at the procedural data flows and the information systems supporting our set of business requirements. It was only then that we could map information systems to the physical infrastructure and engage in the more detailed technical requirements gathering process. These steps were necessary to build a technical infrastructure focused on high availability and data protection, with linear scalability and consistent performance delivered in a flexible, secure way.

This concludes our first wireless application and the methodology followed to build it. We will now move ahead to our next project, a wirelessly enabled sales force automation solution, as an example of a specific application within the larger context of customer relationship management.

Architecting Wireless Customer Relationship Management (CRM) Applications

CHAPTER 12

Overview of Customer Relationship Management

Introduction

Serving a customer is the core tenet of any organization that provides a product, a service, or information. Whether or not it is in business for a profit, the most central business function of any organization is to transfer its offerings to those who desire them. And just as commercial enterprises have customers, so do governments, educational institutions, not-for-profit enterprises, and other structures. The point is that all the mentioned entities will have to interact with their clients in one way or another, sooner or later. However, interacting with a customer requires the allocation of scarce resources. And, especially if you are working at an organization that operates under the profit motive, conserving those resources is one of your critical challenges.

CRM—What Is It?

In an effort to address the challenge of conserving scarce resources, many companies over the years have embarked on cost-cutting missions that resulted in reengineered back offices, improved technology infrastructures, and fine-tuned business processes to increase productivity. Today, however, new revenue growth is challenged by ever increasing market changes, some of which include the commoditization of industry segments, waning customer loyalty in the light of intense competition, and the organizations' legacy systems impeding adaptability and progress. In addition, customers are becoming increasingly demanding. They expect 24×7 access to information, and they expect the same level of quality service no matter what channel they choose to interact with your company, be it in person or via mail, phone, fax, email, or the Web.

What is needed, then, is a way to cost-effectively connect with your customers, using whichever means they prefer. Enter *customer relationship management* (CRM), a business discipline designed to maximize the value of a company's customer portfolio through more efficient and effective marketing, sales, and customer service. CRM is increasingly characterized by the convergence of these three previously separate business functions into a continuous, closed-loop process. Let us briefly examine their definitions from within the context of CRM.

- *Marketing/campaign management* includes the end-to-end management of promotional campaigns including direct mail, advertising, telesales, and Web-based selling. It determines what is the right campaign at the right time for the right customer segment, and measures its effectiveness.

- *Sales force automation* (SFA) can be defined as technologically enabling the selling capabilities of an enterprise. This includes the identification of cross-selling and up-selling opportunities by means of leveraging improved information flows about a customer and the communication of this information across the whole organization. SFA increases the effectiveness as well as the efficiency of a sales organization and its sales efforts across multiple channels.

- *Call/contact center solutions* are evolving traditional call centers from cost to profit centers. This evolution includes the right balance of injecting strategy, business process, and technology into a call center to enable better management of customer relationships through multichannel/point interactions. Such activities can go a long way to facilitate customer acquisition and increase customer retention and value.

The first step in CRM is to define the most meaningful characteristics for each of these business functions and then to establish business rules (via software) to route the customer or information about the customer to the most appropriate channel. For example, a high-value customer who is considering purchasing one of your high-end products should probably be routed to one of your most experienced (highest-cost) agents. Where are those people when you need them?

Operationalizing Customer Knowledge

Knowing your customer enables you to make customer-based decisions. Just being customer led can drive you to bankruptcy. There are banks whose analysis proved that 130 percent of their profit came from 20 percent of their customers. If you are customer led, but most of those customers are unprofitable, then you are throwing good money after bad. Figure 12.1 illustrates the tangible results realized by early adopters of a comprehensive CRM strategy.

Companies that know their customers' behavior and orient their strategy, process, and technology toward the most valuable individuals grow at a disproportionately faster rate than their industry peers. It makes sense intuitively, and it has been proven empirically. These companies also tend to effectively use technology to recognize and deliver value to their customers. Not surprisingly, high-growth companies tend to have more touch points to their customers than those companies still working under outdated business paradigms. High-growth companies apply technology more forcefully, more deliberately, and in the front end of their business, and, whether as cause or effect, these same companies are the ones that are most likely to demonstrate rising satisfaction scores.

Consider the car wash that proactively contacts you at regular intervals to alert you that it has been six months since your last undercarriage wash.

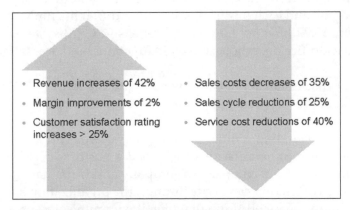

Figure 12.1   Early CRM adopters have seen results. Customer retention, share of wallet, and acquisition strategies are yielding impressive results for the early CRM adopters.

How do they do it? By affixing a bar code label that is scanned each time you use their service. The bar code links to the address you provided to the car wash company the first time you had your car treated. Not only does the company issue this alert, they also offer you the basic wash for free because they know you are a loyal customer—the system shows that you have been there 10 times in the last three months, and it is capable of giving you a detailed receipt that shows your wash history, should you need one. This real-life example illustrates how a company uses its technology to the mutual benefit of the customer—by providing a higher-quality experience—and to the company—by securing you as a more loyal and, hence, profitable customer. In contrast, both parties lose in the scenario in which you go to a hotel that you visit every month and they still ask you to fill out a little white registration card. You are unhappy about having to go through the same routine each time you check in, and the hotel misses the opportunity to develop a closer relationship with you, the customer.

The difference in the performance of these companies is in how they operationalize their customer knowledge. *Customer knowledge management* can be defined as the effective leverage of information and know-how to the acquisition, development, and retention of a profitable customer portfolio. The question is, *What are you doing to harvest the knowledge about your customer in an effort to protect this most valuable asset?*

The Art and Science of Customer Relationship Management

So, keeping all of this in mind, how do we most effectively connect to our customers? As one can imagine from these definitions, there are a number of ways to accomplish this goal, and each business will place a different emphasis on each avenue. Figure 12.2 illustrates how a comprehensive CRM system affects front-office and back-office operations, business functions, and access channels.

Moving forward in our discussion about CRM, we will use the term *customer connections* because it is especially appropriate for the context of mobile business. Today, connecting with the customer occurs via several mobile interactive channels, including cellular phones, PDAs, two-way pagers, and handheld or laptop computers. Yet exactly *how* a business best connects to its customers differs from company to company and from market to market. The primary connection can occur through the organization's brand, its people, the offer, the channels through which it communicates, or even the technology which now expands a company's accessibility, as illustrated in Figure 12.3.

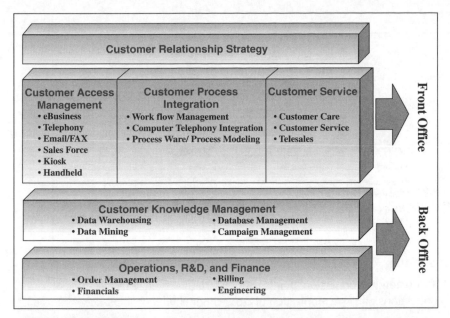

Figure 12.2   Moving to a CRM model. CRM links all front- and back-office processes in a closed loop with synchronized mixed media access channels.

The key is to allocate your resources across multiple connection points, based on your knowledge of your customers, their needs/preferences, and their economic value.

In some businesses, a high proportion of the organization's value provided to customers comes from intangibles associated with the brand. Thus, understanding what exactly attracts customers to your organization and what creates brand equity is absolutely essential. In other cases, a customer may assign a high value to improved access to the company's resources and information. The real differentiation in this case, then, comes from the convenience of how a customer conducts business with the organization. In still other instances, customers do not want a relationship at all. They might just want value for the money and functional utility out of the product. Obviously, in this scenario the key to customer satisfaction lies in the organization's core offerings and efficient transaction management.

Now that we recognize the various connection points, we should start to optimize them. But before this can occur, we need a deeper understanding of who our customers really are. We may have lots of information about them, but have we effectively applied it to the design and management of our business systems, processes, performance measures, and operating behaviors? Has our underlying strategy been the explicit objective of enhancing the customer relationship?

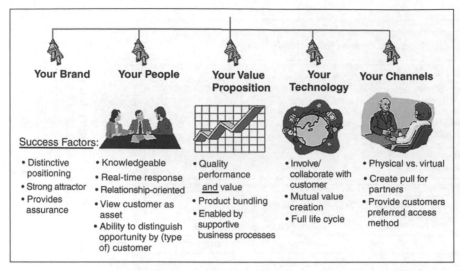

Figure 12.3 Customer connection points. Success requires an effective mix of customer connections and synchronization of cross-channel relationships.

Effective relationship management leads to profitable customers. We are naturally inclined toward focusing on new markets and new customers, yet for many companies there is a tremendous potential lift from just managing current relationships better. We can grow our portfolios simply by knowing who our customers are, identifying their needs and economic value, and then responding accordingly.

Transcending Sales and Marketing

Traditionally the customer language has been spoken by the sales and marketing function within the company. However, increasingly, the value exchange with customers revolves less and less around the core products or service and more around the intangibles, such as speed of delivery and the ability to customize. As a result, nontraditional functions within the organization are becoming instrumental in growing customer relationships. One key area, then, is the introduction of mobile computing, with its ability to capitalize on time- and location-sensitive data. Providing customers with a wireless channel through which to interact with the organization can be extremely effective in relationship management when the technology is integrated across a business enterprise.

In a well-integrated organization, the marketing, sales, and service functions no longer have a monopoly on customer relationships. Instead, the customer is now everyone's responsibility. The most successful companies today are those that manage the customer relationship by learning to follow the customer's life cycle. This means an organization must take a demand-side view rather than following the supply-side approach that got us through the industrial age. Even some of the most monolithic industrial-age companies have taken steps in this direction. For example, at a major international automotive company, there is now a CIO of the customer experience process. The benefits of such a strategy accrue to the organization as a whole. As illustrated in Figure 12.4, completely understanding your customers and following their life cycles can add value to your research and development (R&D) efforts, which will provide your organization the opportunity to improve existing products and/or spawn entirely new offerings that truly reflect customer demand and feedback.

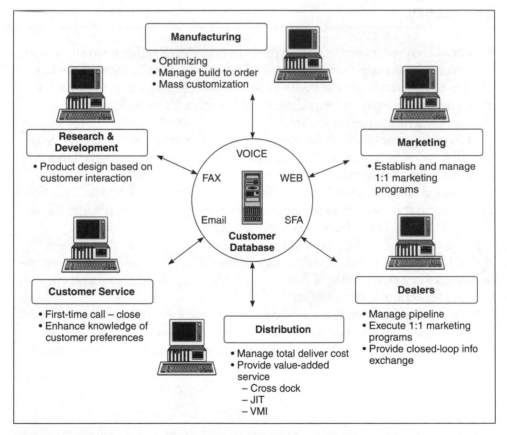

Figure 12.4 A vision of world-class CRM. Properly executed, CRM enables the company to have a 360-degree view of customers, and customers to have a complete view of their experience with the company.

In essence, there is nothing really new about putting the customer first. After all, the customer is the ultimate source of revenue and, hence, profitability. What is different now, however, is that the strategic intent of today's organizations must include technology enablers. Translating this intent into operational reality is what constitutes the real challenge.

Given the recent, rapid advances in technology, mobile computing is accelerating this trend in no small way. The fact that we can now connect directly to our customer information at any time and from anywhere gives us the opportunity for one-to-one relationships that corporations have always dreamed about. By combining mobile computing with other front- and back-end applications and systems, leading companies are realizing the benefits of both improved efficiency (faster, at lower cost) and effectiveness (more positive customer and business impact).

Summary

This brief chapter was intended to provide you with a quick introduction to customer relationship management. The take-away here is that the goal of CRM is to effectively connect with your customers in an effort to grow revenue and profitability by providing superior service through whichever channel today's customers desire.

The concept entails three primary focus areas: marketing/campaign management, sales force automation, and call/contact center solutions. Underlying these areas is a holistic understanding of who your customers are and what they desire. Companies that effectively implement CRM solutions do so by enforcing the concept to permeate the entire organization, not just those areas that traditionally have been associated with customer contact.

Elevating CRM to the next level means extending CRM to the wireless realm. Leveraging wireless technology in the CRM space offers today's organization a very powerful tool that can provide tangible financial and customer satisfaction–related benefits.

Let's now take a look at a specific business case for wireless CRM in the area of sales force automation. We will see how our company was able to realize its strategic intent in a revolutionary way through the application of mobile computing.

CHAPTER 13

Conceptual Business Architecture for a Wireless CRM Application

After giving you the high-level overview of CRM and its key concepts, we are ready to drill down into a specific CRM application. We will explore a case study of a Web-based sales force automation (SFA) application, its functionality, and its benefits. Once we understand the case study's unique business requirements and the corresponding application components, we will then look at the wireless extension, its core feature set, and the special attributes available only in the mobile realm.

The CRM case study we will use throughout Part Three of this book builds upon an SFA application that CGE&Y developed and implemented for a large company in the radio broadcasting industry. The application facilitates the selling of radio spots to local and nationwide advertisers. These time slots are filled with the advertisers' marketing messages and aired over the company's network of radio stations.

We chose to feature a project in this particular industry because selling radio airtime is posing some very unique challenges that are inherent in a product inventory that is constantly changing. While on a call, a sales executive uses the Web-based tool and its wireless version to prepare a quote for airtime on one or multiple radio stations. Using the application, sales representatives are able to rapidly increase turnaround time, resulting in radio stations being able to better manage their inventory and increasing revenue. The specific components of the application allow sales executives to:

- Rapidly create proposals for their clients.
- Filter through all radio stations in the network to find the one(s) that cater to the appropriate listener demographic.
- Route the proposal to superiors and station managers for review and approval purposes.
- Notify other sales executives or managers and receive notifications.
- Manage the proposals by modifying proposal parameters.
- Manage the accounts, such as updating contact information.
- Create reports to review performance metrics or other statistics.

Radio Broadcast Introduction

Although a complete review of the radio broadcast industry would be beyond the scope of this book, we thought it would be useful for us to very briefly look at the industry's past as it relates to advertising sales, point out radio's major revenue sources, and explore some of the challenges encountered in today's competitive environment that we will later address with our wireless SFA application.

Brief History of Radio Advertising

Commercial radio broadcasting can look back on a long tradition of providing news and entertainment to a station's listeners, getting its start around 1920. Back then primarily the realm of hobbyists, inventors, and other tinkerers, radio broadcasting over shortwave was used to transmit original content including musical shows and station commentaries in the public interest. And although there were a handful of radio stations prior to the Queensboro deal that are said to have received barter compensation usually in the form of music records for plugging the names of the companies that donated them, the first truly commercial radio advertisement was probably aired on August 28, 1922, by WEAF in New York City. In return for $100 for a 10-minute spot, paid to the station's owner, AT&T, the Queensboro Corporation real estate firm became the first so-called toll broadcast client by presenting the benefits of a newly built apartment complex over the airways.

Mostly due to word-of-mouth endorsements that radio could be used to plug products, the radio station counted over 30 companies among its clients about one year later. The early advertisers who recognized the mass-market potential of the fledgling medium included American Express, Macy's, Met-

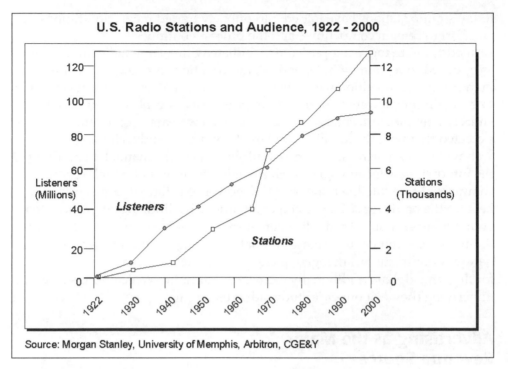

Figure 13.1   The rise of U.S. radio stations.

ropolitan Life, Goldwyn Picture Corporation, and other companies whose names have since vanished into obscurity. Over the coming years, toll broadcasting evolved from soft publicity—usually in the form of program sponsorship—toward full-blown, direct product advertising and sales pitches. By 1995, the number of radio stations in the United States had grown from just about 18 in 1920 to over 12,000 (see Figure 13.1).

Recent Trends in Radio Broadcasting

Jumping ahead to the mid-1990s, we begin to witness significant changes in the industry, the impacts of which are just beginning to play out. Specifically, radio broadcasting is faced with an increasing trend toward consolidation, the threat from alternative media channels in the battle for listeners, and new technologies.

Spurred by the Telecommunications Act of 1996, which effectively deregulated the industry and started the consolidation activity by removing caps on the number of same-market stations a corporation could own, radio operators were striving to rapidly grow bigger in size and reach. Although radio has traditionally been a highly fragmented industry, about 60 percent of today's radio stations have folded into publicly traded megacorporations in an effort to remain competitive, and the process continues.

In addition to the rising pressure to affiliate with a large national network as opposed to remaining independent, radio stations are finding themselves in mounting competition with other media. Not only has television been able to garner more viewers with the rise in popularity of cable channels, but an entirely new medium entered the scene just a few years ago: the Internet. It is expected that with the looming availability of broadband technologies, the Web will play an increasing role as a full-blown media channel. In addition to the Internet, new technologies, such as digital high-definition television, commercial-free satellite radio, and low-power FM stations, are expected to heat up the battle for scarce customer attention. Traditional stations fear that younger generations of radio listeners turn off their boom boxes in favor of finding entertainment over these alternative channels. And this fear is real, as new entertainment technologies, including Internet broadcast, are competing for the same dollars traditionally earmarked for radio. Which brings us to the underlying theme of our case study: advertising sales.

Advertising as the Major Revenue Source

The major revenue source for a radio station is the sale of airtime, or radio spots, scheduled throughout the broadcast day. Figure 13.2 illustrates the breakdown of revenue sources for radio broadcasters.

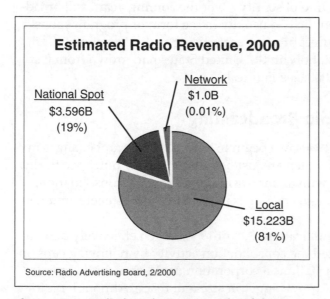

Figure 13.2   Radio broadcast revenue breakdown.

The term *radio spot* refers to a block of seconds during which an advertiser's message is broadcast, interrupting the current radio program in predetermined intervals. Radio sales executives call on their customers—either companies or their advertising agencies—with the goal to sell local spots or nationwide coverage.

There are four primary value propositions that radio offers to advertisers. First, radio spots are a very cost-effective way to expand advertisers' market reach. Radio stations are located in every major city and their signals can cover a wide geographic area. Radio's ubiquity allows an advertiser to announce its message to the entire population within the reach of the radio station's signal, and it catches you wherever you are and whatever you do—as long as you are tuned in to the station's frequency.

Second, radio is very effective at targeting a specific user demographic. A station's format is usually consistent; that is, a station plays either country music or hip-hop, but not both. Listeners are very loyal to their stations, which means radio stations can, to a certain degree, create accurate profiles of their listenership, their age range, their interests, and their lifestyles. Station loyalty is evident in listeners programming their favorite stations' broadcast frequencies into the preset buttons at home, at work, and in the car.

And therein lies the third value proposition: Radio is good at reaching consumers wherever they are—especially while they are on the go. A so-called captive audience, consumers in their daily commute to and from work are the principal targets for advertisers, explaining the primetime status of radio airtime in the morning hours and in the early evening.

Furthermore, because radio waves are everywhere, the medium is capable of delivering a large number of impressions upon a large number of individuals with a very high rate of recurrence. Unlike a magazine that is published only once a month, which requires a reader to pick it up, leaf through the pages, and demand that the reader concentrate on the content, radio delivers marketing messages 24 hours a day, 7 days a week, usually in the background, and it does not require your undivided attention.

Challenges Inherent in the Traditional Model

Now that we have provided you with a brief overview of radio advertising, pointed out the importance of advertising as the stations' major revenue source, and talked about recent industry trends, let's look at some of the challenges associated with selling airtime. We will use a fictional holding company called *Acme Radio*.

Acme Radio owns hundreds of radio stations around the country. Each station is unique in terms of geographic location, format, demographics served, and how the operation is run from a management perspective. Yet all stations are facing the same dilemma: Today's large corporate customers want to purchase airtime slots for advertising purposes in multiple, geographically dispersed markets, and they want to do so with minimal effort and at minimal cost. Especially the large national accounts—major computer hardware/software vendors, automobile manufacturers, insurance providers, and nationwide retailers as represented by Madison Avenue's advertising agencies—desire one-stop-shopping to broadcast their messages from state to state. Although there certainly is a market for, say, small local retailers announcing a holiday promotion or weekly grocery special, it is the large, name accounts that provide the profitable high-volume business every station wishes to attract and retain.

Servicing these accounts, however, is no small feat. Acme Radio's account executives are faced with the challenge of selling inventory that is highly perishable, constantly changing, and targeted at an extremely fragmented market. *Cluster selling*—that is, the practice of packaging multiple stations into one proposal—adds another dimension of difficulty, as we will see.

First, a radio spot is a perishable good. If a spot is not sold in time—that is, before the radio show is aired—it is wasted, without having generated any revenue. Unlike selling hard goods such as lawn mowers, perishable goods cannot be returned into inventory at the end of the day to be offered when the store opens again the next morning. A radio spot is an open time slot that must be filled with a marketing message before its due date. The analogy here comes from the airline and hotel industries: If a seat on the plane is not sold, the plane takes off with an empty space, meaning a lost revenue opportunity that can never be recovered. Similarly, if a hotel room is not booked and remains empty for a night, the revenue opportunity for this particular room during that particular night has vanished. The point is that in order to avoid inventory spoilage in an environment in which you are working with highly volatile goods or services, it is critical for a sales executive to (1) have access to the latest information regarding inventory availability and (2) be able to communicate that availability with as many leads as possible before the product or service expires.

Spoilage aside, a second major challenge for the Acme Radio account executive lies in the fact that the available inventory is constantly in flux. Acme supports a sales staff of several hundred professionals, and each is being compensated to a degree on the number of spots sold. Turning over millions of advertising spots a year, these reps are truly hustling for a living. To do their job effectively, they must be able to access up-to-date information as required by the customers. And that means they need visibility into the latest schedules and pricing for each spot, throughout various times of the day, for each day of the week, for each station in the Acme Radio network. As soon as one open spot is sold to a customer, the station must adjust its available inventory and communicate that change to the field. To make things even more complicated, stations may elect to adjust their spot rates on the fly in a last-ditch effort to sell the perishable item. They accomplish this by reducing a spot's price point as they get closer to the expiration date—the time slot the spot corresponds to. As you can see, access to the latest information, such as that relating to inventory availability and pricing, is absolutely critical for the Acme account executive.

While spoilage and ever changing inventory parameters are two of the major concerns for the Acme account executive, the third headache has to do with selling inventory that is highly fragmented. Although a sales rep may handle only a handful of accounts, these accounts—especially if they are major advertisers—often wish to purchase airtime on many local stations, quite possibly located all across the country. To further complicate matters, these advertisers rarely have the patience to negotiate separately—each spot, each price point, to be broadcast from each station. Instead, Acme account executives are expected to provide a one-stop-shopping type of service that allows the customer to acquire during a single sales call multiple local spots, each priced individually, during different times of the day, in multiple markets, on multiple stations, covering multiple radio formats, and aimed at multiple demographic targets. Not only do advertisers prefer to deal this way, they also want to be able to check availability and compare rates before signing off on their orders. Can you imagine the complexity of being able to check constantly changing availability and price points for thousands of radio spots at hundreds of stations across the country—being sold concurrently by the army of Acme Radio sales reps? The traditional sales model is fraught with major challenges to providing superior customer service and maximizing sales effectiveness.

Web-Based Application Discussion

After reviewing the critical issues surrounding radio advertising sales, we turn to the solution that a CGE&Y project team recently developed for and implemented at a major competitor in this space. For consistency purposes, we will continue to refer to this client as Acme Radio. We will discuss the sales force automation application and how its various components provide strong benefits to Acme sales representatives, station managers, and the company's existing and prospective customers.

The Search for an Internet Business Model

Acme sales executives deal with a perishable inventory that is constantly updated and that consists of radio spots at hundreds of stations all across the nation. Not too long ago, Acme management, upon the urging of the company's board of directors, began to investigate how the Internet could be leveraged in Acme's line of business.

Although management recognized the opportunities associated with reaching out to the stations' listeners by broadcasting over the Web, they also realized that the business model behind such an approach was questionable. Given the extreme fragmentation of the Internet user community, the major value proposition offered to advertisers was lost. Broadcasting over the Web was determined to not be desirable to the stations' advertisers because radio over this medium fails to reach a highly clustered, captive audience.

Although the Internet allows segmentation to the nth degree, Web-based stations fail to reach consumers as a captive audience—the characteristics of radio listeners on their daily commute to and from work, and the prime target for advertisers and their messages. And while the B2C model was quickly rejected, Acme management decided that the Internet could indeed be a valuable technology enabler—but in a different dimension. Instead of focusing on a station's listeners, the technology could be exploited to facilitate the station's internal business operations.

With this major strategic direction in place, the next step for Acme was to determine which of the various internal business functions would benefit the most from the Internet's open architecture and reach. While we assisted the company in exploring several internally focused applications, ranging from Web-based employee self-service all the way to program lineup and content procurement, it quickly became clear that the strongest business case could be created for the application that would directly impact the stations' primary revenue driver—advertising sales. In the next section we will describe the specific functionality of the Web-based media sales application, from here on referred to as *MediaMaven*.

Functional Components of the Web-Based Sales Force Automation Tool

Once the decision was made to pursue opportunities that would streamline business operations as relating to advertising sales, we investigated the various processes that were part of the activities. The following paragraphs describe the processes and functionality offered by the Web-based sales force automation application as relating specifically to media sales.

MediaMaven and its Web-based user interface are hosted in a secure environment. The Web site is accessible only to authorized personnel who are required to submit a valid user ID and password. It should be noted that MediaMaven is available via desktop or laptop PC to Acme reps in the office or in the field, Acme Radio corporate headquarters, and the management of individual radio stations in Acme's network. The following sections examine

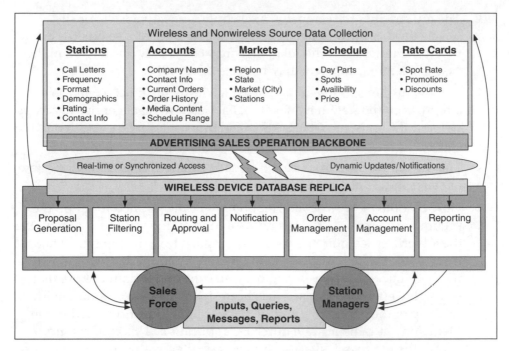

Figure 13.3   Components of the wireless SFA application.

the functionality offered for proposal generation, station filtering, routing and review, notification, order management, account management, and reporting. Figure 13.3 illustrates the MediaMaven application. The following paragraphs describe the user-facing functionality, including Proposal Generation, Station Filtering, Routing and Approval, Notification, Order Management, Account Management, and Reporting.

Proposal Creation

A proposal is a comprehensive offer that the Acme account executive prepares and submits to his or her client, that is, the advertising company or its advertising agency. The proposal contains the details regarding the marketing message, spots purchased, pricing, scheduling, and delivery details, as well as legal terms and conditions. The entire proposal consists of multiple orders, and each order covers an individual radio station at which a spot is purchased. The steps involved in creating a proposal are as follows:

1. The Acme rep logs onto the Web site and begins the proposal creation process by searching the MediaMaven application for, or selecting from a list of all clients, the specific account for which to prepare a quote.

2. The rep specifies the campaign's schedule range; that is, he or she selects the day the message is first broadcast, and then selects the day that it is aired last. A schedule range may cover only one day or, in the case of a larger campaign, may run several months.

3. The rep further specifies a date by which he or she must have received a reply from the station manager(s). Each station manager has to accept or reject the rep's quote before the final proposal can be presented to the client.

4. The rep selects the markets to be reached by specifying which U.S. state or states the client would like to broadcast in.

5. Now that the most basic information has been specified to set up the preliminary proposal, the rep needs to find all the stations that map to the advertiser's requirements. To accomplish this, the rep must choose from a set of station formats, demographics, and markets to be reached. Luckily, this task, requiring an extensive amount of manual research were it paper based, has been significantly accelerated with the help of the online configuration tool that was incorporated in the MediaMaven radio airtime purchase application. This configurator allows the rep to quickly run a three-pass filter against the hundreds of stations located in the geographic market(s) selected in the previous step. First, the rep specifies the desired station formats from a long list of options, including rock, country, jazz, and talk radio. Second, the rep selects the demographic parameters surrounding the desired gender and age ranges. Third, the rep specifies the cities within the markets previously chosen. With all the required parameters selected, the rep clicks on a button that applies the filter. As a result of running the filter to select the desired station formats, demographics, and markets, the application returns a set of radio stations that show the characteristics desired by the advertiser and available inventory. Stations that have no inventory for sale are automatically hidden. If the result set is too large or too small—that is, it contains too many or not enough stations—the rep can retrace his or her steps and modify the configurator to narrow or widen the search.

6. When satisfied with the number of results returned, the rep selects the individual station(s) for which to check the inventory for available time slots. This final set of stations is saved for future reference before the rep moves to the next step.

7. The next step in the media purchase process is to schedule when an advertiser's message is to be broadcast. Using MediaMaven, the rep is presented with a schedule in calendar format for each station selected. Each station divides a day into *day parts:* A.M. drive, midday drive, P.M.

drive, Evening drive, and overnight drive. This division is fairly consistent across all stations in the network. The rep immediately sees which day parts are available and the associated price point per spot. Within the online calendar, the rep now specifies the number of spots to be purchased and the day part during which they are to be aired. The total price is automatically calculated, and additional parameters such as markups, discounts, or comments may be added at this time. Once the rep is satisfied with the scenario, the rep's supervisor approves the quote, and the client has signed off on the total cost, MediaMaven routes the preliminary proposal to the managers of all radio stations named in the document for final approval.

Filtering

You have already seen one application of the station filtering mechanism in the proposal generation feature just described. There, the Acme rep first selected a specific market—say, the state of Florida—and then applied the filter mechanism against all stations in Florida to find the ones that map closest to the advertiser's requirements. Alternatively, the filter could be used from the very beginning of the proposal generation process. Instead of knowing exactly what market an advertiser wants to reach, the Acme sales rep may be asked to provide a list of stations across the nation that show specific characteristics in regard to format, demographics, and/or location. In that case, the rep simply runs the filter before selecting a location, thus being able to rapidly identify the stations most appropriate to carry the advertiser's message.

Routing and Review

Once the preliminary proposal has been created, MediaMaven automatically parses the information. Each order is routed to the applicable station as named in the proposal. For example, let's take a proposal that contains two orders. The first order requests five spots on WXYZ in Orlando, on October 10, for the A.M. drive day part, and, therefore, is sent to the Orlando station manager for review. The second order, targeting the L.A. market, would be sent directly to the manager at KABC in Los Angeles, the specific station named in the order. Each station receives an alert that informs the station's manager of who is the advertiser, what is the media content (i.e., what product or service is to be promoted), what day parts are desired, the spot quantity for each, the unit price, and the date by which a reply is required. Further data includes the total price, markup and/or discounts, and any comments, if applicable. The station manager now uses MediaMaven to approve or reject an order. Why would they reject a quote? Poor inventory management is a challenge not only at a standard retailer selling hard goods, resulting in situations where sales executives quote on airtime already allocated to another advertiser. Or it may happen that all the A.M. drive spots on October 10 have to remain open because the Orlando station is close to finalizing a deal with a local VIP account. Alternatively, the station manager may decide that advertising a special on blowtorches offered by the local hardware store would not be appropriate for a station targeting teenagers between the ages of 10 and 15, potentially putting wrong ideas in the heads of Orlando's younger citizenry. Whatever the reason, the station managers usually have the final authority to approve or reject an order.

Notification

We've just described one application of MediaMaven's notification functionality; it entails parsing a preliminary order, alerting sales directors about a new order, and alerting station managers that the system contains new, pending orders for their station that are in need of approval. The other application, of course, encompasses the reverse of this procedure. Once an order is approved (or rejected), the status must be communicated to the Acme sales executive who initially created the order. Again, MediaMaven employs a notification mechanism that alerts the sales rep of the station manager's decision, in addition to automatically adjusting the order in the system, that is, switching the order status from pending to approved (or rejected).

Order Management

Aside from the functionality critical to creating a proposal, MediaMaven also facilitates standard housekeeping types of activities as they relate to existing proposals and orders. An order may need to be modified, deleted, and/or created anew.

Account Management

Similar to order management activities, Acme sales reps use MediaMaven to update client information, such as contact names, phone numbers, and email addresses. In addition, Acme representatives can create new accounts in the system. Because the application is integrated with Acme Radio's ERP system, MediaMaven also allows Acme reps to view current orders, past orders, customer service issues, payment information, and other customer intelligence. We should note that the reps' access to the information contained in the ERP system is limited by permissions previously established by Acme Radio senior management.

Reporting

Last, MediaMaven provides an extensive menu of reports to choose from. Ad hoc reports can be run to view all pending and/or approved orders. In addition, reports show orders by client, sales rep, market, city, station, format, demographic, and so on. Other reports show current cross-market rates and inventory availability.

Major Benefits of MediaMaven

As you recall, the standard activities of the airtime sales process involve the generation of quotations by Acme sales reps, obtaining approvals from sales directors and station managers, scheduling the broadcast, changing the orders upon request of the advertiser, managing client information, and creating a set of reports that provide operational and strategic insight into station performance, client preferences, and account executive effectiveness. In the offline world, these processes would be performed manually, on paper or with the assistance of spreadsheet programs. Communication would occur in person, via phone, or fax. The whole process would take a long time to complete, and would be extremely tedious due to the scheduling complexity and inventory volatility.

Leveraging Internet technology to streamline the sales process provided Acme sales reps, the company as a whole, and its customers with substantial qualitative and quantitative benefits. For the sales rep, the deployment of the MediaMaven application resulted in the following benefits:

- Faster turnaround of client proposals, from setup through completion, increasing the reps' overall productivity

- Increased number of accounts called upon due to significantly reduced turnaround times

- Reduction of oversold inventory situations, as spot availability was more rapidly updated and communicated to the field

- Faster response times from sales directors and station managers regarding quote approval due to electronic messaging (email)

- Improved accuracy in the quoting process, as sales reps now had access to an online database holding all stations' updated schedules (i.e., available inventory) and pricing information

- Higher average revenue per account due to the reps' ability to quote multiple rates across markets quickly and accurately

Acme Radio, the holding company, benefited from the application by receiving the following:

- Increase in revenue generated by more effective sales representatives

- Reduction of unsold inventory expiring without having generated revenue, due to better inventory yield management

- Reduction in order entry, scheduling, and pricing errors, leading to improved station operations and overall financial performance

Most important, Acme Radio's customers, the advertisers, received added value from the implementation of the Web-based media buying application. An overall improvement in customer service as relating to the advertisement purchase process resulted from the following:

- One-stop-shopping functionality, allowing access to airtime at all Acme Radio stations across the country in a cluster buying arrangement

- More comprehensive and accurate visibility into available airtime and cross-market rates

- Faster notification of order acceptance or rejection, resulting in faster proposal completion

The Wireless Solution

MediaMaven was developed, built, and integrated with Acme Radio's back-end systems, and rolled out to the organization within a nine-month time frame. Because of the large number of radio stations in the Acme network, designing a scalable system and populating the databases with station information, schedules, and pricing data required significant effort. But once the application was deployed, it rapidly became an indispensable tool for sales, station management, and corporate headquarters. However, we need to remember that, out of the box, the application was accessible only via a desktop PC or laptop with a direct connection to the Internet.

A few months after the system had been in operation, it became clear that the solution provided tremendous benefits to all parties involved in a transaction. This meant that Acme's clients received better service, account executives were able to create proposals from any PC connected to the Web-based application, and Acme radio stations could now better manage their inventory. However, even though the process of selling ad space was significantly streamlined, the CGEY analysis identified an additional opportunity to enhance the MediaMaven applications. Some of these enhancements included increased access capability, real-time inventory updates, and updates to the notification process.

During our regular conversations with Acme's director of sales, it became evident that a phase 2 implementation of MediaMaven to a wireless environment aligned perfectly with Acme's future strategy and requirements.

Enter the wireless solution: *MobileMediaMaven* (MMM). Using MMM on their handheld computers, Acme sales executives can check the latest spot availability and pricing in real time, right in front of their customers. The system runs on various platforms, including WAP, PalmOS, and Windows CE. In addition, Acme account representatives and their customers benefit from MMM's immediate notification capability. Let's look at how the wireless solution enhances several business functions introduced in the previous section.

Proposal Generation

The most immediate benefit obtained from MMM, the wireless extension of MediaMaven, is the ability for sales reps to check in real time the availability and pricing of radio spots at all the stations within the Acme Radio network, while on the road or in the advertisers' offices.

Filtering

The filtering feature ported to a wireless device by the MMM works in the same way as in the offline application.

Routing and Review

The routing and review feature within MMM works in the same way as in the offline application; however, both features are also available on a handheld device, freeing the reviewer/approver from the confines of an office with its wired desktop or laptop computers.

Notification

Because Acme sales reps now have real-time access to the Web from a handheld device, proposals generated with MMM can instantly be sent to the reps' sales director for approval, if required, and then on to the individual station manager. Similarly, because MediaMaven is now formatted for mobile applications, both sales directors and station managers can receive alerts while on the go. This means that sales reps no longer have to wait until a station manager accesses his or her desktop-based email inbox, reads the message, and then finds time to review and answer a pending order. Instead, the station manager can be reached wherever he or she might happen to be—and can instantaneously approve or reject a request for specific radio spots. The sales rep using MMM can create a proposal, obtain approval, and close the orders within minutes, reducing the cycle time by hours, if not days.

Order Management

Users can modify orders, amend them, delete them, consolidate them, and so on, via the handheld device.

Account Management

As offered by the Web-based application, the account management feature is extended in full to the wireless device.

Reporting

With MMM, users can run reports while on the fly, presenting customers with their past order history, station format comparisons, pricing reports, and other ad hoc intelligence.

In addition to the extension of the Web-based MediaMaven application previously described, which mostly ports existing functionality to a hand-held wireless device, the mobile version introduces some new functionality not available in the tethered environment. Let's explore some of the additional functionality that really makes MMM stand out as a cutting-edge application.

Interagent Messaging

If you remember, one of the challenges of radio advertising sales was that limited inventory (time slots) was being sold concurrently by a large number of Acme sales reps. Because multiple deals were forged in parallel, the process needed some downtime to allow a station to update its available time-slot inventory. In the offline environment, and even using Media-Maven, it was possible that several sales representatives would propose the exact same time slots to their clients. Whichever rep was able to get the attention of the station manager first usually walked away with a win—that is, an approved proposal—while the other reps had to go back to their clients bearing the bad news. With MMM in place, however, the situation—although not completely eliminated—has dramatically improved. When several sales reps compete for the same radio spots on their clients' behalf, they have the ability to contact each other via MMM's instant messaging capability. Being able to reach a competing sales rep and to come to an agreement instantaneously proves to be a major value-add to the agents as well as their customers, thus giving all parties a means to arrive at a true win-win situation.

Remote Printing

Of course, once a deal is closed, meaning the proposal went through all required channels and was approved, the client would like to receive the agreement in writing. Using the Web-based MediaMaven application, sales executives are able to send the entire agreement to the client contact via regular mail, fax, or email. Using MMM, however, that feature has been ported to the next level. While still in the client meeting, the Acme sales rep can send the entire agreement, consisting of schedules, pricing, terms, and conditions, not only to any email account at the client site, but—even more immediate— to any printer or fax machine at the client's office for remote printing.

Location-Based Profiling

To facilitate the sale of radio spots to local clients, the wireless application features a location-sensitive component that takes the guesswork out of finding the right station in a given geographic location. With the aid of the location sensor, a sales rep is provided with a list of stations in his or her vicinity immediately upon launching the application. Furthermore, the location-sensitive feature allows the sales rep to receive the latest information about the account, which the rep then can use to prepare a more effective sales presentation. Similarly, imagine the sales director of Acme Radio's southeast region traveling to visit the VIP accounts throughout the various states contained in her territory. Upon entering a client's office, the wireless MMM application instantaneously knows where the sales director is located and can instantly provide information such as a history of the client's past orders and listings of the major stations in the area, their format, and listener demographics, or any other last-minute client-specific details such as the latest account aging that would render the sales director's visit more effective. This is clearly a value-added feature for those Acme representatives who find themselves frequently on the road and in unfamiliar territory.

Acme Airtime Auction

The Acme Airtime Auction is a logical extension of Acme's traditional business model. Because MMM provides instant access to airtime schedules and pricing, Acme is developing an auction for unused radio spots to be sold via dynamic pricing algorithms. The idea is to reduce the price of unsold inventory in a dynamic fashion—that is, to progressively lower the price for a spot as expiration nears. Instead of letting inventory go wasted, Acme is expecting to generate incremental revenue from inventory that would have been lost in the offline business model.

Advertising Effectiveness Scoring

This is a feature currently under consideration for future implementation into MMM. Advertising Effectiveness Scoring (AES) is a value-added service that allows a station to generate significant nontraditional revenue by providing its advertisers with statistics surrounding a radio spot's effectiveness. The technology builds upon Acme's ability to know exactly where a listener is located and his or her response to an advertisement. Due to location-identifying technologies to be deployed in mobile devices (cell phones, PDAs, and pocket and laptop computers, as well as portable radios) and modes of transportation (automobiles, trains, city buses, shuttles, and other private or public transportation), Acme will be able to measure the response generated by a particular marketing message—a service highly desirable for advertisers.

Summary

This chapter introduced you to a case study that entailed building a CRM application with the focus on sales force automation. The chapter provided insights into the conceptual business architecture of the wireless solution, outlining the challenges that brought about the need for the application and the type of functionality that would be required to enhance the usefulness of the application. Specifically, the wireless SFA solution encompasses the following critical functionality:

- *Order management*, including wireless proposal generation, routing, review, and approval via actionable alerts
- *Account management*, including wireless access to customer profiles, transaction histories, and the ability to modify such data
- *Reporting*, including periodic and ad hoc report generation, as well as location-aware, real-time, remote access to inventory availability and pricing

Allowing access to critical, rapidly changing data required to successfully create and complete a sales proposal provided all three parties involved in the process—sales representatives, management, and customers—with measurable benefits such as faster proposal turnaround time, reduction of inventory stock-outs, and one-stop-shopping service. Chapter 14 will present in detail how the application and its individual components were built.

Information Architecture for a Wireless CRM Application

If you think back to our model for architecting solutions, the first step was to understand the business needs and articulate them in a business architecture. We did this in Chapter 13 by outlining the major types of functionality that the solution had to provide. In this chapter, we will perform the second step: Define the information architecture using the business architecture as one of several inputs. We will explore how these inputs are synthesized into the information architecture, which in turn will be used as an input for creating the information systems architecture, the third step in the architectural process.

We believe that architecture is a means to a goal, not a goal in and of itself. It is a framework in which the solution is created, and our framework ensures that the business needs dictate the use of technology. What we don't want to do is take a piece of technology and completely realign our business processes—that would be the equivalent of using technology for technology's sake. Realizing that technology is an enabler, we will now build upon our understanding of the business processes within the Acme radio network to take a closer look at their information requirements. We will look at the complexity of data to be moved, the time sensitivity of the data movements, and some strategies for making it all happen.

Overview

As has been mentioned throughout this book, there are several areas in which businesses are concentrating their wireless efforts. This section focuses squarely on one: sales force automation (SFA). Sales force automation refers to automating the ability of a sales representative to perform his or her work in real time while out in the field. The concept is a major contributor to achieving high levels of customer satisfaction since effective SFA can significantly reduce the time a customer has to wait for service. As an important piece in the overall customer relationship management (CRM) process, SFA is not a trivial notion. Figure 14.1 gives an indication as to the relationship between the level of SFA and the complexity of coordinating the efforts of a company's sales representatives.

In the previous two chapters, we provided you with an overview of CRM and the impact of wireless enablement. We also covered the business case for Acme Radio to wirelessly enable its sales force. Now we will begin to delve into the nuts and bolts of how to meet Acme's rather interesting needs.

If you recall, the overall goal of our application is to sell a perishable commodity: radio advertisements. The data architecture facilitates our reaching this goal by defining the data to be exchanged between business processes. The major benefits of creating a data architecture are to uncover the interdependencies between business processes and to create a clean definition of what information is exchanged. The data architecture will serve as a contract between the business components, creating a level of trust in the services pro-

Figure 14.1 The relationship between sales force automation and the complexity of co-ordinating effort.

vided. To quickly reiterate, the high-level business processes outlined in Chapter 13 were account management, reporting, and order management.

In addition to these business processes, we also use Acme Radio's information principles as inputs to create the data architecture. These principles serve as guides in Acme's decision-making process. To illustrate, the following principles were identified:

- Inventory spoils very rapidly.
- Inventory fluctuates very rapidly.
- Inventory is highly fragmented.

These principles need to be considered when creating any part of the general data architecture. In addition, each of the specific application components to be wirelessly enabled will have its own data principles. The data architecture for each application component, therefore, combines general information principles and those specific to that component.

Information Requirements for Order Management

Most benefits to be derived from injecting mobility into Acme's sales force automation efforts accrue in the area of order management because it is the most dynamic and the most time-sensitive piece of the application. In addition, it is the piece most visible to the customers, and thus provides the greatest benefit from improved efficiencies. The following requirements govern the development of a solution that addresses order management:

- Sales reps must be able to rapidly create proposals for Acme's clients.
- There can be several rapid iterations of modified client proposals.
- Proposals must quickly be routed to the appropriate personnel within Acme and its subsidiaries for review and approval. The review and approval process itself is highly time sensitive.
- The proposal creator (sales rep) must be able to receive immediate feedback from the proposal reviewer regarding the order's acceptance or rejection.

Information Requirements for Account Management

- Acme wants to present a unified front to its customers, which requires that the left hand know what the right hand is doing at all times.
- Sales reps can add or modify customer information.

- Sales reps can create new customers.
- A sales rep's access to customer data is limited by policies set by Acme senior management.

Information Requirements for Reporting

- A menu provides access to a fixed set of commonly run reports.
- These reports show orders by client, sales rep, market, city, station, format, demographic, or other data point.
- Other reports show current cross-market rates and inventory availability.
- Ad hoc reports can be run to view all pending and/or approved orders.

We have discussed business processes and information principles as two inputs into the creation of the data architecture. We will now introduce the third input—the business flow—describing interactions between the business processes. Understanding the business flow will require us to look at the business processes more closely. From that analysis, we can derive the data requirements of the general system. Figure 14.2 contains a very high-level view of the processes of the MediaMaven (MM) application.

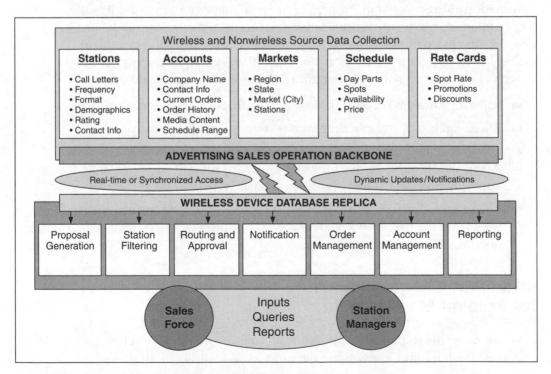

Figure 14.2 Acme Radio MediaMaven's high-level processes.

We will first discuss the as-is picture, describing how order management, account management, and reporting are conducted within the current data architecture. After we understand the wireline application, we will then introduce the to-be picture, known as MobileMediaMaven (MMM). Here we will show the impact of wireless enablement on the data architecture and illustrate the improvements to Acme's business processes.

One important factor in architecting any application is the user base. In our case study, the user base is 100 percent contained within the auspices of the Acme corporate umbrella. As a result, Acme can enforce usage of the application and provide a certain class of client hardware devices on which it is run. Were the application we are building geared toward a public user base—that is, the general customer population—we would have to limit our solution to the lowest common denominator on the client end, restricting functionality and (in many cases) usability. Because private use applications, such as this radio advertising case study, can be more creative in their application of software and hardware technologies, we are facing a significant impact on the complexity and volume of data that can be ferried between components, and on the very nature of the kinds of components that can exist.

We are now ready to discuss the components of the process flow for the wireline MediaMaven application. MM is a traditional Web-based application. Although it seems ludicrous to refer to something as new as a Web application as traditional, alas, such is the pace of change. As a Web-only application, the client will contain a Web browser, whose persistent data capabilities are simple *cookies*, stored attribute-value pairs. This provides enough functionality to create a personalized look and feel, giving the user customized content. It does not, however, provide the ability to store any business data on the client, except in locally cached Web pages. All transactions are conducted synchronously, meaning the user must be online and in contact with the MediaMaven server to use the application.

The following is a description of various processes composing the Media-Maven application. These processes entail components that follow protocols and have contracts with one another. In this way, individual components can be reused and aggregated to form new processes (a.k.a. *services*). This allows for loose coupling between processes and loose coupling between the components that constitute those processes. All of this leads to dynamic and flexible enterprise systems.

In this chapter, we discuss information and information management, not information systems. Therefore, we will keep our references to server-side applications at the level of the MediaMaven Server, even though it may consist of a Web server, an application server, an interface to an ERP system, an interface to a billing system, or the like. For the purpose of looking at the information to be disseminated to our distributed sales force, we will not go into that level of detail here, but will cover it in the next chapter. For now, we want to look at the processes that perform specific tasks, and the data requirements and outputs associated with them.

The As-Is Information Architecture for MediaMaven's Account Management

The account management scheme is a little more complex than it may appear at first glance. Today, companies come into existence and fold up at an unprecedented rate. Mergers and acquisitions are also very commonplace, requiring customer databases to be able to fuse at the drop of a hat. Furthermore, large multinational corporations will have dozens, if not hundreds, of locations around the United States and may own several smaller companies. If Acme wants to achieve its goal of unified, coordinated interactions with all of its customers, it must have solid control over its customer database.

A customer database is more than a record of transactions conducted between a client and a service provider. It is a history of the relationship, and it shows how much the relationship is worth to both parties. If mined properly, it could also show the provider how much the customer *could* be worth. The effective use of this information serves as a major competitive differentiator between average and top CRM organizations. Here are some simple steps that can be employed to make this kind of impact.

1. *Each customer must appear in the database exactly once.* In order for a service provider to understand the value they are providing a customer and the total revenue a customer is providing them, the big-picture view of a customer must be available. For example, the following set of entries was found in a CRM database:

CUSTOMER NAME	CUSTOMER ID
IBM	1001029832
International Business Machines	3742987822
IBM Corp.	7219813201
IBM Corp	9423394224
IBM Corporation	4857486874
International Business Machines, Inc.	8340820100

As human beings, we are able to make the logical assumption that these six entries are parts of a larger whole; our computers, however, cannot make that leap. The fragmentation of the customer's entries will not affect the Acme sales representatives' performing their work or getting credit for it. It does, however, pose a problem for Acme management.

Acme will review the IBM account to find out how much money the company is spending on radio advertising. Without a single, unified picture of the account, however, Acme does not know the total expenditures, and thus cannot compare them to similar clients in an effort to identify how much of IBM's business the company is receiving. Acme doesn't know how big the gap is between revenues realized and revenue potential. The size of this gap dictates Acme's strategic direction: If IBM is spending 90 percent of its radio advertising dollars with Acme, then there is smaller incentive to change the course of action than if IBM were spending only 30 percent. An accurate CRM database is critical to properly make that assessment.

2. *There must be a flexible mechanism to relate customer corporate hierarchies.* The fast pace of mergers, acquisitions, and bankruptcies means that Acme's customer set is constantly changing. For sales representatives to be successful, they need to stay on top of these changes. For example, an Acme sales rep visiting Miller Beer should know that it is owned by Philip Morris, and should know the relationship Acme has with that parent company. The rep should also know the sister companies of Miller, such as Kraft-General Foods, and that company's relationship with Acme. MediaMaven provides assistance with accumulating, storing, maintaining, and presenting this type of corporate intelligence.

3. *Customers can have multiple points of contact in many locations.* This is another aspect of system flexibility. The Acme salesperson must have the ability to work within the framework of the overall customer, yet have the ability to enter and retrieve the localized data important for his or her sales call.

4. *There must be one party accountable for the customer database.* Having one party that is accountable for information integrity is absolutely critical for the customer database. There are basically three categories of stakeholders involved: sales representatives, their managers, and strategists. The sales representatives are the only ones entering data into the system, while managers and strategists read, but do not modify, the information. However, some sales representatives are not very computer literate and thus find using the system challenging. The problem is compounded by Acme management, who wants the company's sales reps out selling, not entering data into a database. To add to this complexity, high turnover rates, with people coming and going, compounds the criticality and challenge of maintaining good data.

Two of the stakeholder categories—sales representatives and their managers—are paid based on the numbers of sales booked in the system. That is the extent of their concern for the data. The strategists, however, are the ones in need of consistency at the national and international level to be able to perform their jobs.

Look at the case of the multiple IBM entries. The first one, "IBM," was created by an Acme sales representative (let's call him Bob). For personal reasons, Bob resigned from his position a week after making the entry. All of Bob's accounts, including IBM, were given to Sally. She searched the database and found nothing under "International Business Machines," so she created an entry using that name (she thought Bob had not done so). Acme salesperson Joe in San Jose has a friend in IBM's local media relations office and strikes a deal to do some local radio ads. He doesn't even bother to check whether another entry exists—he tees off in one hour with his IBM friend and does not want to be late. So he creates the "IBM Corp." entry, feeling happy and secure that he will get the credit he deserves for his work.

And this skew continues over time. Some of the entries become vestigial, some are active, but all reflect uncoordinated activities with the same client. One party (either a person or a group) has to be responsible for imposing the coordination that cannot possibly be accomplished by the salespersons themselves. If this task were to be left to the sales force, they would probably view the requirements as bureaucratic restrictions hampering their sales activities, and as a time drain rather than a time saver. Thus, a dedicated advocate representing the strategists and their requirements surrounding data maintenance should be identified and assigned the responsibility.

Effective maintenance of a CRM client database is not a trivial task; otherwise, every company would do it today. The additional cost of having a data integrity monitor in addition to the traditional database administrator may not be palatable to some organizations. One important note: The data integrity monitor should *not* be involved with transactional data, only with customer data, to ensure that the person can fulfill his or her responsibilities.

Alerts are another way by which account management can be enhanced. This automates part of the task performed by the data integrity monitor. Previously, we alluded to a personalized version of MediaMaven that offers sales reps highly customized views of the application. Customizing the application allows reps to view only the data they deem important, and to tag data elements they would like to watch. For example, suppose you are the Acme sales rep servicing IBM and have configured MediaMaven to reflect this. You meet with an IBM media relations team in their New York office to work out a deal for advertising nationally on Acme's stations. When you connect with MediaMaven that evening, you are alerted to the fact that another Acme rep met with an IBM team in Austin, Texas, that very morning. The alert identifies the Acme rep in Texas, and you are able to contact him to coordinate your IBM efforts.

Without a tool like MediaMaven, it may take days for you to learn of the Texas IBM involvement, by which time it may be too late to coordinate your efforts. The alerting mechanism saves Acme time and money, and provides the ability to present the unified front it requires to meet the company's CRM goals.

The As-Is Information Architecture for MediaMaven's Reporting

The reporting feature of MediaMaven allows sales reps to see how they are performing against their targets and against their peers. It allows Acme's management to perform forecasting, evaluate their employees, and analyze the data in whatever way suits their needs. As a rule, the sales reps will have access to a small number of predictable reports they need, while managers require a broader scope in reporting due to the various performance measurements and operational tasks they have to perform. Because of their extensive reporting needs, managers frequently request new report formats to be provided by IT. Not knowing how these reports are created and not caring about the mechanics of how they work, managers rely on the IT department's outputs. What they do know is that their own performance, and the performance of Acme as a whole, depend on accurate information.

Reporting accuracy, however, is directly related to the validity of the data used to generate the information. If Acme's server in Dallas went down when it was supposed to push its nightly updates to the central server from which the reports are generated, the organization had better be prepared for that eventuality and have implemented strong controls to ensure that the system did not create tainted reports. As a rule, bad data is worse than no data at all. A lack of data is easily explained as a system glitch; bad data makes all the data suspect, because only those with specific knowledge of the data can separate the good from the bad from the ugly.

Because accurate information is critical for the Acme management team to perform its duties, management would be well served by having data with as near zero latency as possible. As we will see in the section on order management, MediaMaven brings Acme close to this goal. Later on, we will see how the wireless version of MM brings Acme even closer.

The As-Is Information Architecture for MediaMaven's Order Management

As we said before, the order management piece of MediaMaven is the most visible to the client. Therefore, it is important to have the process as stream-lined as possible. Figure 14.3 shows the business logic involved in creating an order. We will now describe the process in greater detail.

Client Selection Process

As with every piece of business information in MediaMaven, the set of clients from which a sales rep can choose is housed on the MediaMaven server. Each client record contains at least the following set of information: a name, an ID used for unique identification in the system, and a set of contact information, such as:

- Client contact name(s)
- The positions held by those contacts
- Their phone numbers

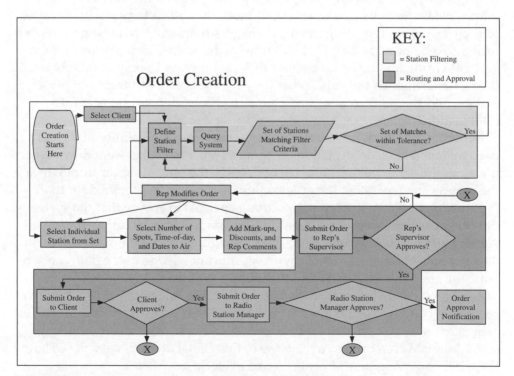

Figure 14.3 Business flows for creating an order in MediaMaven.

- Client address(es)
- History of Acme's interactions with the client, including names of Acme personnel and the names of client personnel with whom they spoke

We briefly touched upon how this data is entered and managed in the section of this chapter that discussed account management. For now, we are only interested in receiving and using existing client data. The set of clients to be presented to the sales rep for selection can be created in one of the three following fashions:

1. Presenting the complete list of clients.
2. Presenting a predefined subset of clients in which the sales rep is interested. This subset can be specified and maintained in the sales rep's preferences as part of the personalized MediaMaven application.
3. Presenting a subset of clients that is created at the time the application is being used. For instance, a sales rep can choose to view all customers who have "IBM" in their name (we saw the reason for this type of filter in the section on account management).

Once the set of clients is created and received at the sales rep's workstation, the rep can then choose one specific account for which he or she would like to create a proposal. Once the account is selected, the rep needs to identify exactly those stations on which the client's ads will run.

Station Filtering Process

The first step in the process to find radio stations in the Acme network that conform to the advertiser's requirements is to define the station filter. In other words, the Acme sales rep has to specify which parameters will drive the query that reduces the long list of stations to only those that are desired by the client. A set of possible parameters is illustrated in Figure 14.4.

These criteria can be used to create a number of interesting station lists. Here are some examples:

- Entering three parameters such as dates, U.S. state, and city, while leaving the format and demographic information blank, will yield the set of all stations in one city.
- Entering two parameters such as dates and demographic information, while leaving the rest blank, will yield all stations targeting a specific demographic across all formats in the entire United States.
- Entering dates and format, while leaving the rest blank, will yield all stations following a specific format across the entire United States.

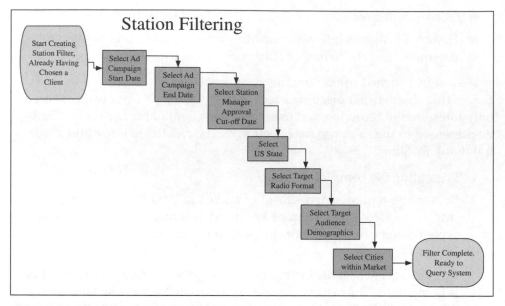

Figure 14.4 A set of possible station filter criteria.

As you can well imagine, there are a number of other possible ways to break down the set of stations. We could even add additional filters, such as the day part and price range for an advertisement spot. What is important to understand here, however, is not so much that specific filters yield specific sets of stations, but rather the concept of empowering the remote sales force to have access to a powerful tool that gives them the flexibility to quickly and effectively find the stations that meet their needs.

Once the rep finds the set of stations matching the advertiser's needs, he or she is ready to find available spots to air the client's advertising message.

Finding Advertising Spots on a Station

The sales rep has several tasks to perform before the proposal can go through its approval channels. First, the rep must select a station to see what the advertising rates and availabilities are. These rates are usually determined by factors that include the station's ranking, demographics, and audience size. Audience demographics indicate spending power, while audience size indicates the exposure the ad receives and the results the ad can be expected to produce.

Radio stations typically use a marketing research firm like Arbitron as an independent source for identifying audience size. Arbitron measures radio audiences in local markets across the United States, while also surveying the retail, media, and product patterns of local market consumers. Since both the advertisers and the radio stations operate using the same Arbitron numbers, both sides can confidently offer and accept payment levels at the proper market value for the spots.

Given this baseline for understanding advertising rates, the sales rep should have confidence that the advertiser will go along with the standard rates. However, because size matters, these rates are not set in stone. A large national advertiser will usually demand and receive discounted rates for airing their ads over the Acme network. Thus, our system must be flexible enough to allow deviations from the standard price while providing feedback on how far the proposed rate can vary from the specified norm.

The sales rep and the advertiser will then agree upon a set of time slots during which the ad will air. Traditionally, this was achieved by the rep printing out a listing with available slots, negotiating them with the advertiser, and then entering the agreed-upon times into the system as soon as connectivity to the MediaMaven server could be established. Of course, there is latency in the data transmission because the data is not entered into the system in real time, but at a later date. This is a reason for conflict, as another sales rep could be selling the very same slot to another customer, unaware that the slot is no longer available.

Because of the potential for two or more sales reps competing for the same advertising spot, Acme must have a set of conventions in place to handle such collisions. The best practice, of course, is to avoid any problems in the first place. The advertisers waste time and effort selecting their desired time slots, only to find out later that they are no longer available, requiring them to go through the selection exercise all over again. Customer satisfaction could certainly increase by making this process more efficient, as we will find when discussing our MobileMediaMaven solution.

After a spot has been selected and the order written up, the rep then places it into the system, where it is routed through the proper channels to get the necessary approvals.

Order Routing and Approval

Once the sales rep has worked out the proposal's details with the client, the orders are routed through the appropriate channels for final review and approval. Chapter 13 detailed the business case for routing approval, including the sales rep's manager, the client, and the station manager. The data requirement here is straightforward—it is the content of the proposal. The challenge lies in configuring the work flow to move the data as efficiently as possible. MediaMaven sends an email to those in need of approving a proposal, but the system is only as good as the approver's access and attentiveness to his or her email inbox. It does the client little good to have the order creation process occurring in real time, when the notification goes unanswered by the station manager for days.

Notification

Notifications in MediaMaven are typically performed by email. It is a cheap, effective way of sending information to a specific party, and email travels well. On the other hand, email can get lost in an ocean of other electronic notifications, clogging a manager's inbox. Latency is a function of the recipient's availability and attentiveness to its arrival. A particularly time-sensitive email can be sent with a "high priority" tag, or some other means of identifying its urgency, but this does little good if the recipient is on a golf course. To bring turnaround times down on urgent notifications, a ubiquitous presence like the wireless MobileMediaMaven is required.

Order Modification

The order creation process is iterative. Clients might change their requirements, a chosen time slot could have been taken, or someone in the approval chain might reject an order for one or more reasons. Because of these and other

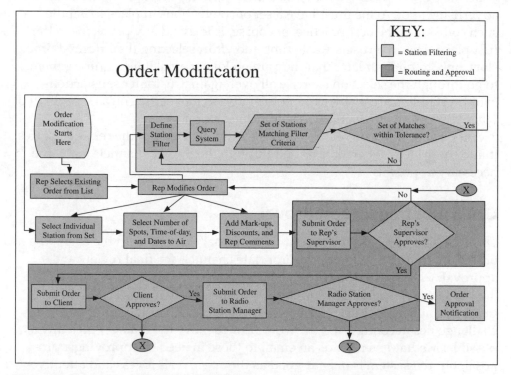

Figure 14.5 Business flows for modifying an order in MediaMaven.

eventualities, orders have to be modified from time to time. The work flow is slightly different from the initial creation process, as shown in Figure 14.5.

Reviewing Figure 14.5, you notice that the sales rep is able to change whatever parameter(s) of the proposal were the cause of its not being approved. The data requirements are the same as for the order creation process, except for the fact that time criticality is of heightened concern, as the customer is sure to want faster results the second time around. The order modification process really magnifies one of the limitations of the wireline version of MediaMaven and provides a very powerful driver for the creation of MediaMaven's mobile extension.

MobileMediaMaven Overview

The MediaMaven application works well, but there are certain areas of the application that would greatly benefit from some improvements. Such areas of improvement include real-time access to perishable inventory (the time slots on a radio station) and immediate turnaround time on notifications. Additionally, MMM provides functionality not possible without the wireless component. Some of all features are listed as follows:

- Order management
- Account management
- Reporting
- Chat
- Calendar integration
- Real-time approvals
- Remote wireless printing
- Location profiling
- Airtime auction

Before diving into the specific enhanced features of MMM, let's discuss the nature of the client/server environment in which it operates. As we said before, MMM is completely contained within the auspices of Acme. This gives Acme management the right to mandate its use and tightly control the technical infrastructure at both the client and server ends. As a result, the architecture can leverage features found in some devices but not others. This gives the architects the choice of going with a thin-client or thick-client model. Let's look at the pros and cons of each approach.

We'll start with the thin-client model, which is currently used by the MediaMaven application. There are advantages and disadvantages of each model, as described here.

ADVANTAGES OF A THIN-CLIENT MODEL

1. Simple to implement and manage.
   a. Latest version of the application is used every time.
   b. Eliminates client version control issues.
2. Synchronous data communication.
   a. All messages are sent and received in real time.
3. All data is located on the server.
   a. No persistent client data.
   b. No issues of data latency.
   c. No issues of data replication.
   d. No problems with data change conflicts.
4. Client can be an inexpensive, small piece of hardware.

DISADVANTAGES OF A THIN-CLIENT MODEL

1. Reinvention of mainframe dumb terminals.
2. Works only when the server is available.
   a. If the server is down, data entry is lost.
   b. If your connection is lost, you cannot perform work.
3. All computation occurs on server.
   a. Bandwidth needed to bring everything to the server for computation and back to the client for presentation.
4. All communication instigated by the client; the server only responds to requests.

ADVANTAGES OF A THICK-CLIENT MODEL

1. Efficient use of bandwidth.
   a. Data is distributed.
   b. Client has computational power.
   c. Saves time in environments with static or rarely changing data.
   d. Server and client can send and receive deltas.
2. Allows asynchronous data communication.
   a. Will work even when the server is not present.
   b. Local client store ensures work will not be lost.
   c. Server can initiate communication with clients.
3. Thick clients have more potential for better user interfaces, better response times (communicating with a local instead of a remote data store), and richer complexity.

DISADVANTAGES OF A THICK-CLIENT MODEL

1. It is more complex.

    a. Requires mechanism for updating clients with latest version of application.

    b. System must handle interactions with a potentially heterogeneous client population.

2. Having to deal with distributed data.

    a. Issues with data latency.

    b. Issues with data replication.

    c. Must develop a strategy to handle data change conflicts.

3. Client must be heavyweight enough to support local persistent data.

4. Since communication is asynchronous, every message must be authenticated.

The model used by an application will depend on the application's data needs and the environment in which it operates. What we will choose for MMM is a *hybrid* client, enabling us to leverage the strengths of both the thick and thin models, while avoiding their pitfalls. We are able to do so for the following two reasons:

- We have control over the client hardware and can specify client capabilities.

- Most of the data required by the client is either relatively static or can be derived without business knowledge.

Let's define what we mean by a *hybrid* client: It's a client that behaves like a thick client (maintaining local data, and computations performed locally) for static or derivable data, and behaves like a thin client (performing synchronous data communication only) for dynamic data. We will now revisit the order management process, to see what a hybrid client looks like in action and how its wireless connectivity can give Acme a competitive advantage.

Figure 14.6 shows the new order creation process. One point to notice is how similar it appears to the MM order creation architecture. That is in keeping with the CGE&Y philosophy that the architecture be less ephemeral than the implementation of the solution. Even though the delivery channel has changed and the client capabilities are quite different, the conceptual image of what is being accomplished is basically unchanged. There are some subtle differences, but they are necessary to fully leverage the additional capabilities brought to bear by the new application.

The first part of order creation is still selecting a client. A sales rep selecting one of his or her clients is not a very dynamic process—sales reps know their clients, and they know which clients they will be visiting on any given day.

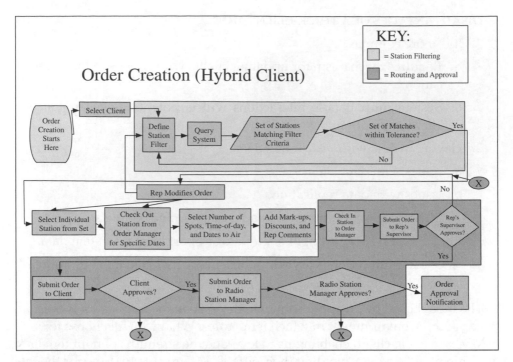

Figure 14.6 Order creation in MobileMediaMaven.

Account management is handled separately from order management, so client information is read-only during the proposal generation process. Additionally, MMM has the previously described built-in alert function that notifies a rep of any changes to a client's data in the accounts database. All these features make it certain that a sales rep can be sure that the local version of the client data is always up to date. Therefore, the client selection process can operate using the thick-client model, requiring no communication with the server.

The next step is station filtering. The set of stations controlled by Acme is fairly stable, and notifications will be built into MMM to give reps fair warning of upcoming changes to the station list. Each station has about half a dozen pieces of information associated with it, as shown in Figure 14.2. To reiterate, the most important information includes call letters, frequency, format, demographics, rating, and contact info.

As stated in Chapter 13, Acme is one of the largest players in the radio business, and there are about 12,000 stations in the United States. Even if Acme owned 50 percent of the stations, that is still only 6,000 stations. If we assume each record in the database containing the stations is 128 bytes long, we would end up with about 768 KB of data, easily contained in a handheld device. Therefore, we can store all of the station information locally, effectively using the thick-client model and, again, requiring no communication with the MMM server to perform this task. In addition to the local data storage, the business logic pertaining to acceptable set size would have to be built into the MMM client.

Then the rep selects a station from the set of matches, exactly as was done in MM. But here is where the shift from thick-client to thin-client models occurs: The next step is to identify spots available on that station and work with the client to select the spots they desire. We have maintained from the beginning that these time slots are volatile, perishable commodities, so no data latency is acceptable. The sales rep must be in synchronous communication with the server to obtain the current set of available spots. Once downloaded, the sales rep is free to peruse the data with the client to find the desired availability.

If the system were really sophisticated, it would classify that station's set of spots as actively being pursued once the request were handled. That way, if another rep tries to reserve advertising time on the same station, conflicts can be prevented by indicating that another rep is already viewing that station's availability. Here's where some of the new functions found in MMM come into play.

Suppose you are visiting your client, Joe's Shoe Company, about advertising their new line of sneakers. They want their ad campaign to be nationally broadcast in the large markets—New York, Los Angeles, Miami, Chicago, and so on—essentially everywhere there is an NBA franchise. You filter the stations and decide that WXXX in New York is the perfect station to run the campaign. You use MMM to find spots on WXXX, but learn that your spouse (who is also an Acme sales rep) is meeting with Phil's Sporting Goods and also looking at ad time on the same station during the same time period. You then use the chat feature of MMM to contact your better half, but your spouse is not currently available (being mobile in a tunnel or a solar flare having knocked him or her offline). MMM then automatically queues up a message to send when your spouse returns online, and searches the calendars of you and your spouse to find the soonest available opportunity in which the two of you can discuss WXXX. In the meantime, MMM has placed you in the queue of those wanting access to the WXXX spot availability information for the time period in question. Should the WXXX record be checked in, you would receive an alert that the record is free, and it would automatically be checked out to you and downloaded so you can continue creating the proposal.

This demonstrates the chat and calendar integration features of MMM. Should your spouse have been available for online chat, chances are that you would not have held your entire conversation by means of that medium. Human factors and usability issues typically prevent most users from being satisfied with real-time data input into most wireless devices, due to the awkwardness of the entry mechanism. Most likely, you would have used the feature to set up an immediate phone call or other such mode of contact that could expedite the process of negotiating for access to the WXXX advertising availability list for the dates in question.

Once the proposal is completely created by the sales rep, it again enters the routing and approval process. The process is essentially unchanged, except for the addition of the order manager service, a part of the MMM server that unlocks the radio station(s) that the proposal affects. The spots chosen in the proposal are put into the "unavailable" category, and will become available again should the proposal be rejected by one of the approvers.

The proposal submission is handled in the thick-client mode. The reasons for this are as follows:

1. Once you check out a record, you want guaranteed return of that record to the order manager.

2. You do not want to have to repeat work you have already completed.

3. You want the server to initiate communications to notify you of status changes as the proposal moves through the approval work flow.

Remember, the thick-client communication model specifies a local server, typically a listening port by which the server can send messages. Another aspect of thick-client communication is a store-and-forward mechanism, so even if the server is unavailable at the time the data was intended to be communicated, the data will be pushed when the mechanism detects network connectivity. From an application standpoint, feedback is given through the user interface about the location of the proposal in the work flow. When the proposal is sent, an icon or some other feedback device is used to indicate the in-transit status. When it arrives at the order manager, the client is notified, and the status bar is updated. Similarly, when the sales rep's manager and the station manager receive and act upon the proposal, the status changes. Should the proposal be rejected, the status change will typically have a message attached indicating the reason for rejection. Should it be approved, the proposal will automatically generate the proper notifications.

For this type of system to be truly effective, real-time access to the station managers is required. Imagine the scene: It's a party celebrating the promotion of a new station manager in the Acme network. The corporate bigwigs come down and say, "Good news and bad news. The good news is you have been promoted to station manager. The bad news is you have to take this wireless PDA and carry it with you wherever you go, and answer it whenever it calls you." A hyperbole to be certain, but not an extreme one. When the sales rep makes a request, she should be able to have it routed through the entire work flow in under half an hour. The biggest rate-determining steps are the availability of the wireless network for all participants, and the responsiveness of the sales rep's manager and the station manager. Should those two parties be available, a 30-minute turnaround time becomes quite reasonable.

One improvement in the postapproval process is getting the client to sign the contract at the time of approval. This can be accomplished by sending the contract (automatically created by the approval of the proposal) to the client immediately. This can be done by several means, such as a wireless connection between the sales rep's MMM device and a local printer, or by sending the contract from the MMM device to an Internet fax server which will forward it to the client's fax machine. By whatever means available, the sales rep is able to leave the client's office with his or her most cherished commodity: a signed contract.

There are two other features of MMM not discussed in the proposal generation process, but they are heavily related. The first one is *location profiling*, which provides data pertinent to the sales reps based on where they are or where they are going. For example, I am from New York and don't know much about Chicago radio. MMM can scan my calendar (again tying in calendar integration), detect that I am flying to Chicago this afternoon, and download all the relevant information on Acme stations in the area. This would allow me to read the information on the plane at my leisure and be as efficient as possible with my time in the Windy City. I could, alternatively, choose to get all pertinent data on my current location, should I take an unexpected trip somewhere or need a refresher on my current location.

The second feature is an Airtime Auction conducted in real time. Again, with the perishable nature of radio ad spots, any unsold spot is lost forever. Therefore, an auction could be held to sell immanent spots before they are gone. Obviously, they would command a lower price than a regular ad spot, because the demander has leverage over the supplier. It would be a win-win situation, as the radio station gets additional revenue otherwise not available, and the advertiser gets discounted rates. Due to the extreme time criticality of the spots involved, real-time access via a wireless channel would certainly be a competitive advantage over tethered access.

Summary

This chapter presented an overview of the data requirements generated by the business processed outlined in Chapter 13. We emphasized that a good solution must follow the previously established business needs—we don't want to implement technology just for technology's sake. Taking these processes, we defined the interactions between them and applied the company's information principles to create a data architecture. We then showed how the MediaMaven application works today, and how its wireless descendent, MobileMediaMaven, enhances the application by overcoming some of the wireline solution's limitations.

The content of this chapter focused especially on the order management piece of the application, since it contains the processes most sensitive to time lapses. A hybrid client scheme was developed, leveraging the strengths of both thin-client and thick-client strategies, while avoiding the pitfalls of both. A locking mechanism was added to the most volatile data elements, eliminating the major problem of colliding advertising spot requests.

Now that we have built the data architecture, we can create the information systems architecture—the focus of Chapter 15.

Information Systems Architecture of a CRM System

Introduction

We've made it through another round of discussions regarding the conceptual business and information architecture, this time focusing on a sales force automation application in the CRM category. As in prior sections, the next logical step is to dig into the information systems architecture of the application. Remember that the information systems architecture covers in detail the technical components of the application, in our case the customer relationship management (CRM) sales force automation (SFA) application for Acme Radio, which we are calling MediaMaven.

As you may recall, the three major areas of any enterprise architecture application are the conceptual, logical, and physical design. We will focus on examining in detail the logical component of the information systems architecture. The format for this chapter will be very similar to the way Chapter 10 was structured. The initial discussion will pertain to the current state architecture of MediaMaven, focusing on the logical IS components and how each of them integrates into a complete solution. MediaMaven started out as a Web-based, nonwireless application. So to make sure you understand the IS architecture completely, the current state will illustrate the application *before* any wireless components have been added. Next, we'll jump into the future state architecture, which will give you an in-depth view of MediaMaven's architecture as it was designed for a wireless environment. This will help you understand what components need to be different, added, or taken away to enable the application for a wireless world. Let's start with a discussion on the current state architecture and what major components the application comprises.

Current State Architecture

To quickly recap Chapter 13, the primary functional components of the MediaMaven sales force automation application are as follows:

Proposal creation. A proposal contains several orders. Each order is a comprehensive offer to sell to a client at a specified price one or more airtime slots, on specific days, and at specific times. The proposal contains all information pertinent to the deal. For a more detailed analysis of the steps involved in proposal creation, please refer to Chapter 13.

Filtering. In selecting the radio station(s) that will broadcast a client's advertising message, the Acme account rep must identify all stations that show certain attributes that are important to the advertiser. The rep accomplishes this task by running a software filter to navigate through all of the stations in Acme's nationwide radio station network. The filter allows irrelevant stations to be sorted out quickly and those stations whose attributes are a good match with what the advertiser desires to be identified. Station attributes include such items as audience size and demographic, format and geographic location.

Routing and review. Once a proposal is created, it needs to be reviewed and approved. In the case of junior sales representatives or in case the proposal contains provisions that vary significantly from a standard contract, the Acme director of sales reviews the document. The next step in the process entails the review of the proposal by each station manager whose inventory is to be sold. The station manager has the final word regarding approval or rejection of a proposal.

Notification. Notifications occur at various steps within the proposal generation process. For example, a director of sales is notified when a junior sales representative needs him or her to review an order. Similarly, station managers are notified when a rep has sold a spot on their station. Sales executives are notified when the station manager approves or rejects an order.

Order management. Orders and proposals may need to be modified. A client may decide to purchase additional spots or change the time a marketing message is to be broadcast.

Account management.   To ensure an accurate client database, sales executives use MediaMaven to maintain their client account information. Account management entails the updating of existing client data or the creation of entirely new accounts within the MediaMaven system.

Reporting.   Acme management, station management, and account representatives require system reports that illustrate their performance. At the corporate level, such reports include financial performance across the entire Acme station network as well as reports for financial planning purposes. Individual stations are mostly concerned about inventory management and station profitability, whereas account reps use reports to track and measure their own performance as it relates to meeting their specific sales targets.

Now that we've reviewed the functional components of the MediaMaven application, let's go through how these processes are turned into a technical solution.

As you probably expect, the design of our application is based on an *n*-tier architecture. In this case we have a three-tier design that is composed of the following logical partitions:

Presentation layer.   This piece is generally concerned with dynamic HTML, Active Server Pages (ASPs), and any other elements used for presentation to the client.

Application layer.   The business logic that makes the system operate is deployed here.

Data access layer.   Contains application components that tell the system how to interact with the data elements and ultimately retrieve information.

Each of these logical partitions will be discussed in greater detail. To start, look at Figure 15.1, which illustrates each of the partitions of the architecture.

Not unlike the standard diagrams that we've used in other chapters, Figure 15.1 takes a high-level approach to a typical three-tier Web-based architecture. Let's go another level deeper, however, and understand what's behind each of these layers to give you a better understanding of our application components. Please note that this application utilizes COM+ components in all layers of the system. The different COM+ components are developed and deployed at each of these tiers. Each tier has a set of standard interfaces that each class of components implements. We will cover each COM+ component in greater detail as related to each tier of the application.

Presentation Layer

As we mentioned, the presentation layer primarily deals with the design components that present the content to the client. In our case study, these components consist of the following:

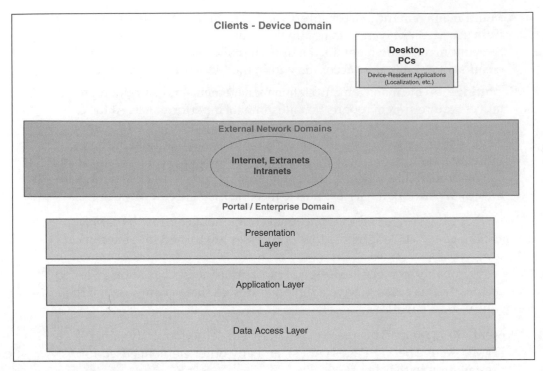

Figure 15.1 Logical partitions of MediaMaven.

Active Server Pages. Usually referred to as ASPs. Active Server Pages are a specification for a dynamically created Web page with an *.ASP* extension that utilizes ActiveX scripting—usually VBScript or Jscript code. When the browser requests an ASP page, the Web server generates a page with HTML code and sends it back to the browser.

DHTML. The same as dynamic HTML. Refers to Web content that changes each time it is viewed. For example, the same URL could result in a different page depending on any number of parameters, such as geographic location of the reader, time of day, previous pages viewed by the reader, and profile of the reader.

There are many technologies for producing dynamic HTML, including CGI scripts, Server-Side Includes (SSI), cookies, Java, JavaScript, and VBScript. In the MediaMaven application, we are primarily using VBScript.

Again, the presentation layer is composed of VBScript-based Active Server Pages, HTML documents, and JavaScript procedures. The ASPs retrieve results from the application layer, which we will cover shortly. These results will be in the form of disconnected ADO (Active X Data Objects) recordsets, or in the case of singleton returns, basic data types. Some of the more complex user interface elements have been built using DHMTL. To reduce the number of trips between the client and the server, all multidimensional and nonsingleton data will be sent as JavaScript arrays to the front end. Thereby, with the help of DHTML, it will be possible to dynamically traverse data using the browser on the client.

The next step is to go into greater detail into each of the COM+ components that are utilized within the presentation.

COM+ Components

The ASP objects that we need to worry about for our MediaMaven application are listed and described in the following paragraphs and detailed in Figure 15.2. You'll see that this illustration breaks down each of the layers we discussed in the previous section and also includes the ASP objects for the application. Let's look at the main ASP object, the ASPProposal.

> **NOTE**
>
> **Com+ terminology and expressions will be used throughout the following sections. These sections are not intended to go into great detail on Com+. For more information on these topic areas, consult *Building N-Tier Applications with COM and Visual Basic* (John Wiley & Sons, Inc.).**

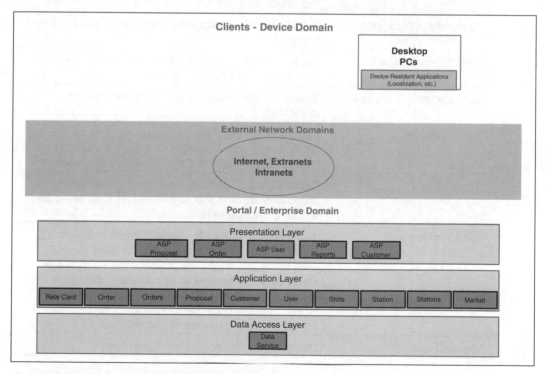

Figure 15.2   MediaMaven ASP objects.

Table 15.1 Com+ ASP Components (Proposal)

METHOD	DESCRIPTION
Construct	Populates a recordset with all the proposal information and returns the same.
Create	Creates a proposal. Every proposal is expected to have, at a minimum, one order. So this method scans through the list of all the orders in the Order Info XML and creates the orders for the current proposal.
	If a schedule is passed, then the schedule gets applied to all the orders of the proposal. On successful creation of the proposal and the orders, a recordset with the proposal information is returned by this method.
	The orders are created by the Proposal.order object
Modify	Updates proposal data for an instance that has already been created.
ApplySchedule	Applies the specified schedule to all the orders of the proposal. The exact same schedule is applied to each of the orders of the proposal.
GetAllOrders	Returns all the orders for the specified proposal as a recordset.
GetFinalizedProposals	Returns all the proposals for a customer that has been finalized.
GetInProgressProposal	Returns all the proposals for customers that are in progress and not yet submitted.
GetSubmittedProposals	Returns all the proposals for customers that are submitted.
GetNewPrices	Calculates the price for the proposal based on the current day's rate. These prices are not persisted as a part of the proposal, but are for display purposes only.
GetSchedule	Returns a schedule for the proposal, based on the proposal start date and end date, and also based on the broadcast calendar.
FilterStations	An extension of the FilterStations method found in cStations. Will filter out those stations that currently exist in the proposal from the returned subset of all stations. This is needed to ensure that a proposal maker does not add the same station more than once when using the proposal modification functionality added in a future release of MediaMaven.
DeleteOrder	Deletes multiple orders from the system.

ASPProposal

This component encapsulates both Proposal and Order entities, such as being able to create a proposal, associate it to a customer, and apply a single broadcast schedule for all the orders. This is more of an envelope to all the orders that are pended to the individual radio stations. The classes/objects supported by this component are:

- Proposal
- Order
- Orders

We will examine each of the methods within each of these objects and understand how they are used in our application. Table 15.1 takes the *Proposal* object to a more detailed level.

The next object in the ASPProposal construct is the Order subsets. These methods are covered in Table 15.2. This component is used for creating orders for a given proposal. A proposal comprises multiple orders for individual radio stations. For example, we are creating a proposal for a customer called Acme Soda, and Acme Soda is buying radio advertisement time at multiple radio stations, such as KXXX-AM in Dallas and KERR-FM in Dal-

Table 15.2   Com+ ASP Components (Order)

METHOD	DESCRIPTION
Construct	Populates a recordset with all the order information and returns the same.
Create	Creates an order and associates it to the specified ProposalID. If a schedule is specified, then a schedule is applied to the order. Returns a recordset populated with all the order information.
Update	Updates the specified order with the specified information. If a schedule exists, then the existing schedule is deleted and a new schedule is applied for the order
GetScheduleASISHeader	Returns a recordset that is used to populate form fields for the order schedule report.
ChangeStatus	Changes of the status of the order and returns a recordset with the order information.
SendOrderContracted	Sends the order.
Notification	A private subroutine that facilitates the email notification of an order being contracted.
GetTotalOrderSpots	Returns the total number of spot requests for an order.

Table 15.3 Com+ ASP Components (Orders)

METHOD	DESCRIPTION
GetAllOrders	Returns all the orders for a proposal.
GetAllPendingOrdersForAStation	Returns all the pending orders for a station.
GetStatusOrdersForAProposal	Returns all the orders for a proposal based on a specified status.
GetAllRejectionReasonsForanOrder	Returns all possible rejection reasons for an order.
AcceptOrders	Changes the status of all of the orders passed in the OrderID string to "Accepted" and returns information about each of the orders as a recordset.
RejectOrders	Changes the status of all of the orders passed in the OrderInfo string to "Rejected," sets the rejection reason ID based on the reason ID supplied in the OrderInfo string, and returns information about each of the orders touched in the form of a recordset.
ContractOrders	Changes the status of all of the orders passed in the OrderID string to "Contracted" and returns information about each of the orders as a recordset.
GetNewPrices	Used to calculate new order prices in the case that prices expire due to fact that a user tries to submit a proposal that was created more than three days ago. This three-day window given to validate prices was created to ensure that a more optimal set of prices is applied to orders based on the principles of revenue management.
ContractOrders	Changes the status of all of the orders passed in the OrderID string to "Contracted" and returns information about each of the orders as a recordset.
GetNewPrices	Used to calculate new order prices in the case that prices expire due to the fact that a user tries to submit a proposal that was created more than three days ago. This three-day window given to validate prices was created to ensure that a more optimal set of prices is applied to orders based on the principles of revenue management.

las. This is called a cross-market buy because it involves buying radio time in multiple markets as well as from multiple radio stations in one market. The entire function is tied by a proposal as one order with regard to the customer. But with regard to the broadcaster, it is one proposal for the customer. This proposal comprises multiple orders for each of the radio stations.

So, in essence, the ASPOrder component has the capability to create multiple orders based on an XML document while also associating a schedule for each of these orders.

The final piece in the ASPProposal object is the Orders subsets. These methods and their corresponding descriptions are covered in Table 15.3. The ASPOrders is essentially a design pattern that is used for the purpose of optimizing the Web site. The object is defined as singular in that it has the capability to get the complete information of the properties of the object, whereas the plural objects do not contain all the information but just the necessary information to display a list of orders associated with the proposal. In this case we are just interested in Order Name, Order ID, and Order Status for a bunch of orders associated with a given proposal.

In the case of a singular object, everything is available, such as the order ID, name, status, start date, end date, schedule (7days * 5 day parts matrix for N number of weeks). In essence the ASPOrders is more of a user interface (UI) list object.

ASPUserManager

The next ASP component we'll cover is the ASPUserManager, which encapsulates the user information of the MediaMaven application, such as adding users, deleting users, and associating passwords. Table 15.4 describes the methods and an understanding of their meaning.

Table 15.4   Com+ ASP Components (UserManager)

METHOD	DESCRIPTION
Construct	Returns all the information about the station.
Create	Creates the user and returns a recordset containing the user information.
Authenticate	Authenticates the user against the password. On successful authentication, returns a user info recordset. If the authentication fails, returns an empty recordset.
Modify	Modifies user information and returns true on success.
ChangePassword	Verifies whether the user has mentioned the correct password and then updates the password with the new password.
SearchUsers	Searches the user table for the specified information. This is a very simple search, based on a SQL search.

ASPCustomer

This component encapsulates Customer information and contains the *Customer ASP* such as adding customers and deleting customers. Table 15.5 describes each of these classes.

ASPReports

The final component of the presentation layer is the Reports ASP. The ASPReports component is a COM+ dynamic link library (DLL) that is used to retrieve all the information used for generating online reports. Some examples of the reports are:

Proposal reports.   Displays a list of proposals with the corresponding orders and also the status of the orders, as in accepted/rejected or where exactly they are in the business process work flow. Also, displays all those orders that have been executed and their corresponding status.

Station reports.   Displays a list of stations categorized by state and markets, and the last time the inventory was pulled and the price updated.

Application Layer

The Application Layer contains the actual business logic that makes the system operate. These business components should be atomic (uninterruptible) program sections. Additionally, each component contains program code required to implement some required business process, but they do not retain information from method call to method call. In this way, each component is stateless.

Each business component should be designed to implement some set of business rules. Business rules should not be used to manage database logic. Instead, database logic controls are delegated to stored procedures found in the Data Access Layer. Each business component communicates with the data store through the stored procedures. The data service is then responsible for collecting and returning requested information to the business service.

Table 15.5   Com+ ASP Components (Customer)

METHOD	DESCRIPTION
Construct	Populates a recordset with all the customer information and returns the same.
Create	Creates a customer and returns a recordset containing the customer information
SearchCustomers	Searches the customer table for the specified information. This is a very simple search, based on a SQL search.

Application objects are called upon by Active Server Pages, found in the presentation layer, to perform desired business operations. The business components accept singleton data in the form of simple/basic data types, whereas all nonsingleton data are passed as XML strings. All singleton results returned by the business components are returned as basic data types, and nonsingleton results are returned as disconnected ActiveX Data Objects (ADO) recordsets.

For the purposes of increased scalability and better system performance, all business service components will be COM+ transaction services-enabled. Each component should be designed so that every discrete process satisfied by a component can be performed with only a single method call. This will limit the number of connections required between the Presentation Layer and the Application Layer, as well as between the Application Layer and the Data Access Layer. Doing so reduces strain on the system by lowering the amount of processing overhead required to establish these connections.

> **NOTE**
>
> **Some of the same methods used in the Presentation Layer are also used in the Application Layer. Refer to Figure 15.2 for the integration points.**

Org

This component provides organization information such as states, markets, and stations. The classes/objects supported by this component are:

- Markets
- Stations
- Station
- States

Table 15.6 covers each method of the Org business layer components.

Data Access Layer

The Data Access Layer contains application elements that understand how to interact with the database to retrieve a logical unit of requested information. There is only one data component used by every business component to retrieve data. The data component follows the Connect—Execute—Disconnect philosophy. It uses a COM+ constructor string for DSN information and is COM+ transaction services enabled. No component in either

Table 15.6 Com+ Business Layer Components (Org)

cMARKETS	
Construct	Returns a list of all the markets in the specified state(s).

cSTATIONS	
Construct	Returns a list of all the stations in the specified market(s).
FilterStations	Used to select only those stations that exist in the user-defined markets given that have the selected formats and the selected primary demographic data.

cSTATION	
Construct	Returns all the information about the station.
Create	Creates a new station in the database and assigns values to the fields (properties) indicated in the method's argument string.
Modify	Updates the fields (properties) indicated in the argument string for the station specified.
GetDayParts	Returns all of the day part information for the station selected.
GetOrdersForaStation	Returns all the pending and accepted orders for a station.

the business or data service layer contains any database SQL queries; rather, all required database queries are facilitated through stored procedures.

All access to the database by the Application and Data Access layers will be performed with stored procedures. This practice results in enhanced scalability of the application, as well as better overall system performance.

The Future State Architecture

We hope you now have a good understanding of MediaMaven and its current functionality. Next we will describe how wireless enhancements and new features can be implemented from an information systems perspective, transforming the application into MobileMediaMaven (MMM). Let's start by looking at a high-level logical diagram of the future state, in Figure 15.3.

Figure 15.3 is based on the current state diagram you saw in Figure 15.1. However, there are some significant differences:

- We now have three different kinds of users: the connected user, the WAP mobile phone user, and the handheld PC user. Connected users will use a Web browser to access the system, WAP mobile phone users

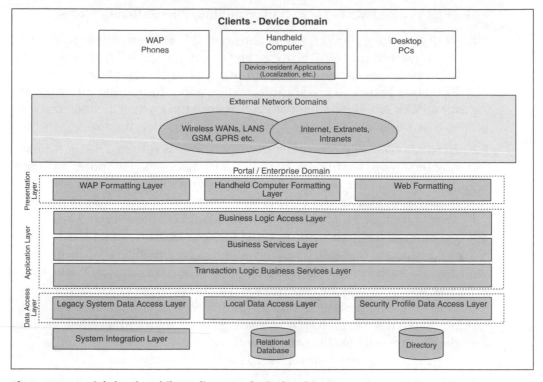

Figure 15.3   High-level MobileMediaMaven logical architecture.

will access a WAP browser, and handheld PC users will use device-resident applications and a scaled-down Web browser.

- All users connect through different networks to get to MMM. The connected user accesses MMM through the Internet or the intranet, the WAP mobile phone user first accesses his or her mobile operator's network and then enters the intranet via a WAP gateway, and the handheld PC user also accesses the mobile network and accesses the Internet directly via HTTP.

- A directory server is now responsible for handling authentication and authorization of the different users. This functionality was previously implemented in the application layer, but we have broken it out logically here to demonstrate a more scalable solution. The application layer is the primary actor using these services and information.

The original application architecture was Microsoft based and we have kept it this way here, even though we describe the applications only in enough detail to demonstrate the general technology direction Microsoft is taking. Figure 15.3 shows how the application is broken down.

The *Presentation layer* has been broken out into three different pieces:

1. *WAP formatting layer.* This layer is responsible for formatting, reading the request from a WAP browser, and creating the WML pages required to present information returned from the general presentation layer. This layer knows all specific information about formatting information for a WAP browser and how to get and use device-specific information (e.g., a Nokia phone model X is being used to access the site).

2. *Handheld computing formatting layer.* This layer is responsible for formatting the task of reading the request from a handheld computer and creating the response required to present information returned from the business logic access layer. This layer knows all specific information about formatting information for a handheld PC browser and how to get and use device-specific information (e.g., a Palm model X is being used to access the site). It is likely that this layer will have to be broken out into individual components for each type of handheld PC due to manufacturer differences. Additionally, this layer will have to handle both online requests and communication with device-resident components. No distinction has to be made between online and offline here, as the device-resident components will communicate with the handheld computing formatting layer over HTTP in both cases.

3. *Web formatting layer.* This layer is responsible for formatting the task of reading the request from a Web browser and creating the HTML pages required to present information returned from the business logic access layer. This layer knows all specific information about formatting information for a Web browser and how to get and use browser-specific information (e.g., Internet Explorer version X is being used to access the site).

The *Application Layer* has been broken down into three different pieces:

1. *Business logic access layer.* This layer is responsible for routing the requests from the client side to the appropriate business logic. We don't want the formatting layers to know much about where the business logic resides and have inserted this layer to ensure that this does not have to be the case.

2. *Business services layer.* This layer is responsible for implementing the business logic. It does not know anything about the presentation layer and where data is located.

3. *Transaction logic business services layer.* This layer knows where to get data, but does not know how to get it. Data retrieval is the responsibility of the Data Access Layer.

The *Data Access Layer* has been broken into three pieces:

1. *Legacy access data access layer.* This layer knows how to access any legacy systems. In this case, the only legacy system is the order management system.

2. *Local data access layer.* This layer knows how to access information in the local database. The local database is used primarily for profiling and personalization, but could also be used for staging data. This is usually done to increase performance.

3. *Security profile data access layer.* This layer knows how to access security information in the directory server.

The future state is a more detailed layering of the application components. The rationale behind this layering is to increase the flexibility and scalability of the solution. You can see in the description that each layer has a specific purpose and has minimal information about the other layers. For example, the business services layer has no knowledge of how to access data or what a WAP phone is. This makes the solution more scalable because the components can be broken down further than in the previous example. It also makes the solution more flexible, mainly because you don't have to change all layers to make modifications to a legacy system or a presentation layer. Going back to our example, we could now implement access via another wireless device without having to change the business services layer.

While reading this, you may ask yourself what these principles have to do with wireless. The answer is that wireless systems must be able to accommodate rapid changes and scalability. As indicated in Chapter 5, we may see much larger volumes of users in a few years than we are experiencing now, and wireless presents us with various different form factors we have to accommodate. For instance, even if we look only at handheld PCs, we could have several different screen sizes, makes, and models. This will force us to change our presentation layers to accommodate these devices in the most optimal way without spending a huge amount of money to do it.

The rest of this chapter will focus on some of the specific applications in MobileMediaMaven. In particular, we will describe the different components needed to implement these applications. The applications covered in the rest of this chapter are as follows:

Proposal generation.   Rapidly creates proposals for Acme's clients.

Filtering.   Filters through all radio stations in the network to find the one(s) that cater to the appropriate listener demographic.

Routing and review.   Routes the proposal to superiors and station managers for review and approval purposes.

Messaging. Notifies other sales executives or managers and receives notifications.

Order management. Manages client orders.

User management. Manages account representative profiles.

Account management. Manages client account information.

Reporting. Creates reports to review performance metrics and other information.

We will also discuss localization features in the context of these applications, though they will not be covered in detail.

Wireless Extensions to Existing MobileMediaMaven (MMM) Functionality

Let's start this section by driving into Figure 15.4 in more detail. Before describing this picture, please note that this picture does not focus on the following:

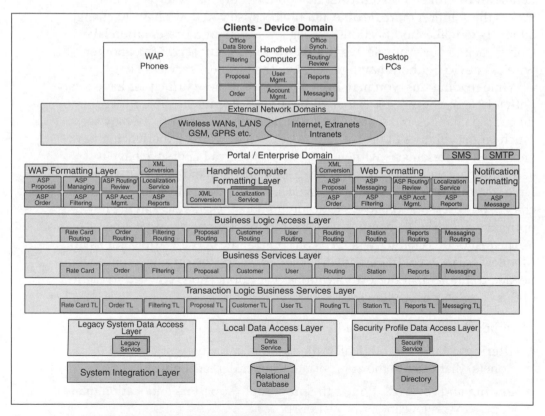

Figure 15.4 Detailed logical application architecture.

- *System services.* This includes services such as session handling, how to provide access to data, and application security. System services are not included in the discussion, because they are not unique to wireless and they can be purchased as part of a product such as an application server.

- *Commerce products.* This includes such things as electronic wallets and auction components. Commerce products are not included, as this functionality can also be purchased.

- *Personalization functionality.* This includes how to customize the front end and so on. Personalization functionality is also not included, as it can be purchased.

It is advisable in all systems implementation and integration efforts to evaluate which functionality should be bought and which should be built. The answer differs from case to case.

Figure 15.4 shows all the major components of MobileMediaMaven. These components are broken out according to the different layers specified in Figure 15.3. A few notes about this figure:

- We have added an SMTP server for email-based alerts. The sole function of this server is to send email messages to the correct physical location.

- We have added an SMS server for SMS-based alerts. The only function of this server is to send SMS messages to the correct physical location.

- Localization components have been added in all of the formatting layers. These localization components are responsible for customizing the information and presenting information according to the preferences of the users. For instance, if an American salesman is working in Mexico, any information presented to him would be in English and only the information relevant to where he is located in Mexico would be presented.

- Offline storage is required and included in the handheld computer client.

- Offline communication and replication are required and included in the handheld computer client. The reason for this is that all information entered in offline mode must be sent back to the back-end system.

- Every business function outside of messaging is structured in the same way. Each has an offline component in the handheld computer; ASP components to generate the WAP and Web pages; an XML generation component in the general presentation layer, which is the same component for all functions in each layer; a business logic component in the business services layer and the transaction logic business services layer.

- Messaging contains a trigger that can initiate a notification from the system. The trigger is responsible for running periodic checks to determine whether a notification is to be sent.

- A common service is used to access legacy systems, a local database, and security information.

The rest of this section will give examples of some of the business functions, including sample interactions between components.

Proposal Generation

We will explain the proposal generation functionality by describing the flow of a transaction launched through a handheld computer.

In the first scenario, let's assume that the handheld computer is online. The Proposal component in the handheld computer allows the user to enter the proposal parameters. These are communicated over HTTP to the XML Conversion component in the handheld computer formatting layer as illustrated in Figure 15.5. This component generates an XML-formatted request and

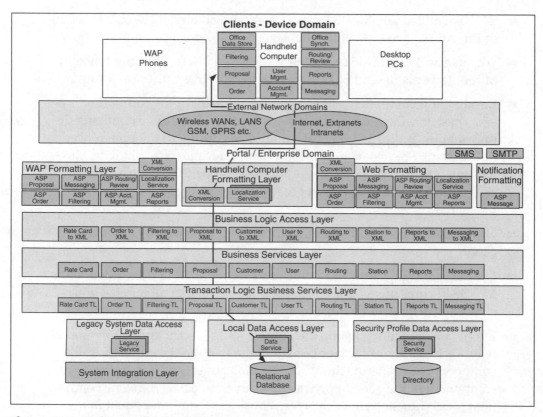

Figure 15.5 Online application flow.

passes it through a Localization component in the handheld computer formatting layer while sending it to the Proposal Routing component in the business logic access layer. The Localization component is responsible for detecting information about the user: where the user is working, what device is being used, and so on.

Once the information is relayed from the handheld computer formatting layer to the business logic access layer, it no longer contains information about the presentation layer. The Proposal Routing component in the business logic access layer translates the incoming information into business logic calls serviced by the business services layer and the business services layer simply executes the appropriate business logic. If data is needed from any back-end systems, the appropriate component in the business services layer sends the request(s) to the transaction logic business services layer. The Proposal component in this layer determines where the information is and how to get it. It sends the appropriate requests to the data access components, which then retrieve the information.

Once the Proposal component in the business services layer is finished executing the business logic, it returns the result as an XML-formatted data stream to the XML Conversion component in the handheld computer formatting layer via the business logic access layer. Here, an XML stream is converted into the format needed to communicate with the handheld computer. Localization parameters may also be applied here. When completed, the response is sent back to the handheld computer.

This kind of flow is standard for all the applications described here. We will describe other aspects of the functionality using other types of flows that exist in the system.

Filtering

The filtering functionality can be executed in exactly the same way as the proposal application described in the previous section. We will not repeat this description. Instead, we will illustrate how this functionality would work offline using a handheld computer.

Figure 15.6 shows the application flow for running the filtering functionality offline. In order to run the filtering functionality, we are assuming that the user has to run the functionality online once to get the information required to run it offline from the back end. Given that this information is there, the user starts the application and runs the Filtering component on the handheld computer. While the user uses the application, information is stored in the Offline Storage component on the handheld computer. When the user exits the application, the information remains there.

After using the application several times during a day, the Acme account executive decides to replicate with the back-end system. While connecting,

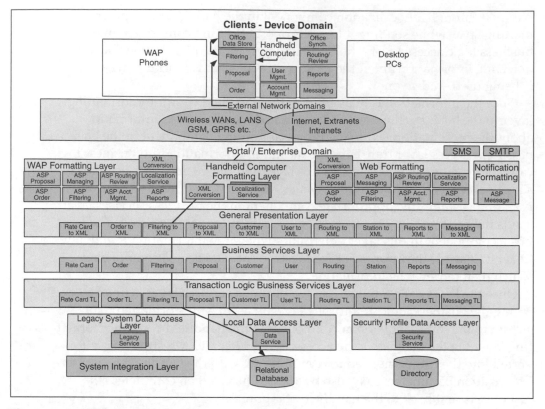

Figure 15.6   Offline application flow.

the Offline Communication and Replication component determines whether any offline data has been stored. If yes, it launches a transaction to the back end to update this information there. This transaction will be the same as described in the proposal generation flow, only using the filtering components in each layer.

Routing and Review

At this point, we have covered how different functionalities can be run using a handheld computer. The next functionality will be demonstrated using a WAP phone.

Figure 15.7 shows how the routing and review process can be executed using a WAP browser. Let's say that one of the Acme executives is giving a talk in London, but that she must still approve several proposals during her stay. Just before her talk, she decided to check, using her WAP phone, whether any outstanding proposals need to be reviewed.

The only difference between this scenario and the proposal generation scenario is that the request from the WAP phone is not sent directly to the handheld computer formatting layer. Instead, a Routing and Review component in

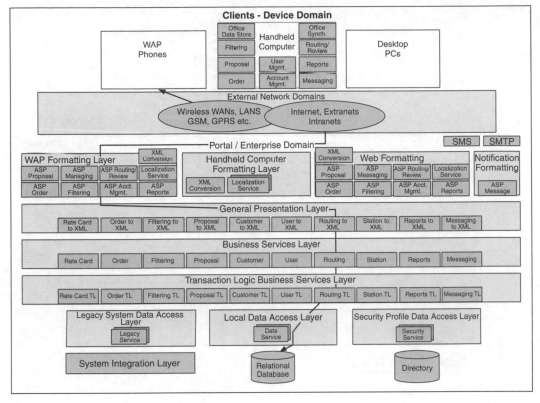

Figure 15.7   WAP browser application flow.

the WAP formatting layer is used to translate the message and, using the localization component, to determine localization parameters for the user. The application now knows that the user is in London, using a Nokia phone model X.

The Routing and Review component in the WAP formatting layer reads and understands the request and forwards the information to the Routing and Review component in the business logic access layer after using the XML Conversion component to convert the request into XML. From here, the process is exactly the same as with the previous examples.

Messaging

The messaging functionality is a little different from what we have previously described. The difference is twofold:

1. The message that is being sent will not be sent back to the source of the request.

2. The system can initiate messages on its own.

We will describe this scenario using a Web browser. Please refer to Figure 15.8 for a detailed description.

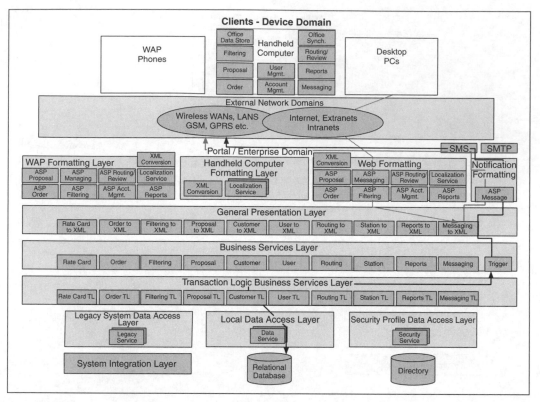

Figure 15.8   Web browser application flow.

An Acme executive using MMM needs to send a message to one of his sales representatives. The executive uses his Web browser to initiate an SMS message. When using the feature, the executive is interacting with the Messaging component in the Web formatting layer. As before, this component understands the request, gathers localization information about the user, converts the request to XML, and forwards the message to the business logic access layer. The Messaging component here calls the appropriate business logic in the business services layer.

The relevant components execute the business logic and then relay the request to send an SMS message to the Messaging component in the business logic access layer. This component forwards the request on to the Message component in the notification formatting Layer. The Message component formats the request appropriately and sends it to the SMS server that is responsible for transmitting a message to the user.

The same process is applied for electronic mail notifications. However, the requested message type will be in the form of an email message being sent via the SMTP server.

Order Management, User Management, Account Management, Reporting, and Other Functionality

The preceding sections described the different scenarios in which components can interact in the system. There is no need to describe the same processes again for the remaining MMM functionality because the component models and interactions are exactly the same. This brings us to the main point of this chapter. The system we described went from one to three channels while also adding new functionality. The components enabling each function were reused to make the system cheaper to build, to attain a faster time to market, and to reduce maintenance cost. This level of reusability was possible because of the application layering and because of how the components were defined. Furthermore, this type of architecture allows you to more easily add support for additional channels or future wireless devices. In essence, our case study illustrates a way to define an effective multichannel application.

New MobileMediaMaven Functionality Implemented with Wireless Technology

The wireless application we described here is relatively simple, yet very effective. In this section, we aim to describe some of the more futuristic ideas on how to extend MMM to be an even more valuable wireless SFA tool.

Interagent Messaging

In Chapter 13, we defined interagent messaging as a chatlike functionality between employees of Acme Radio. How would we implement this in a wireless environment?

Chat has become very popular in the last few years and, as a result, several vendors have created applications for it. Companies are currently using chat servers to provide customers with the ability to chat with service agents and to chat with each other. You can buy chat servers from several vendors and would most likely do so to implement this functionality. One thing MMM wants to accomplish is that the information exchanged via chat be incorporated in what we know about the customer. From a customer relationship perspective, if MMM knows more about its customers, they can target their offerings better and ultimately provide a better service. This implies that the chat application must tie into the CRM capabilities of the site to provide full value.

Remote Printing

Remote printing is a bit more futuristic. Wouldn't it be great to walk into any office and have access to the printing environment via some kind of wireless solution? Some emerging technologies, such as Bluetooth, could make this possible, although there are still several barriers that need to be cleared before this technology can work effectively. This functionality will be explained in more detail later in this book.

Another interesting way to provide customers with information is to send the information to the Acme home server, which then faxes the information to the appropriate number. This is a more tactical solution and could be implemented with today's technology.

In terms of application environment, the future scenario requires our components to be smart enough to know that a printer is available and how to communicate with the printer. The fax-based solution requires some extra messaging components and a fax server to generate and send the faxes. This component should be structured in the same way as the other functionality.

Location-Based Profiling

We have already discussed location-based profiling to a certain degree in the preceding sections. Basically, the advent of wireless gives us the ability to find out more about our clients and where our salespeople are located. With access to this information, we could streamline our operations to become more efficient, while at the same time increasing customer service quality. If the back-end application knows that a salesperson is in San Francisco today, it can filter out all products and services pertaining to other areas of the country and focus only on data relevant to the San Francisco market. This makes the sales rep's life easier by reducing the amount of information that is presented to her and that she needs to sort through during the sales call, allowing her to provide better service to her clients.

Location-based profiling can be implemented using smarter versions of the Localization components in the figures shown previously and by using a location product on the back end. Several of these products are available from established software vendors.

ACME Airtime Auction

The airtime auction feature strives to sell airtime that is about to expire. In an effort not to waste any open time slots, Acme is progressively reducing the price of each unsold advertising spot. In an auction type of setting, Acme uses MMM to sell available inventory to whoever submits the first acceptable bid. The auction runs until close to the very moment the spot is to be aired. This provides some great bargains for those advertisers who are flexible and willing to wait until the last minute, and it provides Acme with additional revenue. The airtime auction feature can be implemented using Auction components much in the same way that other components for the existing functionality have been implemented. In addition, the auction functionality may trigger several notification messages to buyers and sellers of the airtime.

Advertising Effectiveness Scoring

Advertisers and sellers of advertising time are always eager to know how consumers react to their messages. Advertisers want to constantly improve audience targeting to increase the value they receive for their advertising dollars. Media companies such as Acme are always looking for ways to generate nontraditional revenue to supplement advertising income.

MMM could be expanded to include a location-sensitive application that gathers information about a station's audience and its reaction to given marketing messages. A representative group of target users could be tracked over a period of time to better understand their behavior. Acme Radio could accumulate and analyze this information, and then sell it to the company's clients to generate additional revenues. How can this be implemented technically?

Well, first you need to select a sample of listeners that you know will be exposed to the message to be aired and whose effectiveness Acme would like to assess. Asking for volunteers might be one avenue to create a representative list of consumers; paying them a nominal compensation may be a more effective way. After all, we are asking Acme customers to give up a certain degree of privacy to participate in our quest to understand consumer behavior. Once a sample has been identified and enlisted to participate in the study, these users can be tracked using a localization product. This application would ask the consumer for a quick reply to a query about the ad just aired. The query and the user's response could be sent back and forth via SMS. Analysis of all users' responses would provide the value-added component in which Acme's clients are interested.

Summary

As you have seen in this chapter, we have taken an existing system and modified it to make some of its features wireless while we also added new functionality. The important lessons to take away from our exercise include the following:

- We designed our system to be scalable and flexible in anticipation of adding alternative access channels in the future. The result was that we were able to wirelessly enable certain functionality without having to replace the entire solution.

- A layered application architecture enabled us to integrate new channels and services by reusing the architecture and several of the already existing components. The benefits here included more rapid deployment of new functionality, as well as cost-effective implementation and maintenance

- An effective sales force automation tool should be designed to combine the various sources of information required by the solution's users, including senior management, front-line personnel, and third parties, if applicable.

The effort required to architect a flexible and scalable system is considerable at the outset of the project. Yet this approach provides the significant benefits of speed, low cost, and expandability in relation to any required modifications or extensions to the system's functionality in the future. Chapter 16 will explore the technical infrastructure that must be built to support our solution.

Technical Infrastructure Architecture

Introduction

To build our wireless CRM application we previously reviewed the required logical business processes, data architecture, and process and information systems technology. Let's now examine the technical infrastructure that supports the solution. Chapter 13 talked about how wireless functionality added a strategic advantage to the organization by allowing it to manage the customer while transcending the normal limitations of time and location. Our goal has been to map informational concepts from that chapter to the data flow, information systems, and, now, a technical infrastructure. Our task is to integrate the mobile components into the existing physical infrastructure and ensure that all functional, data, and informational requirements are met. We'll look at the entire system, explore the physical infrastructure design, and consider where the mobile elements integrate into the physical environment.

Overview: Technology Infrastructure

In order to secure implementation of this vital sales force automation tool, we need a technical infrastructure that can provide fast and reliable information access and exchange. The enterprise networks must function as an integrated whole between their various components—both central and remote. Our primary concern lies with the managed flow of enterprise information across its tethered and mobile-based communities. This flow needs to be flexible, scalable, and secure. High availability is necessary for our sales force to effectively exploit enterprise information from a multitude of physical locations. The availability of time-critical data leads to faster decision making, improved collaboration, increased sales, and enhanced operational efficiencies.

Similar to the case study that covered the metal exchange example in Chapter 11, we are again using a three-tier architecture of enterprise storage, networking components, and data software supporting the application environment and integrating with existing systems. We need a highly available data center for the radio networks that provides a flexible and secure infrastructure, accommodating data protection, linear scalability, and consistent performance. The general physical infrastructure that supports our network is divided into four main areas: corporate data center and local users, wireless system components and the remote sales force, a disaster recovery center, and individual radio stations with their own LANs and data systems.

Our design focused on a dynamic business environment and was modeled on the concept of a single data center site. This site is connected to a network of local data and enterprise management systems. It has inputs from a variety of carriers and wireless application service providers. The design was provisioned with a redundancy of hardware and software not found in the client's environment. For example, the solution identified a need for a more robust, nondisruptive backup of the database. We did this using business continuance volumes (BCVs), and server-based I/O channel load balancing and failover. Assuring system redundancy was key to achieving the availability targets of our client.

Figure 16.1 illustrates a high-level overview of the technology infrastructure. Here we find the centralized corporate data center that contains the following systems application servers:

- A *rate/yield management system* in which a station manager is given the ability to update system prices automatically or by the click of a button. This package is integrated into the accounting/revenue systems, the enterprise database via communication with the traffic rules database system, and the presentation components.

- A *traffic management system,* based locally at each radio station that feeds back into the rate/yield management systems and into the inventory repository.

- An *inventory management system* that forms an ultimate repository of current state schedules, bookings, rates, station types, and so on. This system is used to create business scenarios to describe the management of the perishable time inventory and the sequence of activities occurring in response to an internal or external event. The system helps the

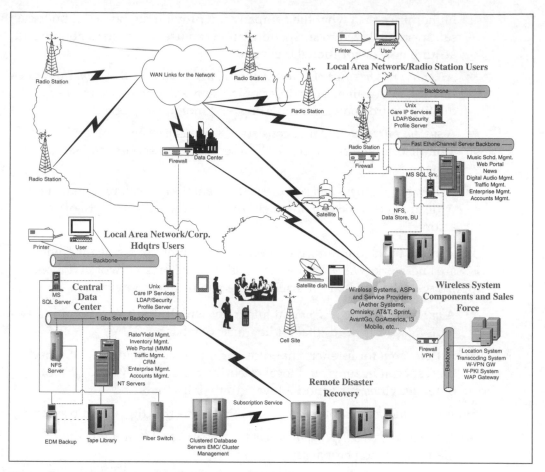

Figure 16.1   Wireless CRM (SFA) infrastructure blueprint.

sales organization and all business entities image and manage the work flow process via a communication program that shares information on current rates, bookings, available time slots, requests, related status reports, and the generation of client quotes. This system allows expectations related to sales quotas, advance bookings, and internal renegotiations to be set up front, prior to the sales call. It enables the Web application with an ability to manage available inventory, produce scenario reports, and prepare customized bid proposals for a client.

- A *CRM system* for managing customer information received via internal and external histories. These are customer demographics related to past airtime purchases collected both from internal transactions and from industry watch groups. Capturing this type of information is intended to alert a given client's sales representative to potential new sales opportunities. For example, information collected from outside agencies reveals that a particular client has been buying radio time at a higher price than what the company can provide the potential buyer. A salesperson alerted to such information can use it to help negotiate and acquire new or additional business.

- An *enterprise resource planning and accounts management system* integrates finance, forecasting, sales order processing, sales analysis, marketing, quality control, powerful reporting, and monitoring tools.

- A *system infrastructure* that consists of various presentation and systems administration servers, application servers, data servers and storage systems.

- *Network connections and security systems* entailing local, regional, and wide area networks, all of which constitute the national network of actors.

In addition to the centralized corporate data center we are dealing with a wide area network (WAN) of several hundred local area networks (LANs) supporting each radio station. Each LAN contains the following:

- A *music scheduling system* that integrates with traffic and accounting systems as well as the digital audio music database

- A *Web portal* for listeners that runs various announcements, DJ schedules, special features, and local celebrity visits, along with a host of other activities to support a loyal group of listeners

- A *news and copy management system* that feeds the digital audio system

- A *digital audio management system* that creates and maintains the scheduled production broadcast

- A *traffic management system* that links to the corporate central inventory

- An *enterprise management system* with bidirectional feeds that is an extension of the corporate headquarters' ERP and accounting systems

- A *system infrastructure* that consists of various presentation and systems administration servers, application servers, data servers, storage systems, networks, and security systems

Wireless System Components

These components constitute the main physical systems of MediaMaven's enterprise. Let's now consider the systems we need to support our mobile application.

Location Systems

It is not uncommon for MediaMaven's sales representatives to cover three different cities in a single day. At other times, they may spend several days in one city, handling multiple accounts and new sales leads. It is important for the sales force to have timely information regarding their customers and their buying patterns in order to manage the accounts. Location systems for the delivery of key information about their customers, like buying patterns or locating other sales reps, are based upon the representative's current geographic location. Tracking is updated continuously. The location component, obtaining its information from the carrier network provider, feeds information into the CRM, ERP, and inventory management systems. As mentioned earlier, the CRM system provides a feedback loop back to the sales representatives, based upon their current position. In the United States, there can be any number of location techniques used by the carriers, based on the particulars of the network and types of devices.

Transcoding Systems

Acme Radio chose not to standardize on any particular type of PDA or wireless WAP phone. Instead, the company decided to pass these costs to their employees and reimburse them for usage. Knowing that devices and technology are continuously evolving, we developed the Mobile MediaMaven application using XML as a metalanguage. This allows our client to translate many sources of content into the various flavors of wireless formatting. To accomplish this, we implemented a solid transcoding system that would supervise information channels, looking at the specific types of devices requesting information and formatting the information accordingly.

Wireless WAP and UPN Gateway

Having a mobile sales force deployed in the field brings with it the need to secure internal systems and mobile nodes from being compromised by unfriendly sources. To do this the company set up VPNs between itself and their wireless network providers. They then deployed a WAP server within their own DMZ for added security. Acme Radio is currently looking at other wireless VPN solutions to support their mobile laptop computers that may require greater throughput in terms of bandwidth.

Wireless PKI and Antivirus Detection System

Finally, our client set up an extensive private key infrastructure (PKI) to maintain confidentiality, authentication, nonrepudiation, and integrity of its data and systems. In addition to this, they also have contracted with virus detection vendors to maintain the security of their mobile devices.

Design Requirements

Whenever we design a technical architecture, we must be clear about infrastructure, integration, and testing implications. Fully comprehending the scope of our project requires us to first develop the answers to questions such as whether the software meets our data and information system requirements. Examples of other questions include how applications are using the network, and what demands new applications would place upon the network and system resources. Answering these questions is critical in assessing the technical requirements for sizing the infrastructure to meet specific application hardware, connectivity, and throughput needs. In our approach to architecting the physical infrastructure, we followed much of the same infrastructure and system documentation that was mentioned in the previous TI chapter for the wireless exchange.

It is important to thoroughly understand the capabilities and integration requirements of the various software and hardware components, both new and old, that make up the client's network. In an effort not to be too repetitive, we are just repeating some key examples of dimensions that need to be investigated:

- The technical network and system requirements for each application
- Vendor-specified capacity requirements and justifications
- The operating system required for a particular function
- Software components required for the application and function
- The number of users who will access the function, at what times, and from where
- Transaction details and bulk data transfer information
- Projected disk space requirements
- How many instances of each database type would be used
- The backup requirements
- Availability requirements
- Requirements for a development test and training environment

Once we had a thorough understanding of the data center and remote stations' computing needs, both old and new, we began compiling the appropriate physical systems components and configurations that would be required. Part of this assessment involved documenting the current network's state.

We needed to know the number of locations, existing WAN and LAN links, the types of networks deployed, and the equipment and wiring supporting the infrastructure. We also needed to understand how existing applications were impacting the network in terms of protocols used, the QoS and types of class issues, and the times when applications were used.

When we began this inquiry, our client provided us with adequate documentation in terms of the physical components and the number of endpoints. However, very little information existed in terms of performance metrics. A few months earlier, Acme embarked on a program to implement a network management strategy across the enterprise but had not completed the project due to several issues related to acquisitions and reorganization. However, the hardware and software had been deployed in the central data center at the corporate headquarters. These components included HP Openview, several LAN sniffers, RMON, and report formatting tools. Ciscoworks and other router/switch management tools had also been deployed.

These tools helped us over the course of a couple of months to gather sufficient data to characterize the network at the corporate headquarters and a few major radio stations. We wanted to know how the applications were impacting the network in terms of bandwidth, peak usage times, the types of protocols running, the types of applications such as data backup, digital audio, WWW, and videoconferencing.

We also conducted a similar survey of the existing security measures implemented by our client and their service providers. For more details about such assessment, see the security requirement section mapped out in Chapter 11 for the wireless exchange.

These activities enabled us to get an accurate view of the existing network at Acme Radio and allowed us to document existing gaps that needed to be addressed in upgrading the existing architecture in terms of systems requirements for network availability, fault tolerance/redundancy, management, operation support, QoS, class of services, types of services, content flow, caching, voice, and video. From here, we mapped out our strategy for bridging these gaps and implementing the network upgrade to support the future state design.

Not only were we able to get an accurate snapshot of the existing network environment, our analysis also supported the underlying need to rely heavily on the utilization and exploitation of uninterrupted, protected, and managed information flows across the enterprise of tethered and untethered actors. It would be impossible to create a well-integrated environment that must function across many different locations without a thorough understanding of all these factors. With this in mind, let's move on and take a closer look at the data center and network components supporting the radio enterprise and our sales force automation application.

Assumptions

Our recommendations to Acme Radio contain a few assumptions.

We expect a certain reliance on external carrier services and wireless application service providers. The recommendation to our client was that these services be collocated within their corporate data center. This provides a higher level of security and knowledge transfer to their own technical staff.

We anticipated the option of an off-site location connected through a VPN as a potentially less expensive solution. Either way, the data center would remain central and not include any geographic load balancing. We recommended the client's partners be chosen carefully, given the volatility of the marketplace for new technology vendors.

Our network and system sizing was based upon the characterization data obtained over the last two months, while anticipating a moderate increase in transaction volume. We marginally oversized the network and data systems to accommodate this assumption.

We performed an extensive analysis of vendor product features, performance abilities and system requirements. This information was used in planning our system integration activities. Our blueprint assumes the vendor-supplied information was accurate and complete.

Our client's definition of high availability means 99.99 percent availability = 1 aggregate hour of unplanned system downtime per year.

Mission-critical data application availability is improved, even when a system fails, because of server clustering and redundant network components. We informed our client that it would likely be a very expensive proposition.

Let's turn our attention now to information systems and other data center components as they are represented in Figure 16.2.

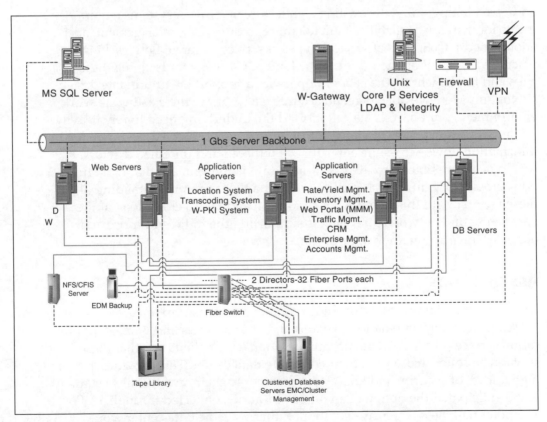

Figure 16.2 MMM data center.

Data Center Architecture

The drawing in Figure 16.2 is almost an exact copy of its counterpart from Chapter 11. There are only minor differences due to the fact that the underlying basic principles of the technical architecture for a highly available data center are the same. We will come across this in the technical infrastructure for Chapter 21 as well. Figure 16.2 shows storage area network (SAN) segmentation designed to move data application traffic away from LAN traffic hitting the Web, application, database servers, and network storage arrays. It's positioned as a dual redundant system. We chose a specific vendor that we believe provides high level of flexibility and scale. The solution provides for database and data software storage and point-in-time mirrors for off-loading testing, backup, and data recovery.

A central storage management solution was deployed to enable planning, configuration, control, monitoring, and tuning of the data infrastructure. Status, configuration, and performance reports are made available on demand to the system administrators. In this environment, software controls dynamic load balancing and access to the disks across multiple paths. These paths are used to route I/O requests on the fastest path, which helps alleviate contentions and delays that might otherwise occur. They also help protect against the failure of a server I/O channel by failing over to a second channel. Because of the client's mission-critical and perishable inventory, it's important to have point-in-time copies of the production database in order to off-load certain tasks. For example, the copy can be used to extract data for data warehouse routines, to test system upgrades, and to provide consistency checks.

We perform both a physical and a logical backup of the database, using the backup host for moving data to tape. This helps to separate and off-load certain routine tasks from the production host. It also enables greater frequency of backups and checks that ensure the integrity of MediaMaven's critical data. Furthermore, the backup software provides administrators with access to independently addressable BCVs for information storage, accomplished independently in background mode. We are doing this to off-load such processes from the main production database. BCVs can be split from the mirrored volumes and used for backup. Once this process is complete, they can then be paired back to the production device.

Oracle is the selected database for the data servers. However, some of Acme's applications run on Microsoft's SQL Server and so they are seated in the design as well. Two Unix machines function as the primary and backup data servers. A second set provides the data warehousing functions. We have configured all of these machines with failover software to guard against the event of a loss node. Ideally, if an application supports preconnected sessions, then almost no delay occurs in switching to the alternative server. Apart from the storage systems, the solution also contains a tape library. The previously mentioned data manager transfers data to an automated tape storage array. We use direct SCSI connection for the backups, which also off-loads this task from the SAN.

Before moving to the application server we should mention that there is also a file server in the SAN provisioned to be an NFS/CIFS server. Each set of application servers consists of four clustered servers. We have provisioned them here as four servers in order to ensure a 75 percent throughput capacity should a single node go down. Further, the NT application servers were set on their virtual LAN (VLAN) so that their intensive traffic demands and non-standard protocols could be kept separate from the main traffic. We connected each server to the network's distribution layer with gigabit Ethernet connections via single-mode fiber. These units are also connected to the SAN via Fibre Channel Arbitrated Loop (FCAL) connections.

A secondary firewall manages access to the application servers and the data environment. A WAP gateway is deployed in front of this firewall. The firewall also functions as a wireless VPN. Behind it are PKI systems. An expectation is that the carrier service and application service providers also represent additional layers to the security equation. As we mentioned and described in more detail earlier, the wireless application servers included the following:

- Location systems for the delivery of timely customer information to the sales representative based upon current geographic location

- Transcoding systems for managing multiple devices and network protocols

- W-VPN and WAP gateway are deployed for cost-effective access into corporate systems

- W-PKI system for security

Because of the transcontinental nature of the Acme Radio enterprise, the use and control of a single wireless network protocol in the United States was neither feasible nor desirable since it would discourage use. It is safe to say that at least two wireless devices (a PDA and a cell phone) and various network protocols were employed. Therefore, an extremely robust transcoding system was required in the systems design.

The ability to locate a remote sales force agent and provide relevant customer data in real time based on physical location was paramount to the agent's success. Again a comprehensive, robust system, comprising several integrated tracking systems, was deployed. This system also tied into other parts of the enterprise, customer profiles and business rules, and knowledge and event managers.

Finally, security was implemented at a variety of levels, including a wireless VPN solution reaching across a number of localized and global carriers. A wireless PKI with the use of digital certificates was mandated.

Network Architecture

Topology

Once again, the drawing in Figure 16.3 is a virtual copy of what we saw in Chapter 11. We've deployed a multilayered topology with an access, distribution, and core layer. Also shown is a DMZ with internal and external firewalls, our Web portals, various system proxy servers, and intrusion detection and security devices. We have a LAN/WAN environment represented in the access and distribution layers. At the core layer we find our server farm and previously discussed data center.

Multilayer Model

We covered the purpose of the multilayer model in Chapter 11. It is aimed at providing a multilayer hierarchical relationship between the three components previously mentioned. When provisioned correctly, such a model can ensure that no device is more than two router hops away from a server across the entire enterprise. In other words, we aim to hold the network's diameter

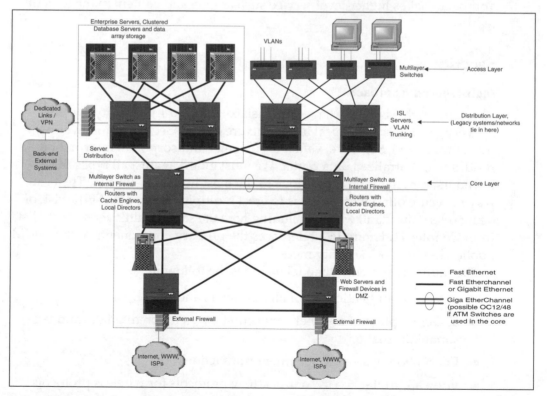

Figure 16.3   MMM network architecture.

to two hops and minimize traffic latency through the deployment of layer 3 switching devices. This allows traffic to essentially span the entire WAN enterprise, giving a global presence to system resources. The core layer can thus be used to accommodate different networks such as asynchronous transfer mode (ATM), switched megabit data service (SMDS), frame relay, and other protocols based upon the quality of services defined by a particular application and services available in a location.

To support these design goals we need to make sure that a contiguous address scheme with a narrow 23- or 24-bit subnet mask is used. Flatter networks with wider masks are not recommended. We also assume in this design an integrated domain name server (DNS) and DHCP server configuration, load balancing, and performance and fault management systems.

Such a topology is scalable. The routers and layer 3 switches use redundant configurations that can easily be maintained and ported across the network. Backbone performance scales at the core layer. More components can be added to the network in a building-block fashion. Finally, the system is highly deterministic, which makes network management and troubleshooting much easier. The model has a highly regulated performance, path determination, and failure recovery component to its design. The multilayer approach makes higher-level service such as QoS and content management feasible.

Security

Multilayered Approach

Throughout this book we have emphasized the need for security systems at every level of the technical infrastructure design. Security can never be overemphasized. Sensitive, valuable customer and corporate information resides in a centralized data center. We must guard against all types of intruders, ranging from the curious to the most hostile of entities. The best way to secure our data systems is to erect multiple barricades. At the risk of being overredundant, we are including here the same multilayer security list from Chapter 11. Figure 16.4 references the primary components of the MMM application network security model.

The multi-layer security model includes the following:

- Password authentication of all customers and participants
- SSL encryption of member connecting pages for credit, data, and personal information
- Encryption of sensitive payment transactions
- Firewall/router access controls (new protocols for wireless protocols must be supported—see Figure 16.4)

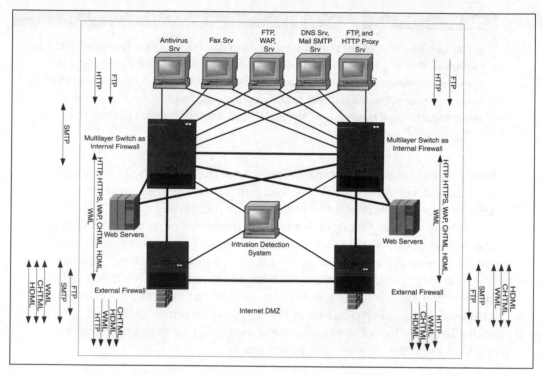

Figure 16.4   MMM network security model.

- OS hardening (system servers as well as clients, both fixed and mobile)
- Malicious content scanning (antivirus)
- Security monitoring
- Limited data storage
- Intrusion detection
- Application sessions and access controls, application password management
- Physical and personnel security controls
- LDAP protocol
- Password authentication of all back-end eCommerce connections, SSL encryption of all credit transactions
- Use of digital signatures, certificates

These specifications will have to be designed and instituted by the organization itself. All personnel within the company need to have an unambiguous understanding of the organization's security policies and procedures, including global procedures and specific ones tailored to each worker's job

> **NOTE**
>
> A site can be secured through the implementation of multiple layers of security. However, this approach will not address all aspects of security architecture such as customized security policies, standards, and procedures; organizational and operational issues relating to the deployment, implementation, and maintenance; and operational, technical, and administrative specification.

function and tasks. We recommended to our client that they take additional steps to test, document the holes, and patch their security systems. This applies especially in the softer areas we just mentioned.

We hope we have brought home the importance of securing your systems and never assuming that they are safe. We talked in more detail about the nature of each security layer in the discussion of our wireless exchange. You can find more details about the various components of a multilayered approach to security in the section on security in Chapter 11. Additionally, please see some of the sources in the References section at the end of the book. They are full of eye-opening examples that have the potential to frighten even the best security administrator.

Summary

In this chapter we looked at the components that make up the technical infrastructure for our wireless CRM application. We looked at all the information system applications deployed across the entire enterprise. We investigated the wireless components that were particular to the sales force application and where they'd fit into the existing space. We saw that our location systems would be tied into the various wireless network providers (carriers). Multiple mobile client devices required a robust transcoding system. They also required a WAP gateway within the corporate DMZ and a VPN to support the roaming laptop computers and PDAs. In order to secure and implement this vital sales force automation tool, we needed a technical infrastructure that could provide fast and reliable information. The enterprise systems had to function as an integrated whole, including mobile clients all the way to back-end systems. Our primary concern was on the managed flow of enterprise information across its tethered and mobile-based community. Our network topology was based on a multilayer topology that was provisioned with throughput, redundancy, and scale in mind. We discussed the components of our application servers and database storage systems deployed with their own SAN. This was in order to keep traffic away from the front-end applications networks. Finally, we emphasized the need to provision these enterprise systems and employ a multilayered approach to security.

This concludes our examination of a wireless CRM application. We will now move ahead to our next project of architecting a wireless Internet application for a financial information and commerce business. This is presented as a specific set of wireless applications within the larger context of the financial services industry.

Architecting Wireless Enterprise Applications for Financial Services and Commerce

Designing Wireless Financial Services Enterprise Applications Overview

Introduction

The financial services industry (FSI) has been a perfect candidate for the implementation of wireless services ever since the first security was traded electronically. Stockbrokers, financial advisors, day traders, insurance agents, and bankers are all prime examples of users that require real-time information to make business decisions. They are ideal candidates for wireless financial services. Traditionally, these financial professionals have relied heavily on data that is real time, whether obtained from a new service such as Reuters or from a computer screen directly connected to a corporate network. Addressing their needs, probably one of the best examples of wirelessly enabling an FSI application, is the mobile brokerage, which also happens to be part of the case study covered in this section. Financial advisors as well as their clients require minute-to-minute information to help them determine their next move. How many times have we seen the stock market shoot upward after the Federal Reserve drops interest rates? Financial services firms rely on this critical data to effectively manage their clients' accounts. The quicker this information is available to the financial advisor and his or her clients, the faster they can react to buy or sell. Wireless technology promotes ubiquity, or anywhere, anytime accessibility to key financial information.

In contrast, there are occasions where real-time information is not a necessity, but a huge advantage just the same. The banking industry is a good example. Financial portals, as operated by Internet-only banks as well as online subsidiaries of established financial institutions, give their customers the ability to wirelessly access accounts—checking, savings, money market, CDs, credit cards, and so on. Even though these services can be provided in real time, they are not as time sensitive as trading a security whose price is rapidly fluctuating.

Where it gets interesting is when you combine both types of financial services, banking and securities trading, and throw in insurance products and services for good measure. Due to the recent deregulation of the financial services industry in the United States, we are beginning to see various new types of integrated offerings that allow customers to bank, trade, and purchase insurance policies online, all from the same provider. And all of these services can be provided on a wireless device, whether it's a cell phone or a PDA.

This chapter and the rest of this section will focus on three major areas within the financial services vertical sector:

Mobile banking. Areas within mobile banking may include wireless access to checking, savings, and cash management accounts; credit cards; foreign exchange services; loan management; and trust services. Also, online bill presentment and payment applications offered via wireless devices provide opportunities to pay your invoices while away from a PC.

Mobile brokerage. Financial brokerage houses and financial services firms deal with the buying and selling of various financial instruments such as stocks, bonds, derivatives, mutual funds, and various other services. The mobile brokerage application enables customers to obtain real-time news services, access their portfolio, make buy and sell decisions, receive alerts, screen stocks, and so forth. Financial advisors are able to better service their clients by having access to their portfolios, reach them while they are on the go, and build a closer relationship through improved communications.

Insurance. Virtually every area of insurance can benefit from wireless services. Insurance agents who manage auto, home, life, and health insurance can have the ability to access a client's book of business on a PDA while away from the office. Insurance rating applications and financial analysis tools can be carried in a briefcase and brought to client meetings. Policy due alerts can be sent to clients who subsequently can pay them using the same wireless device.

Let's look at the role of financial applications and the hurdles that must be overcome to allow such services to enter the mainstream.

Role of Financial Applications on the Internet

Based on wireless devices such as mobile phones, personal digital assistants (PDAs), and pagers, mobile commerce (m-commerce) will have a larger impact than e-commerce has had during the past few years. Giving people an untethered means of connecting to the Internet offers tremendous opportunities. As technology becomes smaller, faster, cheaper, and better, the possibilities seem endless.

"Electronic Commerce, A Need to Change Perspective," Cap Gemini Ernst & Young Special Report on the Financial Services Industry

Financial services firms are racing faster than ever to make wireless applications available to their customers. Part of the rush is that firms see this as the second coming of the Internet wave and they don't want to miss it. They also see the wireless Internet as a second opportunity to extend their services and offer new services via wireless devices. Figure 17.1 illustrates a compelling study done by Forrester Research, Inc., on the types of services that some of the leading financial services firms are offering.

No doubt financial services firms are making a significant investment in servicing their customers through wireless offerings. This response is driven by a significant demand for financial mCommerce services. Figure 17.2 illustrates a recent study by Ovum on the dramatic projections for mCommerce revenue in the coming years.

To add to the overwhelming statistics in Figure 17.2, other studies indicate that:

- By 2002, there will be over 1 billion potential global users of the wireless Internet.

- By 2003, more than 70 percent of major financial institutions will offer mobile banking.

- By 2005, the number of users of wireless financial services will grow to 83.7 million in Asia, 76.6 million in Western Europe, and 35 million in North America.

Convergence, Consolidation, and Globalization

In our discussion of the role of the Internet in financial services we must also consider the industry trends that are forming and shaping the sector as we

Financial instituttions	Trades and quotes	Alerts and notifications	Access account information	Funds transfers and bill pay	View portfolio	Access research and charts	News, weather, and horoscopes	Service availability	Partners	Devices
Ameritrade	✔	✔	✔					Available	In-house	Phone
Bank of America			✔					Available	724 Solutions	PDA
Bank of Montreal	✔	✔	✔	✔	✔			Available	724 Solutions	Phone, PDA
Charles Schwab	✔	✔	✔		✔	✔		Available	Aether	Phone,PDA, pager
Citibank		✔	✔	✔				Available	724 Solutions	Phone,PDA, pager
Claritybank.com	✔	✔	✔	✔			✔	Available	724 Solutions	Phone, PDA
DLI direct	✔	✔	✔					Available	In-house	Phone,PDA, pager
Dreyfus Brokerage	✔		✔	✔			✔	Available	w-Trade	Phone,PDA, pager
Fidelity	✔	✔	✔		✔			Available	In-house	PDA, pager
Harris Bank	✔	✔	✔	✔	✔		✔	In trial	724 Solutions	Phone
MasterCard			✔	✔	✔			Planned	724 Solutions	Phone, PDA
Merrill Lynch	✔		✔	✔		✔	✔	Available	w-Trade	Phone,PDA, pager
Morgan Stanley Dean Witter	✔	✔	✔			✔	✔	Available	Aether	PDA
National Discount Brokers	✔	✔	✔				✔	Planned	Aether	Phone, PDA
Suretrade	✔		✔		✔	✔	✔	Available	w-Trade	Phone,PDA, pager
Visa		✔	✔					Planned	Aether	Phone, PDA
Wells Fargo (to be announced)								Planned	724 Solutions	TBD
1800DAYTRADE.COM	✔		✔			✔		Planned	w-Trade	Phone,PDA, pager

Source Forrester Research, Inc.

Figure 17.1 The wireless FSI landscape.

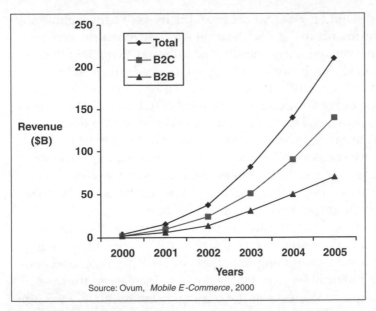

Figure 17.2   Worldwide mCommerce revenue 2000–2005.

speak. Three megatrends that we are observing right now deal with convergence of various financial offerings due to deregulation, consolidation of the provider base, and the race for global expansion. Let's look at the convergence opportunity first. For those readers who are familiar with the European financial environment, the commingling of various financial products and services is nothing new. For years, European financial institutions have been able to offer their customers a comprehensive portfolio of financial services, including brokerage, banking, and insurance products. Consumers in Europe are accustomed to developing close relationships with one universal financial services provider, as this provider usually addresses all of their financial needs. The aggregation of financial services has always spelled a win-win situation for providers and consumers alike. Consumers received the convenience of one-stop-shopping, whereas the providers were able to establish close, long-term relationships with their account holders, often seeing them from their first financial steps of, say, opening a savings account when they were teenagers, all the way to providing assistance with home loans and insurance needs as the customer matured.

In the United States, on the other hand, the financial services industry has been very regulated. Banks could offer banking services and security firms dealt in equity trading, whereas insurance carriers were limited to offering policies. This separation of business was established by the Glass-Steagall Act of 1933. The Act, preventing full affiliations among banks, brokerages, and insurance carriers, attempted to prevent a repeat of the Great Depression.

The thought process behind it was that the risky business of banks' gambling with misguided investments in the stock market was one of the reasons for the collapse of the market and the economy. And although economic theory in later years questioned this line of reasoning, the Glass-Steagall Act remained in effect until very recently. In November of 1999, finance history was made with the release of the Gramm-Leach-Bliley Act (GLBA). This act effectively deregulated the financial services industry and opened the doors for financial institutions to expand their offerings. A so-called financial holding company may now own banks, brokerages, and insurance carriers, which places them on equal footing as the integrated financial services companies that are commonplace in Europe. Finally, American institutions will be able to offer a full range of financial services to their clientele.

In addition to the aggregation of different financial services, the second major trend in the industry concerns consolidation. Although GLBA is still very new, the U.S. market is beginning to witness increased merger and consolidation activity in financial services that is going to increase as the quest for financial leadership accelerates over the coming years. In the race to align with the largest players in this space, we will witness the emergence of financial powerhouses that will dominate the U.S. market, mirroring the financial behemoths of Europe. Those companies will strive to provide all financial products and services under one roof in an effort to establish as close a relationship with their customers as possible. Customer retention will become the paramount objective as companies now have the regulatory freedom to service all of their customer financial needs. Once a customer chooses a provider, that customer can effectively be locked in by catering to his or her every financial need. Large institutions see the opportunity to significantly increase revenue by obtaining a larger share of their customers' wallets.

Remember the unrestricted environment of Europe? And the opening of the American market due to GLBA? The large European institutions have had years to refine their service offerings to closely address consumer demand, understand consumer behavior in light of multiple financial services, and learn how to attract and retain a customer base from a customer service perspective. Yet the European institutions were prohibited from fully expanding in the United States not so much because of differences in consumer behavior or geographic boundaries, but because of Glass-Steagall, which restricted the aggregation of financial services—the primary value proposition to the organizations' customers. By removing this barrier, the United States is now becoming a target for large, multinational financial corporations who understandably view the American market as a huge opportunity to pursue in their globalization efforts. On the flip side of the coin, American institutions may look overseas to expand their reach and presence on a global scale. Whether it is the United States or Europe in the wake of the European Monetary Union, large financial conglomerates will become even bigger via intrasector mergers, cross-industry mergers, global expansion, acquisitions, or strategic alliances. The objectives behind this drive are clear: Own the customer, increase revenue via expansion of the customer base, increase your share of the customer's wallet, and reduce operating expenses from economies of scale and shared infrastructures. In the end, those who will be successful stand to gain tremendous benefits from a worldwide market presence.

Taking this brief detour, you now understand the tremendous pressure facing financial institutions. Especially the larger players who are continually evaluating how technology and the Internet can be leveraged to gain that competitive edge. After adding Web access to their customers' accounts, the financial providers are now looking to take their services to the next level by adding wireless features to their offerings. The next section illustrates some of the applications and benefits provided by mobile technologies.

A Wireless Situation

One of the unique opportunities that mobility brings to the table is the fact that wireless devices are generally associated with a person as opposed to a fixed location. In addition, mobile devices, such as cell phones, are usually not shared with other members of a household or business. This provides unique opportunities for companies to target very specific users according to their demographics, likes, and dislikes. As we discussed in previous chapters, localization changes the way business can attract the attention of potential customers simply by knowing what they are interested in and where they are located. Here's a small example. Michael Moneybags, an extremely busy executive at a high-profile law firm in downtown Chicago, is an affluent client of the Dewey, Cheetem, & Howe Financial Advisory Firm. Typically, Mike talks to his financial advisor, Larry Dewey, a couple times a week to discuss his portfolio and potential buy/sell decisions based on current market conditions. Currently, Larry's firm provides access to client accounts through a PC connected to the Web. Services include stock quotes, buying/selling of financial instruments, and research. Mike spends a great deal of his time traveling, which limits his opportunities to work in front of a connected PC. Thus, Mike must speak to Larry more often to obtain financial data about his accounts. Recently, Larry's firm launched the Dewey, Cheetem, & Howe wireless FSI portal for their clients. Some of the new features that are available to the company's clients are hot alerts, buy/sell stocks, financial research, and even payment systems. With these handy new features, Mike can now use his PDA or cell phone to receive real-time information based on his location and preferences. Mike can also react to market conditions by buying or selling stock through his wireless device without even once contacting Larry. Finally, when Mike travels to other cities, or other countries, he can receive financial advice anywhere and anytime.

At any rate, you get the basic idea. Financial firms such as Dewey, Cheetem, & Howe are making waves in the FSI sector by offering services to customers that were never available before. Other opportunities to service clients have opened up through wireless technology, such as financial advice based on real-time market conditions. This information availability anywhere and anytime provides the consumer with the power to react immediately instead of waiting to speak to an advisor. However, there are still some precautions that firms need to consider before investing wildly in wireless technology.

Barriers to Overcome

With all the advantages that wireless technology brings to the table, what could possibly be a factor in slowing its continued growth? Some of the issues that you may want to consider as potential barriers to mobile commerce include the following:

- The United States is generally known as a wired society. In other words, PCs, phones, TVs, and so on, are wired devices, and consumers and business alike have made significant investments in these areas. However, change is on the horizon, and, as we've seen, the number of wireless and Internet users is increasing dramatically due to a more mobile lifestyle. With 30 percent of the U.S. population (a figure that continues to grow) owning a cell phone, mobility will continue to make its presence known ["Half the Country Is Mobile Mad," *The London Times,* July 4, 2000].

- The payment structure for wireless services is complicated and inconsistent. Frankly, there seem to be just as many service plans as there are cell phones today! That is getting a little carried away, but certainly the consumer is presented with a great deal of confusion when it comes to pricing plans and usage of wireless services. Wireless service providers are definitely making strides by decreasing fees and offering simpler plans for Internet access. AT&T, Sprint, and Verizon are just a few that are bundling their voice services with Internet usage.

- We've made several references in this book to the current fragmented and slow bandwidth of wireless technology experienced in most countries around the globe. However, consistent, fast, and reliable always-on service is one of the keys to the success of future wireless services. The development of new technologies in second- and third- (2.5G, 3G) generation infrastructures will answer the call.

- Finally, wireless applications have been accused of having a poor user interface and providing a limited user experience. The key to overcoming this barrier will be for firms to establish a sound wireless strategy. Frankly, cell phone screens will never be the size of a PC monitor. In addition, the speed of transactions may never catch up to wired Internet environments. Firms must determine what the value proposition is that they will be providing their customers. Is simply offering stock quotes and financial trading over a device enough? Probably not, since that type of offering already exists on the Web sites of several financial firms. Alternatively, a firm could offer a complete analysis of a client's position based on current market conditions and bundle that offer with comprehensive online transaction functionality. So if I'm Mike Moneybags and I receive an alert on my phone that IBM will be offering new chips smaller than the size of an ant, take a guess what could happen to the stock price. If Mike's financial advisor can offer the analysis and advice while Mike is mobile, and Mike in turn is able to immediately act upon that advice and buy IBM securities, the possibilities are endless. The bottom line is that FSI firms must offer more than access to existing Web sites and hot alerts. Hopefully, the next few chapters will show you how it's done.

In the following sections of this chapter, we'll cover some of the wireless FSI applications in two primary business markets: the B2C (business-to-consumer) and the B2B (business-to-business) areas.

Consumer Financial Applications (B2C)

The focus of this section as well as the following chapters in the FSI portion of this book is on consumer markets—if not for any other reason than (you guessed it) demand and growth is in this area. Figure 17.3 indicates that the number of online investors is projected to increase 37 percent CAGR (compound annual growth rate).

The two primary areas we will cover within the B2C space are mobile brokerage and mobile banking. Each of these markets is a separate and distinct business area, but you'll see that wireless services complement each very nicely.

Mobile Brokerage

As we've mentioned throughout this chapter, a mobile brokerage is one of the primary wireless services offered by several financial institutions today. There are two main building blocks associated with mobile brokerages: wireless financial transactions and financial knowledge networks.

Wireless financial transactions.   This entails the ability to engage in wireless online trading of stocks, bonds, mutual funds, and other financial instruments such as Treasury bonds, currency, and commodities, using mobile devices. This also includes the ability to transfer funds between individual accounts, to deposit or withdraw currency.

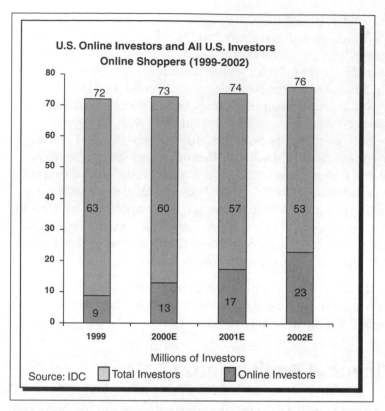

Figure 17.3 The number of online investors to increase at 37% Compound Annual Growth Rate (CAGR).

Financial knowledge networks. These are communities that consist of financial subject matter experts, such as financial advisors who service brokerage customers. Financial knowledge networks are built around financial intelligence as represented by financial news, research and analysis, financial planning tools, and communication enablers such as online chat and real-time actionable alerts.

Let's go through both of these building blocks in greater detail.

Wireless Financial Transactions

While trading on the Internet is nothing new to most of us, wireless trading is quickly becoming a natural add-on service for clients of online brokerages. However, wireless financial transactions entail much more than the ability to trade Microsoft stock on a cell phone or a PDA. The mobile brokerage opportunity represents an entirely new business model, of which wireless trading is one segment. Let's refer back to Mike Moneybags, our Chicago attorney. A typical scenario might go like this:

1. Mike goes to work at 8:00 A.M., riding the train from Naperville to downtown Chicago on the train. While traveling to work, Mike receives an alert on his mobile phone that Cisco stock has reached a trigger level.

2. Mike checks the Cisco real-time quote and news through the wireless financial portal provided by his brokerage firm. He accesses the site through his mobile phone or PDA.

3. He next checks the portal for real-time analyst ratings and research reports on Cisco. Using all of this information, Mike can then make an intelligent decision on whether to buy or sell elements of his portfolio.

4. Subsequently, Mike identifies stock he wants to buy or sell and checks the funds that are available in his account. If funds are running low, he can transfer money out of his savings and into the trading account. He can then place the order online and immediately receive a confirmation that the order was processed accurately.

These four points cover the main services enabled by the wireless transactions building block. These services include:

- *Wireless access to market and investment information.* Steps 1 to 3 illustrate what we mean by this. Basically, it's the ability to receive financial information wirelessly to make informed investment decisions. The user can access such information by initiating contact with the portal, or the user can receive the information via a server push that automatically transmits the data to the device in the form of an alert.

- *Execute transactions.* Step 4 illustrates the ability to buy or sell financial instruments such as stocks on a wireless device. Other transactions include transferring funds between accounts and withdrawing or depositing cash.

- *Maintain and manage personal accounts.* The preceding steps show how Mike was able to check existing account information that helped him make the decision on whether to buy additional stock. In addition, Mike can peruse financial analysis tools such as stock screeners or financial calculators offered on the portal that will provide him with further direction regarding a financial transaction.

Another exciting feature that will become commonplace among wireless phones is bill presentment and payment, which is covered in greater deal in the B2B section of this chapter. Now, instead of having to use a credit card or cash, clients can make payments through their mobile devices.

Financial Knowledge Networks

Information, real-time or not, is key in making intelligent decisions about one's financial investments. Stock prices vary by the minute, sometimes even by the second. Still, unless you are a day trader or are extremely bored with your normal career, most individuals are happy with data that is accurate to the nearest 15 or so minute range. Financial data accessible from a wireless device is driving wireless applications to become critical components in the area of electronic financial research and knowledge management. The primary areas within financial information management are:

- **Research.** The typical wireless applications covering financial research include:

  - *Market indices.* Clients should have the ability to see major market indices, such as those of New York, London, Tokyo, and Frankfurt.

  - *News agency headlines and articles.* Major news services such as the *Wall Street Journal*, Reuters, and *New York Times* provide up-to-the-minute information for investors.

  - *Financial Web site links,* The Internet provides links to Web sites that help clients research companies or securities. Wireless Internet devices can display those links and allow users to follow them— just as they would in the tethered Internet experience. Popular Web sites include www.cbsmarketwatch.com, www.cnnfn.com, and www.morningstar.com.

  - *News search via customizable filters.* Ability to search by country, language, date, publication, industry, keywords, company name, stock symbol, or other parameters.

  - *Stock screen/mutual fund screen.* Allows filtering of a large database of equities via customizable filters, such as SIC code, EPS, P/E, revenue, market cap, annual return, or other parameters.

  - *Alerts.* Messages that are deposited in an email account or sent directly to the device, triggered by changes in market indices, stock prices, company profiles, or the release of news content, analyst reports, or other key events.

- **Account holder forum.** Similar to an online community, but accessible through wireless devices. Topic areas include:

  - Popularity lists, such as the top 10 traded stocks of the day/week/month/year, top 10 Gainers, top 10 Losers, and so on.

  - FAQs.

  - "Dear Advisor" weekly column.

- User forum/threaded discussion board.
- Real-time chat.
- Investment tips from financial advisors.
- New product announcements.

Wireless transactions and financial knowledge networks complement each other very well, since the investment decision-making process is well served by the ability to, first, retrieve real-time information and, second, to utilize that information to perform a transaction.

Another key service provided by the financial services portal we are using as our case study within the B2C wireless FSI space is the mobile banking solution.

Mobile Banking

Similar to online and wireless trading, mobile banking is also becoming an increasingly popular tool for the wirelessly enabled mobile customer. The traditional banking products and services that a bricks-and-mortar bank offers today, such as checking, savings, money market, CDs, and automatic teller machines (ATMs), are now beginning to crop up on mobile devices. The concept of the online financial portal as an extension of the physical banking branch is increasingly becoming a mainstay for banking clients throughout the world. Essentially, the Web portal provides all the services your bank would normally offer, except for physically dispensing cash. But more on that later. The two categories we will cover within mobile banking are account access and bill presentment and payment.

Let's go through each of these areas separately.

Account Access (Checking, Savings)

Similar to brokerage houses, banks are providing their customers the ability to access their accounts through mobile and nonmobile devices. Almost all of us are customers of traditional banks. Typically, we have at least one major account, be it a checking account or a savings/money market account. In the old days—a couple of years ago, that is—the only way to access your accounts with a bank was to either visit one of the bank's branch locations or use an ATM. Once banks began discovering the power of the Internet, they started to offer bank-online services that gave their customers the ability to use a computer with an Internet connection to access their accounts, view balances, access customer service, and even download account information into Quicken or MS Money. Soon, however, customers demanded more, which brought about the Internet FSI portal concept. Now not only could customers access their accounts online, they could also perform real transactions such as moving funds between accounts and paying bills using the portal's online bill presentment and payment function. They could even access other services such as online brokerage, mortgage lending, loans, and credit cards in a one-stop-shopping fashion.

In addition to receiving access to traditional banking products and services via the new, wireless channel, mobile customers are offered some new services that didn't exist before.

Hot alerts.   While on the go, the bank's customers can receive actionable notifications about balances, notifications that checks have cleared, warnings of insufficient funds or bounced checks, and other email alerts from customer service.

Wireless funds transfer.   This function works well with hot alerts. For instance, a customer receives a low balance alert, and subsequently transfers money from his money market to his checking account, or vice versa.

Mobile bill presentment and payment.   The service that brings it all together is bill payment online, providing a customer with flexibility never experienced before. While on the go, the system alerts the customer that a new bill was received and allows the customer to specify the payment of this bill, including the amount (full or minimum payment) and the exact date of the transaction (i.e., the date when the funds are transferred to the payee).

Wireless ATM.   As personalized services and electronic wallets are becoming more and more evident in the wireless world, the ability to obtain cash through your mobile device will also be here soon. Secure e-cash, which is nothing more than your personal account information stored in embedded devices, is not far off. Customers will have the ability to walk to an ATM or any other wireless-enabled merchant and obtain goods and services simply by punching codes into the phone that performs the transaction.

Business-to-Business (B2B) Financial Applications

The B2B world is also making waves in the wireless realm. As you might recall, Part 2 covered a case study on the MetX Exchange concerning how businesses can perform secure transactions within an exchange type of environment. In the financial services world, wireless applications are hitting big with brokerage houses, insurance firms, and financial advisors. The two areas we'll cover in the B2B wireless FSI space are as follows:

> **Member direct trading.**   Financial advisors are becoming quick adopters of wireless technology. Mirroring services available to the customer, financial advisors, and brokers can also perform buy/sell transactions, usually as part of managing their clients' accounts.

> **Insurance.**   Insurance agents today are less dependent on real-time information, but they are more concerned about establishing and maintaining positive customer relations. Wireless CRM applications (see Part 3) are one example of how wireless services make inroads into the insurance world.

Let's cover each of these areas in more depth.

Member Direct Trading (Stocks, Futures/Options)

Remember Mike Moneybags from earlier in this chapter? Mike relied heavily on information that Larry, his financial advisor, provided to him when he needed it. Larry, on the other hand, is very much concerned about providing his clients, including Mike, with valuable, up-to-date information, analyses, and financial insights. Being a financial advisor, one of Larry's activities in the office is to routinely check his customer's accounts and keep them informed about market conditions. That might mean performing a simple financial analysis or risk assessment, and then counseling his client about the implications of his analysis. Larry relies mainly on his company intranet site to access his clients' accounts, on financial news services, as well as on his financial advisor peers within his firm. Larry also happens to travel quite a bit to visit his clients to review their portfolios and accounts in one-on-one sessions. In the old days (a few months ago), Larry had to bring loads of files with him to those client meetings. Today, in the wireless world, Larry leaves the bulk of his files in the office and takes with him only the minimum number of documents and his mobile device. The new services that are now available to Larry include the following:

> **Wireless account access.**   With his new PDA, Larry can access his customers' accounts in real time. So as market conditions change, Larry can provide immediate advice based on his financial know-how and skills. He can also transfer funds between accounts, which he previously could do only over the phone or at the office.

> **Wireless alerts/messages.**   Larry carries an SMS-enabled mobile device that allows him to receive messages from his staff and other sources for financial news. In addition to receiving information, the SMS feature allows him to stay in close contact with his clients since he can send out communications from wherever, whenever.

Wireless CRM tools. Larry can access several wireless CRM tools that help him to manage his schedule with clients, coordinate his marketing activities, and track his own performance against targets, among other activities.

Insurance

The insurance industry is one of the oldest professions out there. Insurance has traditionally been a very paper-based environment, as insurance agents used to fill out applications on paper, send/receive claims on paper, fax/send customer information on paper—paper, paper, and more paper. That's where wireless services come in. An interesting study by CGE&Y discovered that, "For major insurance carriers, a 5% migration of paper-based processing to mobile technology could produce savings of $70 million or more." Wireless or any other application will probably never be a complete replacement for paper, but let's see how it can help streamline the processes and reduce costs.

Generally, insurance products and services come in many flavors, but the most prevalent might be auto, home, life, and health policies. To a certain extent these products are somewhat similar in nature in that they all provide some sort of payout if injuries or accidents occur. The process behind selling these products is also fairly consistent. A customer can benefit from wireless access to these products and services in the following ways:

Insurance concierge. A fancy term for a service that integrates auto insurance with telematics to automate accident reporting, or combines homeowners' insurance with home alarm systems to notify the agent of a potential claim.

Policy additions or renewals. After making new, expensive purchases, insurance customers can use their mobile devices to check current coverage levels and increase limits to cover new items. Similarly, customers can renew expiring policies using a wireless device.

Microadditions. Mobile devices allow customers to purchase small, discrete insurance policies to cover them for a specific period (trip to South America) or event (roller coaster ride).

Location-based offerings. Location-specific offerings are delivered to mobile users who have chosen to opt in and receive messages, presenting nearby opportunities to speak with agents, learn about policies, or find the nearest bank or ATM location.

Customized pricing. A mobile device in an automobile can record key data on the driver's location, driving habits, and mileage. This information is then used to provide more accurate pricing of an automobile insurance policy, for example, tailored to a customer-specific driver profile.

A DAY IN THE LIFE OF AN INSURANCE AGENT

Having reviewed these customer-facing applications, let's take a look at a sample day-in-the-life scenario of an insurance agent and how wireless technology facilitates her daily job responsibilities:

START OF THE DAY—9:00 A.M.

Agent arrives at office.

Checks her voice mail and email, responds to time-critical messages, and then reviews her activities for the day.

Client arrives at 10:00 A.M. to discuss her life insurance needs.

Conducts in-depth client interview, capturing data on her PC.

Recommends policy.

Generates a price quote and comparison of competing products on her PC.

Client accepts policy terms.

Fills out policy application and submits to carrier's policy processing system.

Meeting ends and agent synchs wireless PDA with her PC.

WITH THE CUSTOMER—12:00 P.M.

Agent checks her wireless PDA for directions to the next appointment, while eating her lunch.

Arrives for her 12:30 meeting to discuss homeowners' insurance with an existing client.

Pulls up client's policy history and notices that client's life insurance policy needs to be amended.

Client requests the change and agent updates the policy in real time.

Performs price and product comparisons using real-time information from wireless PDA.

Customer accepts policy terms and gives credit card for payment.

Submits complete policy application through wireless connection to carrier's proprietary policy processing system.

Meeting ends.

Sends thank-you email through PDA along with electronic version of policy.

WITH THE CUSTOMER 3:00 P.M.

Agent gets referral from 12:30 appointment for new client interested in auto insurance.

Conducts in-depth client interview, capturing data on her wireless PDA.

Displays auto insurance products to client with aid of PDA.

Generates a real-time quote and client likes terms.

Client accepts policy terms and gives credit card for payment.

Fills out policy application and submits to carrier's policy processing system.

In addition, agent realizes that one of her annuity products would be perfect for the client.

Displays annuity product description on her PDA.

Client likes the offering and purchases it.

Meeting ends.

END OF THE DAY—4:30 P.M.

Agent returns to office to clean up administrative tasks.

Sends thank-you email through PDA along with electronic version of policy to her 3:00 appointment.

Synchs her PDA with her PC.

5:00 P.M.

Leaves for daughter's soccer game.

Checks PDA for directions to the game.

Self-service quote estimator. A customer is considering a new auto or home purchase and wants to get an insurance quote. A mobile application lets the user enter key data and receive a real-time quote from the insurance company.

As you can see, there are numerous benefits that accrue to a customer who uses the wireless service. Additional benefits for the agent include the following.

CUSTOMER INFORMATION:

- Avoids unnecessary reentry of information from paper-based forms, ensuring accuracy
- Faster turnaround time on quotes
- Comprehensive price and product comparisons to assist in the selling process
- Reduction/elimination of unnecessary paperwork
- Increased sales agent efficiency—reduced cycle time and increased sales call potential
- Increased sales agent effectiveness—real-time access to information

CROSS-SELLING:

- Increased revenue due to more effective cross-selling
- Agent is always connected and has access to critical information
- Increased sales per agent

CUSTOMER EXPERIENCE:

- Client meetings can be shorter and more informative
- Provides targeted information
- Improves customer experience

Summary

This chapter has given you an idea about the background and trends in the industry, characterized by deregulation, service convergence, consolidation, and global expansion. The overarching goals, again, included customer attraction and retention as well as improvements in financial performance due to revenue expansion and cost reductions. We then covered some of the advantages that wireless services can bring to the financial services industry, including benefits that accrue to providers and customers in both B2C and B2B environments. The rest of Part 4 will dive deeper into a very specific CGE&Y case study to show how, in particular, customers are able to benefit from a full-service wireless financial services application.

CHAPTER 18

Conceptual Business Architecture

Now that we have received an introduction to the financial services industry, let's turn our attention to the case study that will be the foundation for this and the following chapters. The case study was provided by CGE&Y's financial services practice. The client, a diversified financial services institution, commissioned CGE&Y to design, develop, and integrate a Web-based solution that would allow the company's customers to access their accounts from their homes, as opposed to having to visit the local banking branch or brokerage. We selected this case as it is quite different from the radio advertising project described in the Part 3 of this book. Whereas the previous application aimed at providing a company's account representatives with wireless functionality to streamline advertising sales, this study focuses on leveraging wireless technologies that touch and provide significant benefits to a company's consumers.

The primary business drivers that propelled our financial institution to explore Internet-based customer self-service before offering financial services over the Net had gone mainstream, included the company's innovative corporate culture that looked favorably upon experimentation with new technologies. Even more important was the increasing desire of the company's customers for self-sufficiency. Furthermore, the company had some very quantifiable objectives in mind. These included enlarging the customer base and reducing operating expenses by shifting account support of smaller clients toward a Web-based self-service model.

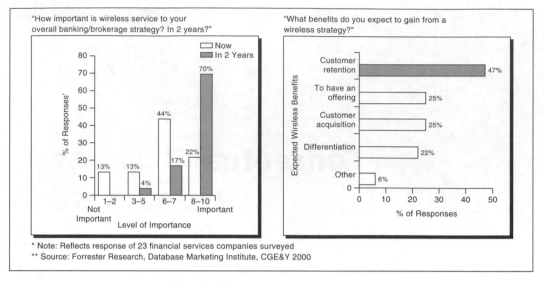

Figure 18.1 Planning for wireless in the financial services industry.

However, our company was not alone in its cutting-edge thinking. Figure 18.1 illustrates that already in early 2000 several players in this industry assigned a high degree of importance to offering wireless capabilities, predominantly in an effort to retain clients.

Aware of its competitors' intentions, CGE&Y did not sit still after it successfully launched its Web-based financial services portal. Instead, the institution quickly turned toward incrementally adding wireless capabilities to take the company and its service offerings to the next level. But we are getting ahead of ourselves. Before exploring the wireless solution, let's first understand the business drivers behind building the Web-based application.

Web-Based Application Discussion

In light of the financial service industry's recently surfaced megatrends of convergence and globalization, it comes as no surprise that especially the larger institutions were eyeing Internet technologies to help them in their scramble to remain strong contenders, if not leaders, in their sector. When CGE&Y was approached by a major provider of financial services—let's call the institution Acme Financial—a competitive bidding process was already in full swing. The company was in the early stages of soliciting proposals from the top five consulting firms to assist with building its Web-based application.

Before we jump into discussing the specific functionality of the portal, let's take a step back to the early stages of the project. Even before any technology architecture was contemplated, before systems and applications were developed, selected, or integrated, the project team had to gather the business requirements that would drive the application's functionality. However, different groups within Acme had different expectations regarding the most pressing business needs. Acme Financial's board of directors, for example, expressed its long-term strategic vision for the portal. Senior management considered the application an opportunity to increase the performance of the revenue centers they were leading. Line management expected the application to assist in service delivery, while front-line personnel weren't exactly sure *what* to think about the Web site: Would it make their jobs easier, or would the technology pose a threat to their continued employment with Acme?

All groups of Acme constituents considered the portal as a vehicle that would take the company into the future, yet each had a different viewpoint of how that would be accomplished and what the impact might be. To illustrate the disparate opinions that the project team had to take into consideration, look at Figure 18.2, which illustrates some of the business requirements as expressed by the various internal stakeholders. While the illustration is an oversimplification of the comprehensive business requirements analysis conducted, it serves as a nice means in this context to show that each interest group was concerned about different aspects of the overall enterprise.

To begin, let's look at some of the expectations voiced by Acme's customers. Desiring access to a one-stop shop for various financial products, the financial portal, in essence, had to present itself as an easy-to-navigate supermarket for securities, bank accounts, loans, and insurance products. Access had to be secure, and data had to be handled in strict confidentiality. The

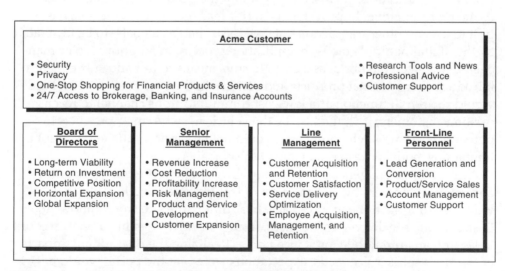

Figure 18.2   Diverse business requirements at Acme Financial.

supermarket had to be open 24 hours a day, seven days a week, while offering its shoppers all the tools required to make informed decisions. Should customers not be able to answer their questions by helping themselves, they wanted the option of obtaining advice from professional financial advisors or receiving answers about nonfinancial support issues from trained Acme customer service representatives. This was a tall order to fill.

On the enterprise-internal side, Acme's board viewed the portal as a strategic differentiator that would propel the company forward in terms of its competitive position in the industry, thus ensuring that the company would remain in business, and profitably so. In addition, the board viewed the portal as a critical stepping stone in expanding horizontally into additional financial service areas to be offered to the current and potential customer base. Also on the board's mind had been a push toward global expansion, and how the portal could become the vehicle to enter overseas markets. With a close eye toward the financial community and the company's ownership—that is, its shareholders—the return on the investment had to be positive, satisfying stringent profitability requirements as established in detailed cost/benefit analyses and a financial business case.

The most important goals for Acme's senior management were to improve their profit centers' financial performance. Thus, this group envisioned the site to increase Acme's bottom line via an increase in commission revenues from opening the trading desk to customers around the clock, which would enable them to trade more frequently than ever before. Attacking profitability from the middle section of the income statement, the self-service portal would reduce operating expenses, as customers could check balances, open and close accounts, find answers to frequently asked questions, and complete all loan or insurance applications themselves—without the need for assistance by an Acme representative. Improvements in risk management via the analysis of past loans and other transaction data was also cited as a goal—to be facilitated by the digital nature of the Web application process. In addition, senior management viewed the portal as a cost-effective means to test and subsequently introduce new financial products and services to the customer base. Not only would the portal function as a low-cost communications channel to reach existing customers, it would also be the vehicle to sell Acme's products to potential customers who previously could not be serviced due to geographic limitations, both at the national and at the global level.

Acme's line managers hoped the portal would help the company attract and retain customers by offering a set of features and functionality that would be unique in the marketplace. In addition, the site would allow Acme to streamline the delivery of services, from trading to opening a bank account to issuing an automobile insurance policy. Realizing the operational impacts of the site's self-service focus, expectations were that the portal would become a cornerstone to improved customer service, resulting in a boost in customer satisfaction, thus warranting immediate focus on implementing support-type functionality. Communicating to the labor pool that Acme was at the cutting edge of using technology in its industry, line management intended to leverage Acme's new image during the attraction, motivation, and retention of highly skilled employees. Accordingly, both senior and line management championed the suggestion that the portal should feature an employment section that would not only act as a recruiting tool, but could also be used for internal processes such as training, time and expense reporting, or benefits administration.

Probably the employee group that would be most impacted by the portal from an operational perspective were Acme's front-line employees, including brokers, financial advisors, and customer support personnel. Whereas most brokers considered the low-cost trading enabled by the portal as a clear threat to their traditional way of earning a living, many financial advisors welcomed the new technology because it would serve as a tool to bring financial instruments to the masses and allow the advisor to form a closer relationship with his or her customer base via online portfolio management and analysis as well as instant communication capabilities. Support personnel viewed the portal as more of a mixed blessing than anything else since it was unclear whether the ability to answer a customer's own questions by perusing an FAQ database or other research tools and the associated reduction in workload would be outweighed by the increase in users who had to be helped along the way.

As you can see, even though the business objectives among these groups were not diametrically opposed, finding common ground required finding an acceptable compromise that would satisfy each group and ensure their commitment to and support for the task at hand. Without such support and buy-in at all levels within Acme, the undertaking would not have been successful.

Being fully aware of the different expectations held by Acme's constituents, the next step was to think about what the portal would look like from a functionality perspective. The two major activities during this process included functionality scope definition and sequencing.

The first step was to identify all the features that were to be offered online. As the portal was supposed to cover the three areas of brokerage, banking, and insurance services, the process of distilling each area's key functions was a demanding one. To satisfy customer expectations, all services offered in the bricks-and-mortar world had to be replicated online. To exceed expectations, additional functionality had to be designed that took advantage of the unique Internet medium. The scope of the fully loaded application was tremendous, and it quickly became obvious that trying to release all features at once would significantly delay the launch of the site. The decision was made to roll out the site in stages—which meant functionality had to be prioritized.

Because the application would be built in various releases, the first version of the site would have a subset of capabilities, while the second version would be enhanced with additional functionality, and so on, until all features were rolled out. Adding functionality incrementally and in stages meant the team had to be very clear about what type of features were absolutely required for release one, and in what sequence additional features should be made available. As you can imagine, different owners of a particular application had different opinions about which features were to be included in the first iteration of the site, versus those that could wait. After identifying a cross section of Acme's client base, the team accelerated the prioritization exercise by conducting in-depth interviews and focus groups, resulting in a rank-ordered list of features from "absolutely critical" to "nice-to-have," timed along the lines of three major releases staged over an 18-month horizon.

The following section will explore some of the functionality included in the final version of the portal. You will notice that some of the aforementioned functionality, such as the employee administration piece desired by senior and line management or the specific components assisting Acme's financial advisors in their daily work, is not covered here. The reason is that we wanted to limit our discussion in this section to the consumer-facing applications as opposed to those that facilitated enterprise-internal management aspects.

Functional Components of the Portal

Building a highly secure, Web-based financial supermarket accessible via desktop computer was a massive undertaking. The project required the efforts of a team consisting of CGE&Y business and technology architects, Web designers, and system integrators, as well as dedicated Acme financial representatives. Because of the ever-increasing advances in technology such as the addition of a wireless channel and the unique functionality it enables, Acme knew from the project's early beginnings that the effort would require an ongoing commitment.

Because of the portal's complexity and the limitations of this publication, we can focus on only a small but, we think, interesting subset of all features. If you already are a customer of any larger financial institution that has a significant online presence, you might be familiar with some of the applications we discuss here. For those who are new to the personal financial management tools offered via the Web, we hope that this section serves as an introduction that might pique your interest, not only because you are a technology architect, but also because you may soon join the online financial customer crowd.

Let's start with a look at the overall application. Figure 18.3 illustrates the system's data sources (user profile, accounts, news, products, community) and application components (preference setup, research and analysis, products/services, quote generation, transaction, notifications/alerts, community). We will describe the functionality of each component in detail, both from a Web-only and from a wireless perspective.

Preference Setup

Whether it's the first time users interact with the application or they adjust their settings from time to time, preference setup readies the application for custom usage. It is here that users specify parameters that the application will check before allowing access or displaying data and information. Specifically, users enter residency, language, currency, data presentation, and security/privacy parameters.

Residency settings are stored by the system to satisfy regulatory requirements associated with brokerage and banking applications, such as the calcu-

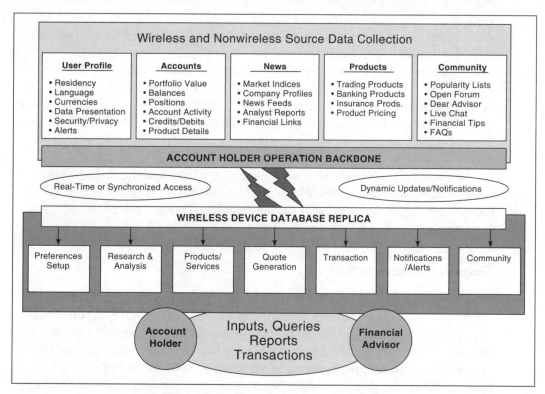

Figure 18.3   Wireless financial services application.

lation of prices, fees, and tariffs, or the reporting of transactions to tax authorities. These settings are required in addition to the basic user profile data that is commonly used for paper-based communications.

Although you may be the legal resident of one country for tax purposes, you still may prefer that your account communicate with you in a *language* different from that country's major tongue. The application allows you to specify a language in which all information is displayed, regardless of origin. Or, for the truly multilingual, we may instruct the system to show information in the language in which it originated. Thus, a news editorial in *Le Figaro* is shown in French, whereas an article published by *Der Spiegel* will be shown in its original German.

Similar to language preferences, the system allows you to either select one particular *currency setting* of your choice or have all securities in your portfolio displayed in their original format. This feature is of special importance for the global investor community.

In addition to specifying the *presentation settings* of data and information at the micro level, the application allows you to select your preferred views at the macro level. By this we mean that the system is fully customizable in terms of starting screens and data/information layout. If you so choose, the system after login may jump directly to the securities portfolio screen and within that area display your holdings in alphabetical order, or by industry, or by top gainers/top losers, and so on.

Last, but certainly not least, the financial portal allows you to customize *security and privacy settings*, such as your user ID and the password used to log in to the system. Although you are willing to disclose highly confidential information to the online brokerage and banking application to be able to fully use these services, you may not want the insurance branch to have access to this data.

After setting your preferences, you are ready to access the financial services portal and explore its various sections. To assist you with selecting the financial products best suited for your individual needs, the portal offers an extensive array of research and analysis tools.

Research and Analysis

When it comes to making informed decisions about trading in a security, opening a bank account, applying for a credit card, or purchasing an insurance policy, a consumer usually first collects a certain amount of information—to the extent required to make him or her feel comfortable about the impending deal. Before engaging in a transaction, we want to know our options. Thus, the portal provides its users with static and dynamic information as well as a set of analysis tools.

Static research sources encompass company profiles (company description, address, key financial data, stock price chart, senior management, annual reports, etc.), the market tickers indicating the activity at large exchanges in major financial centers around the world, financial analyst reports as published by major brokerages and other institutions, articles from a slew of business publications, links to financial Web sites, comparisons of interest rates, and so on.

Dynamic research tools include customizable filters that let you search static sources for company names, major events such as earnings releases, or keywords. Or you could screen all securities for those with a certain earnings-per-share ratio, level of institutional holdings, and other key metrics you might consider when deciding whether you want to trade a stock. Also included here are alerts, which we will describe in more detail later.

For the *analysis* part, the application provides a set of tools that facilitate decision making before and after making a transaction. For example, the site offers several views of your securities portfolio. These views allow you to answer questions such as: Which are my biggest gainers/losers? What is my portfolio's ratio of securities to bonds to cash, and what should it be, given my attitude toward risk? How does my portfolio compare to a sample portfolio reflecting high/medium/low risk levels? In addition to these little helper applications, the site offers access to a variety of financial calculators to run what-if scenarios relating to stocks, mutual funds, bonds, retirement planning, IRAs, insurance, budgeting, and personal savings.

Having researched the plethora of investment choices, banking products, and insurance offerings, you are ready to have a closer look at the products and services provided by the operators of the financial services portal.

Products/Services

The purpose here is obviously to describe the offerings that the operator of the portal makes available so that you can familiarize yourself with their particular attributes. The core offerings of the portal include brokerage, banking, and insurance products and services. Whereas the *brokerage* offers debt and equity securities, mutual funds, and other financial instruments, the *banking* section deals in savings and checking accounts, certificates of deposit, money market accounts, business accounts, credit cards, and loan origination, to name a few. *Insurance* products and services include life insurance, medical, dental, vision, disability, automobile, home, and other offerings.

The next step in selecting a financial product is to obtain a quote, or price, for a product or service you want to acquire.

Quote Generation

Obtaining a quote is straightforward for the *brokerage* application; all that is required for you to make a buy/sell decision is the security's current price. However, obtaining a quote for a car loan in the *banking* section of the portal is slightly more complex, as the price of the loan—in other words, the interest rate and the loan's duration—may depend on several factors, including the car's make, model, year, and mileage, as well as your personal information including age, income, and collateral. The most complex quote generally concerns the *insurance* offerings, as the development of a price for home insurance, for example, depends on a very large number of parameters that establish the value of your property—location, age of the house, square footage, construction, proximity to a fire station, and so on.

With a quote in hand, you are ready to make the final purchase (or selling) decision. Making a financial decision in this context means conducting a financial transaction.

Transaction

Using the *brokerage* services, such a transaction involves the buying or selling of a financial instrument. The complexity here is medium, especially if you are a more experienced trader who uses various parameters such as buy to cover, sell short, market, limit, stop, stop limit, day, good till canceled, all or none, do not reduce, and fill or kill. Within the *banking* section, conducting a transaction can range from simple, such as funding a new account, to complex, such as when paying your phone bill using the online bill presentment and payment feature. Probably the most straightforward transaction takes place in conjunction with *insurance* products. While they are difficult to quote, once the price is established you go ahead and pay the quoted premium, either in full or in installments.

As a value-added feature, the portal provides notifications sent to your mailbox established within the portal or to an external email address. As described in more detail elsewhere, real-time alerts can also be sent to your handheld mobile device.

Notifications/Alerts

In the *brokerage* section, you can set up a notification to be sent to you whenever a stock, for example, reaches a certain level you specify. So if your favorite equity reaches a certain price, the system sends you a message so that you may add to or subtract from your current holdings. Within the *banking* application, and if you subscribe to the online bill presentment and payment feature of the site, the system can be configured to send you a notification when you receive a new invoice—say, from your utilities company—or when a payment is overdue. The insurance application can send you a note that alerts you to the fact that your car insurance is about to expire and provide you with a link to the area within the portal that allows you to renew the policy.

Finally, the financial services portal offers a community area where users can help themselves or seek the assistance of other account holders. This area provides part of the stickiness that attracts users back to the site over and over again, adding value to the overall offering via the integration of the individual account holder in a group of similar-minded individuals for the purposes of information exchange, advice, and even entertainment.

Community

Specific community features include popularity lists, FAQs, questions to and tips from your financial advisor, threaded discussion groups, and real-time chat. *Popularity lists,* such as the top 10 most heavily traded securities of the day, week, month, or year, provide insights into the attractiveness of certain securities or mutual funds to the trading community. Other top 10 lists may include the top gainers or losers during a specific time frame. Frequently Asked Question lists (FAQs) allow newcomers to the community in particular to quickly obtain answers to those questions that have been asked—and answered—many times before. In addition to the FAQ self-help mechanism, the community section may also offer the latest tips from your financial advisor. In a Dear Advisor type of format, the site selects one question a day as the Query of the Day and posts an exhaustive answer, usually written by a financial services professional. In addition, *financial tips* are offered on a regular basis, covering basic through advanced financial topics that do not require one-to-one interaction with a specific account holder. An *open forum*— that is, a threaded discussion group—allows users to post questions to a monitored forum, while other users can post answers or ask follow-up questions. A *live chat* room provides a venue for those account holders who have especially time-critical questions or who simply enjoy the live interaction with like-minded portal members.

Now that we have explored some of the specific functionality offered by the Web-based application, let's look at some of the benefits provided to account holders and financial advisors, the two primary users of the financial services portal.

Major Benefits

Recalling that customer attraction, retention, and service were a few of the primary business drivers behind building the portal, we are ready to review the major benefits that accrue to the application's user groups, including account holders, financial advisors, and Acme Financial in general.

Benefits to account holders include the following:

ONLINE BROKERAGE:

- Low-cost trading, often at fixed rates, versus percentage commissions charged by full-service brokers
- Access to market information such as (near) real-time prices of securities from the customer's own desktop
- Convenience of 24 × 7 access to the trading desk without the limitations of physical brokerage business hours
- Rapid executions, as a customer doesn't have to rely on other intermediaries to service trade requests
- Access to an extensive list of research content, including analyst reports and market, company, and industry news
- Powerful tools to evaluate a customer's portfolio and set up shadow portfolios, tracking the securities the customer is interested in before making a transaction
- Access to positions, balances, recent transactions, statements, and so on, at any time
- Notifications of significant events from intelligent agents scouring research and metrics on the customer's behalf

ONLINE BANKING:

- Free or low-priced banking products and services, such as interest-bearing checking accounts, that can be set up, funded, and managed entirely online without the need to see a banker
- Access to account balances, online registers, funds transfers, and recent transaction history around the clock and from any computer connected to the Internet, regardless of geographic location
- Ability to comparison shop for interest rates on bank accounts, credit cards, and automobile and home loans before settling on the offering of the best provider
- Paying bills online via online bill presentation and payment services, providing the ultimate in convenience for the busy professional
- Importing online statements into personal finance software such as Quicken or MS Money for further analysis

ONLINE INSURANCE:

- Comparison shopping, 24 × 7, for the best deal based upon various scenarios

- Online transaction and order fulfillment—no need to talk with or visit an insurance agent

- Access to all documentation online versus having to wait until the insurance policy, for example, arrives in the mailbox

- Online renewals, modifications, and cancellations of policies at the customer's convenience, rather than during the regular business hours of the insurance company

Benefits to financial advisors include the following:

- Management of client accounts and their portfolios 24×7, with full access to all portfolio balances, positions, past transactions, and other account activity detail

- Direct communication access to clients via notifications, alerts, or real-time chat

- Online sales and marketing tools to assist the advisor in customer relationship management and with the customer life cycle

- Planning and monitoring tools to benchmark the advisor's performance against established goals

Aside from the benefits realized by account holders and Acme Financial advisors, the company itself realized several of the goals it set for itself. While the battle for leadership in the global financial services world is still being fought, Acme was able to very quickly increase its customer base and reduce operating expenses. The reduction in expenses was primarily the result of shifting the costs associated with servicing individual accounts to the customer, who willingly picked them up in the name of convenient self-service. Studies have shown that the cost per transaction could be cut in half by having a customer check her balance over the phone, as opposed to visiting a banking branch. This already reduced cost could be further cut by a factor of 10 by shifting customers to use their computers or wireless devices, thus eliminating the need for any human intervention from the bank's side. Instead, customer service representatives could be freed from such administrative, low-value-added tasks as repeating a balance in order to assist clients with more challenging questions.

A similarly strong business case can be made for using wireless services to improve customer retention, shown to have a direct impact on profitability. CGE&Y research indicates that a 5-percent increase in retention can increase customer lifetime value by 75 percent in both banking and insurance sectors. Thus, wireless services spell a true win-win situation for the customer as well as the company that makes its services available via the new channel to consumers who value the freedom associated with remote personal financial management.

Having reviewed the Web-based application and some of the portal's components, let's look at the impacts the wireless extension had on the site's functionality and service offered to the customer.

The Wireless Solution

The benefits of the Web-based application we just explored were significant—as confirmed by the several hundreds of thousands of users (existing and new) that signed up with the portal within months after launch. These users now regularly visit the portal for news and advice regarding their financial needs and to monitor and maintain their current accounts, preferring to do business with Acme on their own terms and at their own convenience. And although the Web-based solution was the first step toward increasing the company's value proposition to current account holders and potential customers alike, Internet-enabled financial services quickly ceased to be viewed as a novelty. Racing to catch up with Acme, several of the company's larger competitors implemented Web-based self-service capabilities. In an effort to reach the next level in customer retention, acquisition, and competitive differentiation, Acme Financial commissioned CGE&Y to investigate the possibility of making certain portions of the portal accessible to wireless devices.

Being able to offer wireless access to account holders in all three sections of the portal—brokerage, banking, and insurance—not only was envisioned to serve once again as a differentiator in the battle for the customer, but also would again propel the company to cutting-edge status. Investing in this new technology, in incremental steps, would allow the company to learn those important lessons associated with new technologies, while at the same time it would provide customers with a wireless access channel to their accounts. Within Acme Financial's corporate culture, it is the company's express mandate to explore new technologies and selectively invest in those that hold the biggest potential. By following this approach, Acme has been able to gain a level of implementation and evaluation expertise that allows the company to quickly incorporate select solutions and exploit them long before the company's competitors discover their value.

The following is a discussion of the various portal components and how the wireless dimension enhances their functionality. Later, we will introduce a couple of features under consideration at Acme that are unique to the wireless financial environment, such as transforming a handheld gizmo into a universal payment device.

Wireless-Enabled Functional Components

Let's first inspect the financial portal's components and how wireless technology enhances their value. We will follow the same order as was used during the previous section.

Preference Setup

All system settings are controllable while on the go. As you are using your wireless device to access all functionality, it is important that you are able to adjust the settings in real time, from anywhere at any time, so as not to impair your user experience. And although you may not adjust your security and privacy, residency, data presentation, or language settings on a regular basis, there are other settings that you now have to consider. Specifically, these settings are associated with features unique to the wireless environment, such as the enabling of real-time alerts and location-based services.

Alerts are one of the truly revolutionary features of this system. In addition to setting specific *triggers* that launch an alert, such as a security reaching a certain price, the portal allows you to also specify *alert settings* at the macro level. At the macro level, you may want to instruct the handheld device to vibrate only (versus alerting you with a tone, which can be annoying during a meeting), or you may want to establish that no alerts will be accepted after 11:00 P.M. (for those of us who sleep at night).

Besides alerts, wireless technologies enable *location-based services*, for which settings must be specified. You can instruct the system to disclose your physical location always, never, or only upon request. So, while at times you may not want the insurance branch to know where your handheld device and, presumably, you are located (what would they say if they knew about your frequent trips to the local sky-diving grounds?), you most certainly would want to receive location-specific information such as the nearest bank or ATM around the sky-diving center to load up on cash to pay for the next jump or any first-aid necessities to patch up those skin abrasions.

Similarly, if you are on a business trip in a foreign city or country, you might welcome location-based services that address your financial needs while on the go. Imagine yourself traveling abroad and in need of local currency. Your wireless device will allow you to quickly identify which institution offers the best exchange rates, where they are located, and how to get there.

Exceeding the financial realm, there are many other location-based features that are worth mentioning, even though we will not discuss them in more detail. For example, while on the same business trip, your device could be configured to also present you with localized weather, travel, and other reports or alerts, effectively extending the portal's functionality to include concierge services. Acme Financial might decide to incorporate such services into its financial portal via partnerships with appropriate content providers. The aim here is to increase the portal's stickiness and to lock in customers by increasing switching costs. Once you have invested a good amount of time and effort into personalizing the content to be delivered to your device, it is very inconvenient to switch to a different service provider. Thus, in the name of customer retention, the financial services might be augmented with ancillary, value-added features in the future.

Research and Analysis

All of the sources, such as analyst reports and news feeds, are available over the wireless channel. But because the user experience on the mobile device is not exactly the same as the one delivered by your 21-inch color monitor, we rely heavily on alerts. So, although working with a PC connected to the Internet via a broadband connection makes perusing a long list of financial articles for the information that you are really interested in a breeze, using a mobile device over relatively slow connections does not allow us that luxury. Instead, we need a mechanism that allows us to home in on the information we desire, and only that information. Alerts to the rescue!

Instead of surfing for information, we can have it delivered to our devices. Enter a few keywords so the portal's intelligent search agent knows what to look for, or enter a few financial metrics that will be your triggers, and presto! Here come the results, delivered as actionable alerts to a wireless device near you. Note that the operative word here is *actionable.* It is nice to receive an alert while on the go, yet it is even more important that the alert allows you to act upon the news you just received. You want to be able to immediately buy or sell a stock, not just hear that the price is on its way into the cellar or going up through the roof.

Products/Services

The feature of being able to review and evaluate all offerings and their attributes is extended straight to the wireless device.

Quote Generation

While shopping for a mortgage or signing up for automobile insurance, using your cell phone may not be feasible due to the sheer amount of required input parameters. Having access to stock quotes—especially for those day traders among you—surely is a strong value proposition. Coupled with alerts, being able to obtain a real-time quote and to act on that quote (i.e., to conduct a transaction) is a powerful extension of the wireline brokerage functionality. Similarly, having access to your trusted online banking branch and its car loan rates may come in handy when sitting in front of that intense car salesperson who is trying to set you up with financing through his or her own organization.

Transaction

Conducting a transaction as the next step after obtaining a quote was a feature that we also transferred to the mobile device. Stock trades are a relatively simple application, as previously described. Another feature that provides the ultimate in convenience for those of us who are on the road more often than not is the ability to pay bills from anywhere at any time. Soon, many of us will pay our bills using an online bill presentment and payment service. Instead of receiving a paper bill in the mail, the invoice is routed to our online bank. Once there, we can specify whether we want to pay the amount in full, or make the minimum payment, and we can specify the date on which the bank is supposed to electronically transfer the amount to the payee. Using our wireless devices, we can now pay bills while waiting to board our next flight.

Notifications/Alerts

We already reviewed the benefits of receiving an alert from your brokerage as soon as a particular stock hits a set target price. But there is a plethora of other situations where alerts come in handy. Wouldn't it be nice if you received an alert on your wireless device that you are close to reaching the limit on your credit card *before* you offer to pick up that dinner tab you and your VIP client ran up? And wouldn't it be even nicer if the alert were actionable, meaning it offered you the opportunity to apply for a credit increase right there on the spot? And wouldn't it be the ultimate if your application for a credit increase could be instantaneously routed to whoever or whatever system had to authorize this request within the bank, so that you received an answer within seconds? This is now possible, as our system is capable of sending these alerts not only to our desktops, but also to a multitude of handheld gizmos.

Other alerts include those that notify you when funds that were deposited to your account via wire transfer or check become available. There are alerts that let you know that a new bill has been received and needs to be paid, or when your insurance policy is about to run out and needs renewal. Actionable alerts in general constitute a tremendously effective application that expands the financial portal's value. Not only do they reach mobile customers with time-critical information wherever the customers might be located, but they also allow them to act upon that information immediately, using the wireless device.

Community

All community features, including popularity lists, FAQs, tips, discussion groups, and chat, are also accessible via the wireless device. However, the added wireless dimension again enhances this feature. For example, when you post a question to a discussion group, you can opt to receive an alert to your device as soon as your question is answered. Again, time-critical information is relayed to you faster than ever before. On a different note, you may find yourself in the airport, waiting to board your delayed flight. With your wireless device, you can quickly check your financial advisor's latest tips or engage with other members of the community in a real-time discussion about how the market was doing that day, and ask them about what recourse you have against airlines making you late as a result of mechanical problems.

Additional Functionality

As of this writing (in late 2001), wireless technologies are becoming more and more engrained in the financial services industry. Wireless trading, for example, is slowly but surely being rolled out by all major brokerages, establishing some sort of level playing field in the industry. What was cutting edge last month becomes run of the mill today. And so, in its continued quest for technological leadership in the sector, Acme Financial is currently working on taking the financial portal to the next level. The main features under consideration include the following.

Session Management

One of today's realities is that our network connections are not 100 percent reliable. All of us have experienced dropped calls or lost connections, mostly due to network issues, sometimes because our batteries ran out. While such inconveniences are more of a nuisance during a person-to-person telephone conversation, they can be truly detrimental during a complex financial transaction. Imagine yourself in the middle of selling a security when you get disconnected. Did the brokerage receive your request to sell? Did the entire transaction go through, or only half? Do you need to submit a second request? Or what if you are in the middle of applying for automobile insurance, when the screen goes blank. All that time you spent on the device filling in the forms, all the data you previously entered—gone. With wireless session management technologies in place, dropped connections are no longer an issue. If you get disconnected using your cell phone, simply reconnect with your phone switch over to your laptop computer and pick up the session exactly where you left off before the signal was dropped, using the other device.

Cashless Vending

Since you are running most of your financial management through Acme's portal, including credit cards and bank accounts, why not extend the reach of your cash to your wireless device to enable small dollar transactions. Already a standard functionality in Europe, a cashless vending feature would allow you to use your cell phone to connect to a soda machine, stamp dispenser, or parking meter via your handheld gizmo. Various methods of connecting with the vending machines are currently being evaluated, including dialing an 800 number displayed on the machine, linking via an infrared beam, or connecting using the Bluetooth protocol. The processing of these micropayments could be handled by the telecom operator or the bank. In our case, Acme Financial is the sole processor of any and all transactions conducted by the company's customers. At the same time, however, Acme is considering providing its payment processing services to major operators in an alliance type of arrangement, which would enable the operator to add these payments to their customer's phone bill at the end of each month.

Wireless Wallet

Taking the concept of cashless vending to the next level, a true wireless wallet would be able to handle small dollar amounts in addition to large transactions. Because Acme is the issuer of credit card services, the wireless device could become the true universal payment system, replacing not only the loose change in your pockets, but also your checkbook, credit/debit cards, and any ATM cards you carry. Acme envisions its customers paying for groceries, gas, movie tickets, restaurant visits, dry cleaning, and other merchants' services by using their handheld devices, consolidating all transactions into one seamless financial network.

Not only would this service enhance the consumer experience, it would also allow Acme to generate additional revenues from accumulating and analyzing such shopping data in the aggregate and selling it to various parties interested in such research. The information gathered via the wireless wallet feature could be used to support buyers' preference portfolios, to track previous purchases, to browse selections, and to offer suggestions to the consumer regarding similar types of products. The system is buyer determined. Thus, information regarding a buyer's purchasing patterns or habits is stored in the system and is available for future use to suggest other products to the buyers, based on their purchasing patterns.

Once the wireless wallet is deployed, several additional features would enhance its basic function as a payment device. The wireless wallet can be extended to offer consumers a way to purchase products that adapts to their busy and mobile lifestyles, meaning at the exact time the need first occurs or when they encounter an advertised product. Imagine yourself in a mall, when you see that big-screen TV you always wanted, and today it is on sale at 20 percent off. Because you are using a handheld device with an integrated bar code scanner connected to the Net, you can instantly pull up the TV's detailed specification by scanning the UPC symbol attached to the product or the box. Those of you who are ready to jump at a bargain when you see one may actually purchase that TV right then and there. Scanning the bar code not only provided you with details about the product, but also allowed you to run a quick price comparison against the model's standard pricing at the major online electronics stores. That search revealed that the price of the TV, at 20 percent off, was indeed a true bargain. In fact, Ericsson, Symbol Technology, Motorola, Sprint, and others have heavily invested in applications that will add bar code scanning functionality to wireless devices.

The point is that there is a large, untapped market of consumers that can be reached for impulse purchases by leveraging wireless technologies. Think about a song you hear on the radio as you are driving the kids to school. You like the song and want it, but don't recognize the artist. And even if you knew the band, what would be the probability that you would take the time to look it up on your PC? Or even go to a record store to find it? With a wireless wallet, you might simply push a button and the song or album would be immediately available for purchase—from your handheld device. The true power of this technology is that it allows consumers to make a purchase at the very time an advertisement or special deal is presented to them, wherever they are. And all parties to the transaction win: the retailer who realized additional revenues, Acme Financial who handled the transaction, and the consumer who received the convenience of value-added shopping, independent of location.

Summary

In this chapter we described how Acme Financial leveraged Internet technologies to achieve specific business goals including customer attraction and retention, cost reduction, and competitive positioning. We learned that adding a wireless dimension not only ported the applications' functionality to Acme's consumers on the go, but, more important, added to the value received by the company's client base. Chapter 19 will take you through some of the steps required in the actual build-out of the system.

Information Architecture for a Wireless Financial Application

Chapter 18 described the business aspects of our wireless financial case study, Acme Financial. This chapter will take the business requirements and use them as a foundation upon which the data architecture will be created. Since the business case for Acme Financial is quite different from that of Acme Radio, the data architecture will be quite different, too.

The disparity we are observing is consistent with the architecture metaphor we use in application development: When compared to architecting buildings in the real world, the architectural plans for a cathedral are quite different from those of a high-rise office building, and quite different still from a center hall colonial residential house. So, too, the data architecture for a wireless application built to facilitate the tasks performed by sales representatives and their managers (a business-to-employee, or B2E, application) will be significantly different from the data architecture for an application used by John Q. Public to access financial data and perform financial transactions (a business-to-consumer, or B2C, application). Let's take a look at that architecture now.

Overview

Financial transactions is one of the key areas in which many companies are concentrating their wireless efforts, which is one of the reasons the application developed for Acme Financial was selected as our case study. The application was originally designed as an Internet portal for all the financial services that Acme offered via the company's traditional channels. Acme realized that the Internet service could be exploited as a tactical competitive differentiator, yet within months of launch the company's competition had caught up by offering comparable services. To keep their advantage, Acme realized that the offering could significantly be enhanced by providing wireless access as yet another channel and to enable entirely new value propositions to the company's account holders.

In this chapter, we will create a data architecture to help us achieve that goal. We start by defining the data to be exchanged between business processes, which will help us to both uncover the interdependencies between these processes and create a clean definition of what information is exchanged. The data architecture will provide the contract between various business components, which helps to ensure that the services perform adequately. The functionality contained within each component will be covered later in this chapter.

Additionally, the business, information, and technical principles of Acme Financial are used as inputs to create the data architecture. They were touched upon in Chapter 18:

- Acme Financial looks favorably upon experimentation with new technologies.

- The company wants their customers to perform as many tasks as possible without direct involvement of Acme personnel.

- Acme would like to expand its customer base.

Let's look at each of these principles in greater detail to understand how they apply to the data architecture.

Acme Has a Favorable View of Experimentation with New Technologies

Corporations that experiment with new technologies can have far-reaching successes or devastating failures. Just imagine: You could have been the first company to embrace LANs and email capabilities, or you might have been the company that, after VHS became the standard, found itself sitting on a warehouse full of Betamax players. Figure 19.1 shows what is known as the *technology adoption curve*.

Let's look at each point on the curve.

- *Early adopters* find themselves on the bleeding edge of technology and, as a result, often get cut. They are true risk takers, which allows them to find the next big thing *if* all goes according to plan. In the best-case

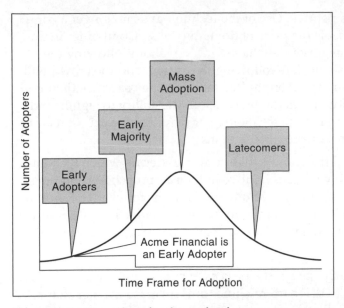

Figure 19.1   Curve for adopting technology.

scenario, they are the ones who stand to reap the tremendous rewards of betting on new, unproven technologies before anyone else does. In the worst case, they find themselves joining the technology-of-the-month club. At any rate, early adopters often fund the R&D of their entire industry, as others follow their lead. They have the opportunity to gain mind share and market share with innovative breakthroughs in ideas and technologies, but run the risk of being burned by failure. As long as these companies understand and accept the attrition required to be on the front of the adoption curve, they should not become gun shy—missing an opportunity they should be able to seize. They have the advantage of being standards creators and can leverage their position as innovators to force others to play on their terms.

- *Early majority* members are always on the lookout for new technologies and ideas, but want to temper their forward thinking with prudence. They obtain a tactical advantage by adopting new technologies before the majority, but wait until some critical mass of acceptance has been reached before fully jumping in. The technology is quite expensive to purchase at this point, because users must subsidize the R&D performed by the inventors.

- *Mass acceptance* comes when standards have gained widespread adoption, and it becomes obvious to all that a particular technology must be used to remain competitive. Use of the technology is no longer a competitive advantage, it is the price of doing business. Companies in the mass acceptance category rarely have to worry about following the wrong path, because they are following the crowd, and the crowd will not ignore the dollars spent on this technology. They sacrifice their own innovation for safety, and in the process lose the ability to significantly influence the direction of the technology or reap the rewards that early adopters are able to secure by being first.

■ *Latecomers* wait until the game is all but over to decide which technology to use. They are very conservative in their approach to technology and probably do not have very strong motivations for change. They are typically the haves in an industry—resistant to change because it threatens their position of strength. They may even try to squelch innovation in an effort to keep the status quo.

Acme Financial is clearly part of the *early adopter* community. It reflects the intense pressure felt to drive up revenues and increase market share. Being an early adopter will allow Acme to leverage new data technologies, such as XML and LDAP, before their competitors catch the drift. Acme will have to make a strategic as well as a financial investment to adopt these technologies before they have gone mainstream, but the investment will position the company to meet its business goals and provide it with competitive, quantitative, and qualitative returns many times over the original outlay.

Customers to Perform as Many Tasks as Possible without Involvement of Acme Personnel

Acme is trying to create a knowledge vending machine capable of automatically servicing a wide variety of requests. It's in keeping with the old IT 90/10 rule, whereby 90 percent of the enterprise's efforts are used to solve 10 percent of the problems. Conversely, this means that 10 percent of the effort solves 90 percent of the problems. When applied to the Web-based self-service model, this means that Acme personnel can be freed up from performing the mundane tasks to be automated by the portal. Instead of answering routine inquiries or performing low-value-added tasks, customer service reps and financial advisors can focus all of their attention on their customers' nonstandard problems.

Another old IT rule is that there is nothing new under the sun. This also applies to the Acme self-service portal, because it is exceedingly rare for a small investor to have a question, concern, or problem unique to him or her.

On a similar note, different customer groups will require different services. For example, high-net-worth customers will have different issues (estate concerns, for example) that never trouble small investors. The Web-based portal allows the majority of customers to find answers to their most frequently asked questions in a self-service fashion, while Acme resources are freed up to provide highly personalized service to the company's most profitable high-net-worth customers, without sacrificing service to either group.

When we discuss the functional components of the portal, we will demonstrate how an effective data mining strategy on the customer support database can enable this level of support.

Acme Would Like to Increase Its Customer Base

Who wouldn't? But the portal provides a strategy and path to make this goal a reality. From a data architecture perspective, the portal must reach as many customers as possible. Advanced technology that may render the portal a cool application from a technology perspective may block some current or potential clients from accessing the system and, therefore, is not an option. Targeting as wide a range of technologically sophisticated customers as possible places some limitations on just how leading edge the application can become.

All current Web browsers are compliant with the HTML 3.2 standard. Thus, we can safely use this standard as a baseline for the type of content that the portal can provide. Anything beyond that standard runs the risk of blocking customers from effectively using the application. Security is another matter that we must closely watch. Banks regularly require a browser with a minimum of 128-bit encryption to protect the customers' data.

Remember, the financial services portal is not like the B2E sales force automation application—usage of the technology is neither compulsive nor monitored. Users are free to come and go as they choose, and they may use all kinds of wireless access devices. Something as simple as slow server response time could turn off a new customer and send that customer straight into the arms of your competition, regardless of how superior your portal may be from a functionality perspective. These issues will be discussed at length in the technical infrastructure chapter (Chapter 21), but are mentioned now to demonstrate the tenuous grasp an enterprise has on the users of its Web application. For the application to succeed, numerous factors, including availability, reliability, security, and functionality, must be working absolutely flawlessly.

In an effort to reach the broadest possible customer base, then, we must make sure that we are adopting the lowest common denominator of client hardware and software requirements. That means we must use a thin-client model. Let's review the issues involved with thin and thick clients.

ADVANTAGES OF A THIN-CLIENT MODEL:

- It is simple to implement and manage: The latest version of the application is used every time, and it eliminates client version control issues.

- It uses synchronous data communication; all messages are sent and received in real time.

- All data is located on the server; there is no persistent client data, no issues of data latency, no issues of data replication, and no problems with data change conflicts.

- The client can be an inexpensive, small piece of hardware.

DISADVANTAGES OF A THIN-CLIENT MODEL:

- It involves the reinvention of mainframe dumb terminals.

- It works only when the server is available. If the server is down, data entry is lost. If your connection is lost, you cannot perform work.

- All computation occurs on the server. Bandwidth is needed to bring everything to the server for computation and back to the client for presentation.

- All communication is instigated by the client; the server only responds to requests.

ADVANTAGES OF A THICK-CLIENT MODEL:

- There is efficient use of bandwidth; data is distributed and the client has computational power. It saves time in environments with static or rarely changing data. The server and the client can send and receive deltas.

- It allows asynchronous data communication. It will work even when the server is not present, the local client store ensures that work will not be lost, and the server can initiate communication with clients.

- Thick clients have more potential for better user interfaces, better response times (communicating with a local instead of a remote data store), and richer complexity.

DISADVANTAGES OF A THICK-CLIENT MODEL:

- A thick client is more complex: The model requires strategies for updating clients with the latest version of the application. The system must handle interactions with a potentially heterogeneous client population.

- You have to deal with distributed data: There are issues with data latency and data replication, and you must develop a strategy to handle data change conflicts.

- The client must be heavyweight enough to support local persistent data.

- Since communication is asynchronous, every message must be authenticated.

Acme Financial must use a thin client for its portal, because a thick client would be contrary to the goal of enlarging the customer base. We will talk about how this impacts the application in the next section.

Data Requirements for the Functional Components of the Web Portal

Chapter 18 outlined the data sources (user profile, accounts, news, products, community) and application components (preference setup, research and analysis, products/services, quote generation, transaction, notifications/ alerts, community) of the Acme Web portal, as shown in Figure 19.2. We will use that framework to build the data architecture, first of the Web portal, then of the wireless application. First, we'll look at the Web portal functionality.

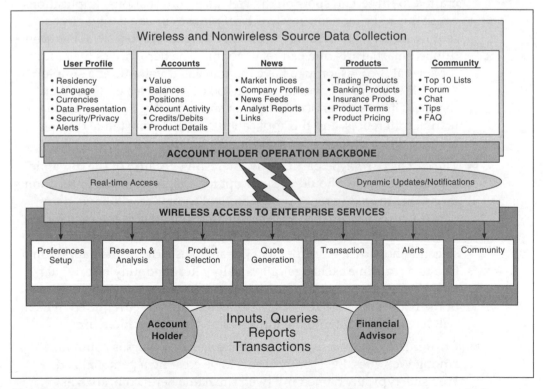

Figure 19.2 The functional components of the Acme Web portal.

Preference Setup

The goal of preference setup is to provide persistent knowledge of the user's likes, dislikes, needs, and requirements. Typically, the preferences are stored on the Web client (even a thin client can store such simple attribute-value settings as cookies). The overhead of storing the preferences of millions of users that can potentially be on the system is too great—it could adversely affect system performance and cost. The information is not very sensitive, so client-stored preference data seems to be the most reasonable path. The downside is lack of mobility: Since the preferences are stored on the client, a user accessing the system from a different computer will not have the cookies needed to render the customized content. This limitation would apply only on a borrowed Web browser, since cookies can be ported from machine to machine (so a user with both desktop and laptop computers could copy the cookies from one machine to the other without the overhead of reestablishing the preferences).

- The *residency settings* stored by the system can have value in many ways, as mentioned in Chapter 18.

- The *language* settings pose an interesting challenge to application developers. Typical Web user interface design has flowing text, carefully crafted to fill certain spots on the Web page. International applications cause this to be more difficult, as brief phrases in one language can require much more verbiage in another tongue. Anyone who has seen a dubbed martial arts film can relate to this: The hero moves his mouth for several seconds, only to have the translated voice utter a simple "yes," leaving you to wonder why it took 25 syllables in its mother tongue. The user interface design can suffer greatly from the analogous textual differences, and the application designers must take this into account when dealing with multilingual data.

- The *currency setting* is particularly interesting because of the dynamic aspect it can have. Should the user opt for a representation of the monetary value to be different from its original format, it will have to undergo a currency conversion. The dynamics element is the exchange rate, which changes constantly. The user should have the option of selecting how dynamic an exchange rate to use: the nearest approximation to a real-time exchange rate, a daily rate, a monthly rate, a quarterly rate, or a yearly rate. The data requirement is significant because of the real-time access to Acme's exchange rate server required, which will have to be built into the information systems architecture.

- Customizing the macro and micro *presentation settings* is common among Web sites. Everybody seems to be creating a personalized Web site, typically named my [web site name here].com, such as myYahoo.com, myAmericanExpress.com, and myAcmeFinancial.com (of course). Typically, this is accomplished via a presentation settings Web page, with checkboxes indicating a closed set of areas in which you can claim interest. Order is important, as your settings appear on the page in the sequence in which they are selected, further customizing the user experience. For most users, this is one of the most powerful techniques for providing a positive experience—most novice computer users are amazed when they log in to a Web site, and the page greets them by name.

- The last customized preferences are *security and privacy settings*. Don't be alarmed—nothing secure in and of itself is stored here, only the permissions for transactions and content providing.

One interesting side note on the security of online transactions: Everyone using an online Internet system to conduct financial transactions is very concerned with security, and with good cause. But the primary source of danger is not in transmitting credit card information on the Web; it's transmitting that information to a dubious partner in an online transaction. The transmission of the information is safe; it is the merchant who may compromise the data—willingly (a fraudulent merchant) or unwillingly (loose security on the server that stores your confidential information). Another threat comes from a holder of sensitive data being hacked, but that can happen any time a credit card is used. It is safe to say that your credit card information sent via the Web is more secure than when you are using your card at the small restaurant down the street.

It's amazing that the security of online transactions receive so much scrutiny, when ATM transactions do not. One of banking's dirty little secrets is the way that ATM transactions are conducted: Account and PIN information is sent over the various banking networks in clear text to perform the account access and transactions. Granted, the networks are not as accessible as the World Wide Web, but the only reason that people are not panicking about using ATMs is the fact that these machines were accepted long before the potential security holes became known to insiders who understood the technology. Imagine trying to sell such an unsecure model in today's highly conscious environment!

The next area to be explored covers the portal's array of research and analysis tools.

Research and Analysis

Due to the thin-client nature of our portal, all of the computational power used in research and analysis is housed on Acme's server. This means that every piece of data upon which an operation can be performed must pass over the wire to the server and return to the client for presentation. This burden can be eased to some degree by knowing some of the questions the users are going to ask and having the answer already formulated.

- *Static research* essentially means using some filters predefined by the Acme portal. The portal is designed to provide easy access to the most commonly performed data queries, so the user does not have to formulate them. Typically, the results of static research queries are cached for optimal access from clients.

- *Dynamic research* provides greater flexibility than static research. Since there is a direct correlation between complexity and flexibility, the challenge of the application designers is to keep the dynamic research functionality rich enough to provide value without complicating the interface beyond the capacity of the novice computer user. Essentially, the user has to formulate a query without ever knowing the mechanics of how to do so. Since the dynamic query must be run in real time, it may take somewhat longer to complete than its cached, static counterpart.

- The *analysis* part is simpler than might be imagined, because the set of functions performed by the user is closed. Don't misunderstand—there is a rich set of tools available, but the server performs all of the calculations (per our thin-client model), so it must know a priori the set of tasks it can be asked to perform. The data transmitted during analysis is usually more sensitive than in research, which is typically a customized set of publicly available information. Therefore, the utmost in security is used when analysis is performed.

We will now look at the products and services Acme provides in its portal from a data needs perspective.

Products/Services

Just as the user sets limits in the security and privacy preferences, Acme is able to filter the products and services seen by the individual users. This is handled by profiling the user upon login to the portal. The user preferences are fairly low security, hence the willingness to store that data statically on the client. The user information pertaining to the products and services they use (credit card numbers, account numbers, etc.) is highly sensitive and is stored only in a very well protected server behind Acme's firewall. It is certainly worth the effort and cost to provide database servers to store and protect this information; otherwise, the system would not be trustworthy. Without trust, user acceptance would be low, and system use would be low. The careful protection of sensitive data is vital to the portal's success.

The product/service bucket on the back end is not just for static information about service offerings only, so the user would be able to review terms and conditions for a savings account, for example, to learn about minimum balances, how to open and fund the account, the interest rate paid, how to transfer funds, how to close the account, and so on.

Another aspect of the portal is to obtain a quote, for a product or service you want to acquire.

Quote Generation

As was stated in Chapter 18, obtaining a quote is straightforward for the *brokerage* application; all that is required for you to make a buy/sell decision is the security's current price. Obtaining any secured loan in the *banking* section of the portal requires more data: the loan-to-value ratio of the item against which the loan is borrowed, the debt-to-income ratio of the borrower, and the credit score of the borrower, as provided by one of several credit bureaus (there are three nationally in the United States). All of these factors combine to set the level of how attractive the lender finds the proposition. Suppose you want to buy a reasonably priced new car and you make lots of money. However, you have had your last two cars repossessed for nonpayment on the liens. Acme may approve your application, but generate a quote with a slightly higher interest rate than you would have received if you had perfect credit. That reflects the higher probability that the loan will default.

Quote generation for the *insurance* offerings are typically the most complex, because there can be dozens of factors involved in setting the premiums required. The online application will be similarly complex, because there can be cascading data needs. For example, suppose you are filling out a medical questionnaire for an Acme disability insurance policy. One question is, "Have you ever had any heart difficulties?" If the user answers no, then the application moves on. If the user answers yes, then a whole slew of follow-on questions are asked to specify the nature of the heart trouble. It is this increasing complexity of both user interface and data needs that makes online brokerage applications much more common than those providing insurance policies.

The culmination of this work is to conduct financial transactions.

Transaction

As was pointed out in Chapter 18, the level of complexity in generating a quote for the various services is roughly inversely proportional to the level of complexity in performing a transaction, from a user interface and data collection perspective. In fact, all transactional Web tasks must be handled carefully to ensure the event is recorded exactly one time. Imagine an order to buy 200 shares of Sprint stock that is executed repeatedly or a bill-paying service that pays your water bill twice—or not at all! The complexity of Web portal transactions has to do with the synchronous nature of Web data communications. Let's take our purchaser of Sprint stock. He researches the stock, likes what he sees, checks the price, and decides to buy. He goes into the brokerage section, fills out the form so that 200 shares of Sprint will be purchased, then hits the Submit button. But he's not sure he hit the button properly, so he hits it again. Does the system recognize this as one transaction submitted twice? Or are they viewed as two separate transactions? Let's suppose the client's system freezes after the button is pushed—how will he know if it went through? Will the transaction be recorded quickly enough so if the user reboots his computer and gets back into the portal, he will see the request to purchase the stock? These are the questions that need to be answered when designing and implementing the application, and the system must have safeguards to prevent these kinds of holes. For example, a transaction ID can be associated with the form upon entering the transaction page. If the Submit button is pressed several times, the system recognizes the duplicates and takes the appropriate action (probably discarding the duplicates and sending notification to the user of the situation).

Let's look at the data requirements for alerts in our Web portal.

Notifications/Alerts

Web-based alerts are quite simple from a data perspective—there is a closed set of alerts from which the user can choose. The user then provides the email address at which he or she should be notified when the chosen event occurs. It will become more challenging and interesting when using wireless devices, however.

Finally, let's look at the data requirements for the Web-based community area where users can help themselves or seek the assistance of other account holders.

Community

As was mentioned in the beginning of this chapter, an effective data mining strategy can optimize the automated help provided by Acme's portal. By monitoring the questions, concerns, and issues of the user base as expressed in the online community, Acme personnel can identify the most frequent problems and formulate the *FAQs* to answer them. In addition, this feedback can be incorporated into future releases. Frequently asked questions can also be used to guide the *Dear Advisor* and *financial tips* content. The *live chats* and *open forums* require real-time server access, as found in such common Web applications as AOL Instant Messenger and Microsoft Chat. They require the ability to move large quantities of small pieces of data, such as a million people sending one sentence per minute. The applications will have to be carefully geared to handle such scaling issues.

Now that we have explored some of the data requirements of the Web-based application, let's look at the wireless version of the portal.

Information Architecture for the Wireless Solution

The data architecture of Acme's wireless Web portal has many common characteristics with its wired sibling. As with Chapter 18, we will keep the same order for consistency's sake.

Functionality Common across Both Portals

There are seven functional areas common to the wireless and wired portals. Let's look at each one, to help better understand their data requirements.

Preference Setup

The nature of the wireless client device will impact the preference setup. A PDA or other such wireless device has the ability to store state in the client, so the cookie model will work. A WAP phone has no such persistent client storage, so it will not. This leaves three possible solutions. First, require a client device with persistent client storage (not a good choice—it goes against the principle of ecumenical access). The second choice is to disallow WAP client customization (again, not a good choice—it seems prejudicial against the WAP user). Finally, the third option is to store the WAP user's preferences on the server, which seems to be the best idea.

Though many users on the system may configure their WAP phones for the portal, the relatively poor user interface and low bandwidth will ensure that usage via WAP is limited. WAP will most likely be used in the *alert/response action* motif and for *location-based services.* Since most WAP phone users are certain to sleep at night or be in business meetings during the day, at which times they do not want a default beeping sound emanating from their phones, providing them the same level of customization as other users is important. Should the use of WAP phones become so extensive that storage of client customization data becomes problematic, steps can be taken to alleviate the crunch, such as offering users incentives to switch to a PDA.

Alerts have to be set up to reach the wireless device in real time. The Web is not equipped to push alerts out to clients—that's why the wired version of the portal has to send them via email. It defeats the purpose of having wireless alerts if they cannot reach the client device immediately. Whereas WAP phones have a disadvantage relative to other wireless devices in client storage capabilities, they have an advantage in their ability to be located. Companies such as Kenemea are developing strategies, such as client harnesses wedged in between the application and IP layers, to provide client visibility and guaranteed transmission of data services for handheld computers. This will solve many of the wireless device shortcomings currently found.

Research and Analysis

Since the data required for alert generation must reside on the server to be of use, the triggers for research and analysis will be stored in the user's server-based preferences (they will be kept to a minimum for server cost considerations, but all users will have some server-based information in the system).

Products/Services

The data requirements for products and services are the same, regardless of wired or wireless network access.

Quote Generation

It's a given that filling out a form on a WAP phone is not feasible for most orders. The exception would be alert responses (for a stock alert, you simply have to say, "Yes, I want to buy/sell," in response to the alert, and a quantity of the commodity to be traded). Other quotes are simply too complex to be reasonable, given the state of wireless client devices. In the future, however, the limited presentation abilities are sure to be overcome.

Transaction

Chapter 18 described wireless stock trading and bill paying, both useful wireless services. Their data requirements are the same for wired and wireless clients. They are quite light and capable of being handled on a limited wireless device. Let's explore a possible future transaction, when the aforementioned wireless client limitations have been overcome. At such a time, it does not seem unreasonable that the pressure received from a car dealer to finance through their organization be met not with your request for a lower rate, but with an approval from Acme. This is possible with a minimal set of data: The bank already knows all about me, so I don't need to enter any personal data. The bank only needs authority to seek a credit bureau rating and the value of the car, obtainable by the year, make, model, and submodel. Since there is a finite set of car types, it is a closed set, and can be provided in drop-down style lists, making data entry even easier. The only other data required is the requested loan amount. The bank now has all the information it needs to make an evaluation. Like most banks, Acme has an existing autoscoring system, which means that this loan data can come in through the wireless portal and be funneled directly to that back-end system. Approvals can be generated in seconds. The car buyer's negotiating position has gone from "I can go to my bank and apply at such-and-such a rate" to "I have an approval at this rate, so my monthly payment will be X. Can your dealership beat that?"

Notifications/Alerts

We already reviewed the information requirements for wireless alerts.

Community

The only limitations placed on community participation have to do with wireless client device limitations, not wireless access in and of itself. The throughput needed for real-time chatting would not exceed the capacity of even a WAP phone, since they are only simple text messages being transferred. The real usability issue is real-time data entry—even a skilled practitioner will not be able to enter data rapidly enough into a palm-held computer, let alone a WAP phone.

Wireless-Only Functionality

Let's discuss the data requirements for the additional functionality that are only reasonable from a wireless client, as outlined in Chapter 18.

Session Management

This is a critical factor in multichannel enterprises. It requires all session data to be stored persistently on the server, since transactions are to be performed from multiple clients. A balance has to be struck between transmitting every keystroke to the server and losing large amounts of data (the former has huge implications for system scalability; the latter has user satisfaction concerns). A possible solution is to have the wired infrastructure remain the same (most people have reliable wired connections to the Internet) and introduce something akin to the Kenemea solution on the wireless clients, to ensure that transactions can be conducted asynchronously. This will prevent data loss and remove uncertainty about transaction completion because of the guaranteed delivery of the transaction exactly one time.

Cashless Vending

The biggest data concern for cashless vending is not the quantity or ability to transmit the data to the vending machine, but the transmission's reliability.

There has to be a mechanism in place to prevent a malfunctioning vending device from draining my account while freely dispensing Cokes without my knowledge, or a hacker from assuming my identity and performing the same mischief.

Wireless Wallet

In general, the data issues of the wireless wallet are similar to the cashless vending concerns. The data mining possibilities, of course, are significantly higher with the integrated wireless wallet. The data requirements to facilitate impulse purchases are all in place once the wireless wallet is in use. In fact, it is the natural evolution of the technology. Acme wins with its nontraditional revenue source. Companies buying the consumer data win by knowing exactly how much of your spending goes to their competition, so they can segment you properly. And you, the customer, win because companies will try to woo your business by offering you incentives to use their products and services.

Summary

This chapter presented an overview of the data requirements generated by the business processes outlined in Chapter 18. We took those processes, defined the interactions among them, and applied the company's information principles to create a data architecture. We showed how the Acme Financial Web portal works today, and how its wireless counterpart enhances the offering.

The B2C nature of the portal requires us to use a thin-client model. This not only ensures the largest possible set of users, but also allows a wide set of client hardware devices to be used. The thin-client model also makes sense because of a large percentage of dynamic data. Since dynamic data needs to be retrieved from the server in real time anyway, it does not make sense to have thick-client capabilities.

The wireless portal has most of the same functionality as does the wired version, except for the data-intensive research and analysis components. They simply do not make sense, given the nature of wireless client devices. But the wireless functionality is not a subset of the wired version—it has location-based services and real-time alerts possible only with the wireless infrastructure.

We are now ready to discuss the information systems architecture, using the data requirements described here as source material.

Information Systems Architecture for Wireless FSI (Mobile Brokerage)

Introduction

Again, we've reached the information systems architecture analysis, this time for the design of our financial services wireless application. Remember, the key points of any information systems architecture are to understand the underlying aspects of the logical layers and application services that must be built so that all system components can work together as one unit. As you know, the question of *how* the application is built is answered in the information systems architecture. We don't want to seem like we are reinventing the wheel in each IS section, but it seems to be worthwhile to at least go through a quick review of the information systems architecture layers:

Conceptual. Answers the question of what the overall system is supposed to accomplish. The conceptual business architecture of the Acme Mobile Brokerage (AMB) system was covered in greater detail in Chapter 18.

Logical. Answers the question of how the system is built. As architects, you'll want a design that reflects the conceptual business requirements as completely as possible. Generally, logical architectures don't go into great detail about the physical components, such as hardware and software services. Those areas are left for the physical layer.

Physical. Essentially, the physical architecture is the implementation of your design. In other words, the physical dimension will tell you with what you are going to build your architecture. This book will not go into great detail on the physical layer, for the simple reason of longevity of design. Software and hardware versions are constantly changing. It's not the job of the architect to necessarily recommend or design a specific version of software or hardware components. However, it *is* the architect's job to design a system and components that logically work together in a collaborative fashion. Physical design is a critical component of any system design, but the minute we recommend a particular version or software release, it seems to become outdated by a new version.

Perhaps we should begin our discussion with a brief review of the functionality of the Acme mobile brokerage application. Remember from Chapter 18 that the Acme Mobile Brokerage application started out much like our other case study examples, as a nonwireless, Web-enabled system. This chapter will cover the current state and future state logical architecture of the initial Web-based release and subsequently the future wireless services pieces. Figure 20.1 illustrates the key functionality for the Acme Mobile Brokerage application.

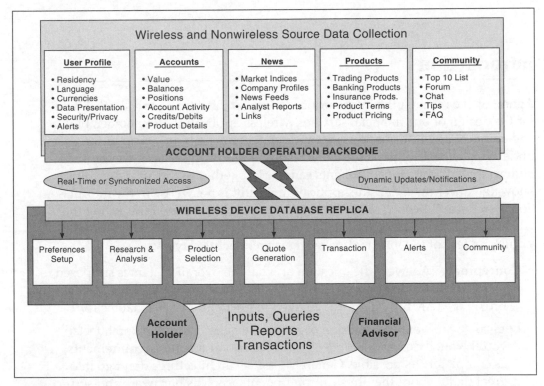

Figure 20.1 Acme FSI portal conceptual business flows.

As a review, the key functions of the Acme application are as follows:

Preference setup.   Preference setup readies the application for custom usage. It is here that the user specifies parameters that the application will check before allowing access or displaying data and information. Specifically, the user enters residency, language, currency, data presentation, and security/privacy parameters.

Subfunctions of the preferences setup are:

- Residency settings
- Language
- Currency setting
- Presentation settings
- Security and privacy settings

Functions of the financial services portal of the Acme application are as follows:

Research and analysis.   It's important for customers to have the ability to make informed, intelligent decisions regarding their portfolios. The Acme FSI application portal provides its users with static and dynamic information as well as a set of analysis tools to help with that decision-making process.

Subfunctions of the FSI portal are:

- Static research
- Dynamic research
- Analysis

The ability to actively select financial products and services is an important piece of functionality within the Acme application. These functions include the following:

Products/services.   Describes the offerings that the operator of the portal makes available so that you can familiarize yourself with their particular attributes. The core offerings of the portal include:

- Brokerage
- Banking
- Insurance

Another step in making a financial transaction is to generate a quote for a product or service that you would like to purchase. Quote generation functionality includes the following:

Quote generation. Most of the quote generation functionality is essentially selecting various fields required for the system to spit out a quote. For instance, brokerages may simply require the stock/bond symbol and off you go, whereas the banking component requires several other factors to be considered, such as loan duration and interest rates, depending upon the type of banking service. Finally, insurance may also become rather detailed, depending upon the insurance product. Life insurance quotes, for instance, may require personal information such as medical data, age, and coverage amount.

Transaction. Essentially, the ability to perform financial transactions such as buy, sell, transfer funds, bill presentment and payment, and premium installment payments.

Notifications/alerts. There are a number of ways of leveraging notifications/alerts in the Acme application. Examples are a notification that a stock has reached a certain price or receiving an alert from your bank that an account balance is low or a payment is due.

Finally, the Acme financial services portal offers an online community where similar-minded individuals can gather electronically for the purposes of information exchange, advice, and even entertainment.

Community. The popularity of online chat sessions and online communities has been increasing daily ever since the dawn of the Internet. The value that an online community in our Acme case study brings is the ability for clients and advisors to communicate directly at any time of the day. Additional functions within the community portal are:

- Popularity lists
- Frequently Asked Questions (FAQs)
- Dear Advisor column
- Financial tips
- Open forum
- Live chat

We wanted to provide a quick review of the conceptual business architecture of the Acme application to give us a framework for our information systems technology. Please refer to Chapter 18 for an in-depth overview of the conceptual business architecture.

The next step is to get a better understanding of how the functionality transfers into logical application services. As in the preceding parts of this book, let's start with an understanding of the logical current state design of the application.

Current State Logical Architecture

In the prewireless state, the Acme financial services application was designed to be accessible through any Web browser, such as Netscape Navigator or Internet Explorer. As in any Web-based architecture, several fairly common components must exist. Figure 20.2 provides a bird's-eye view of each layer in the current state logical model for Acme's mobile brokerage application.

This layered *n*-tier architecture should look familiar. Let's touch upon each of the layers in greater detail.

Web formatting layer.   One fairly consistent component of our layered architecture approach has been the Web formatting layer. This important component is required so that the application can read requests and generate HTML pages. Without these services, the client (Web browser) would have no way of viewing the presentation information. The business logic access layer and the Web formatting layer work hand in hand in generating the content that is presented to the user. This layer is also

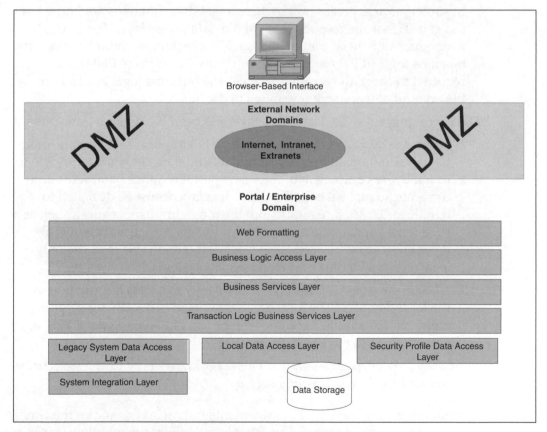

Figure 20.2   Acme FSI portal current state logical architecture.

responsible for obtaining very specific information about the format of the HTML pages and determines how that information is used on the page. For instance, if Michael Moneybags is a client accessing the Acme FSI portal to manage his accounts, he must first access a browser and type in the HTTP address for the site. The Web formatting layer generates the pages to be presented to Mike.

Business logic access layer. The business logic access layer, as related to the Acme FSI mobile brokerage application, is the application elements that know where the business logic resides and what needs to be called to fulfill certain application requests. The purpose of this layer is to ensure that the Web formatting layer does not have business logic and that it doesn't need to know how to find it.

Business services layer transaction logic. The business services layer has been broken down into the business services layer and the transaction logic business services layer. The business services layer is responsible for implementing the business logic. It does not know anything about the presentation layer and where data is located. This transaction logic business services layer knows where to get data, but does not know how to get it. This is the responsibility of the data access layer. For example, to conduct a financial transaction, such as checking account balances, the business logic utilizes application program interfaces (APIs) that are required to check the account balance. The business logic needs to know the account balance and how much to deduct.

Data access layer. The data access layer has been broken into three pieces:

Legacy access data access layer. The Acme FSI application would be of little value to its customers without data and data elements to present to users. Every component of the Acme portal requires data, static and dynamic, to make it complete. The data layer must be designed to handle application requests from the most simplistic request, such as a preference type, to the most complex financial analysis report. This layer knows how to access any legacy system. In this case, the only legacy system is the order management system.

Local data access layer. This layer knows how to access information in the local database. The local database is used primarily for profiling and personalization, but could also be used for staging data. This is usually done to increase performance.

Security profile data access layer. This layer knows how to access security information in the directory server.

System integration layer. The system integration layer handles the way you access legacy systems. This layer is essentially the middleware service for the application. The middleware service can be custom built or package based, depending upon the complexity of the application. It all depends on what and how much data needs to be accessed.

Each of the layers is important to the ultimate success of the application. Within each of the layers are various components and application services that we'll look at in closer detail. Let's start with Figure 20.3, which illustrates the Acme logical current state architecture.

Web Services

Most Web services refer to the application components that deliver the presentation to the user. The Acme FSI portal is very similar to the traditional Web-enabled application. Figure 20.3 illustrates the Web services logic within the DMZ zone.

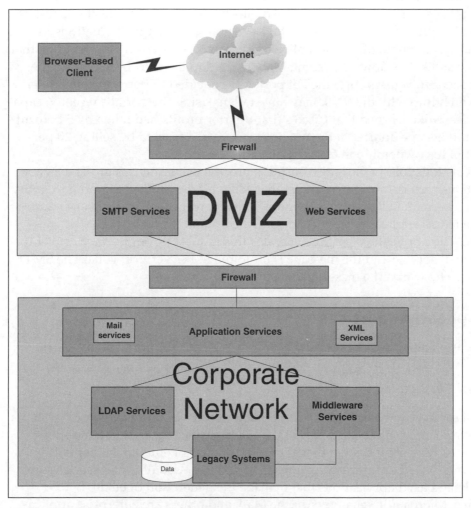

Figure 20.3 Acme FSI logical current state (detail view).

Traditionally, Web services consist of a Web server that translates HTTP requests from the client and subsequently makes requests to the application services based on the type of request.

SMTP Services

Simple Mail Transfer Protocol (SMTP) is the mechanism used to push email messages to their proper physical location. The Acme FSI portal allows for notifications and alerts that are sent using email messages. The SMTP service receives a client request specifying that an email is to be sent/received. The SMTP service then communicates with the mail server that actually generates the message. If required, the application services may come into play if a transaction is requested.

Let's again use Mike Moneybags, our Acme client, in an example. Mike owns several hundred shares of Cisco stock. He uses the notifications/alerts functionality of the Acme FSI portal to be notified of large fluctuations in the security. Recently, Mike was online and received an alert in his email box that Cisco stock was down 3 percent. Mike reacted to the alert by accessing the transaction page through the FSI portal. He decided to perform some quick research into why the stock may have fallen. Using the portal's research capabilities, Mike learned that Cisco's first-quarter profits had fallen by 5 percent from the same quarter in the previous year. Mike reacted by selling 20 percent of his ownership in Cisco stock.

Now let's follow the entire business process using our design. The legacy systems services react to a trigger that Mike placed on the FSI portal to notify him by email if Cisco drops by more than 2 percent. The trigger sends a message to the application service to generate a standard formatted XML message. The application service generates this request through a Java applet to the mail service and the message is off to the SMTP service. Voila! The SMTP service forwards the message to its proper location.

Application Services

The application services are the heart of the Acme FSI portal architecture.

Mail Services

Mail services have been a key component of most eCommerce-based systems. Electronic mail has been a part of the Internet since the first byte was sent across a wire. Any system that allows for the use of email, whether it's Internet-based or internal, requires mail services. Usually, mail services are package-based applications that manage the generation of email. Lotus Notes, Microsoft Exchange, Apache Mail, and others are enterprise applications that manage mail for clients. The way the process works is that a client initiates a mail message through any browser-based client. The SMTP services forward the message to the mail service where the message is generated. The required data for the message, such as address and authentication, are obtained from the LDAP services and, if necessary, from legacy systems. Finally, the message is sent back to the SMTP service for delivery.

XML Services

The XML service is used to transfer information between different layers of the application. The reason for doing this is that it makes the data independent from the application that is processing it. The XML file is originally generated in a specific format. The format specification is also used when the XML is interpreted to make sure that the data has not been corrupted.

LDAP Services

The Lightweight Directory Access Protocol (LDAP) services are used to store addressing information and authentication data about the users of the Acme portal. Again, in an n-tiered application like the Acme FSI portal, it makes sense to break the LDAP service into its own separate component for scalability and usability.

Middleware Services

The middleware components are basically the code needed to integrate the legacy systems, or other existing systems into our design. Middleware components can take various forms, from small, custom Java or C++ applications that simply send data back and forth from the legacy system, to complex packages like EAI WebMethods, IBM MQ Series, or SeeBeyond.

Legacy Systems/Data

The grandfather of our current state family is the legacy systems and database systems. As you know, legacy systems can be almost any application that may be used for database access, accounting, or an enterprise function that has been deemed critical to the business. Usually, we think of legacy systems as monolith-sized mainframe computers with terabytes of data storage. Since it's very expensive and impractical to try to convert a legacy system to a more current technology, why not integrate the old with the new? If you can't beat 'em, why not join 'em? That's where the middleware components come into play. Middleware joins the old with the new. Since protocols, services, hardware, and software are usually directly incompatible, middleware must be used for integration purposes. Again, a middleware application can be almost any component that solves a particular technical problem. For instance, XML and Web services are not directly compatible with an IBM mainframe. Thus, a middleware component is used to convert the XML to a format that can be read by the legacy system. Alternatively, the XML engine can't read native legacy throughput. So the middleware is used to convert legacy data to XML and subsequently to the presentation to the user.

The Future State Architecture

Now that you have an understanding of the ACME Mobile Brokerage current state architecture, let's focus on how we would make it wireless. Take a look at Figure 20.4.

There are many similarities between Figures 20.3 and 20.4, but also some significant differences:

- As in Chapters 10 and 15, we have three different users: WAP, handheld computer, and Web browser users.

- SMS services have been added to implement alert functionality.

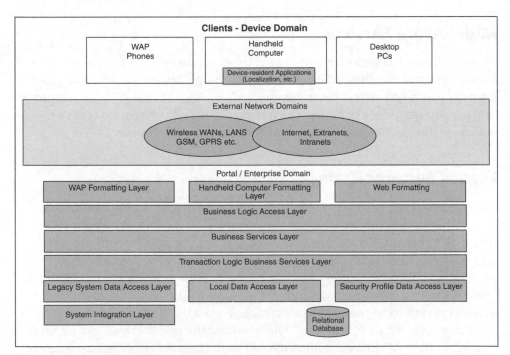

Figure 20.4 High-level Acme Mobile Brokerage logical architecture.

The underlying application is capable of handling multiple channels and therefore has not been changed.

Please note that Figure 20.5 does not focus on the following:

- *System services.* This includes services such as session handling, how to provide access to data, and application security. System services are not included in the discussion, as they are not unique to wireless and since these can be purchased as part of a product such as an application server.

- *Commerce products.* This covers electronic wallets, auction components, and so on. Commerce products are not included, as this functionality can also be purchased.

- *Personalization functionality.* This covers how to customize the front-end, and so on. Personalization functionality is also not included, as it can be purchased.

It is advisable in all systems implementation and integration efforts to evaluate which functionality should be bought and which should be built. The answer differs from case to case.

Please note that at the logical level, the layering is exactly the same in this chapter as was described in Chapter 15. The reason for this is that we want to provide a more general way to look at application structure in this book. The

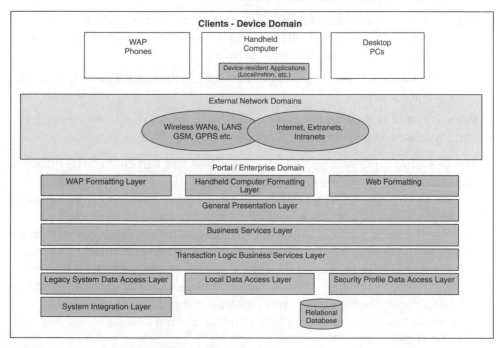

Figure 20.5   Logical application architecture.

main differences between the application architectures in Chapter 15 and those in this chapter will lie in the component architecture since the system described in Chapter 15 is Microsoft-based and the one in this chapter is Java-based.

To recap, here are the major layers in the architecture. Refer to Chapter 15 for a more detailed description. To start, please note that we have the same users and networks as in Figure 20.4. In addition, we have the following application layers:

WAP formatting layer.   Responsible for creating WML pages and handling WAP requests.

Handheld computing formatting layer.   Responsible for creating handheld computer pages and handling requests from handheld computers.

Web formatting layer.   Responsible for creating HTML pages and handling Web requests from desktops and laptops.

Business logic access layer.   Responsible for routing requests to the appropriate business objects.

Business services layer.   Responsible for running business logic. All business components and business logic are located in this layer.

Transaction logic business services.   Responsible for routing data requests to the appropriate place. The purpose of this layer is to make sure that the business components don't have to know where to find information.

Legacy access data access layer.   Gets data from legacy systems. This layer knows the function calls and/or data structure of the legacy systems and can access all relevant information.

Local data access layer.   Gets data from a local RDBMS. This is focused on local data access.

Security profile data access layer.   Gets data from a directory server. In many cases, this will be an LDAP server.

The basic principle behind this layering is that you can change the information in each layer without affecting all subsequent layers. For instance, if you want to change the presentation layer for WAP, you don't have to change the business services layer unless some specific business logic needs to be added. By the same token, if you want to change a legacy system, you don't have to change the business services layer or any of the presentation layers because you have encapsulated the legacy system access with the data access layer. The benefit of making different layers independent of each other is that you have a more flexible architecture. It's easier and faster to make changes or additions. Relating this to wireless, you can see that adding a wireless channel is thus easier.

The rest of this chapter will focus on some of the specific applications in the Acme Mobile Brokerage. Let's recap some of the business components described in Chapter 18:

Preference setup.   All system settings can be modified through wireless means. This includes language, location, and alert preferences.

Alerts.   This topic covers a host of different types of alerts:

- *Market condition alerts.* For instance, a customer may want to know when a certain security reaches a certain price.

- *Location-based alerts.* Alerts based on where you are. An example is finding a local branch of a bank when you are climbing in the Himalayas.

- *News feeds and analysis alerts.* It is difficult to access a large article on certain wireless devices, but alerts about the article's availability or a short excerpt of the article can effectively be sent.

- *Information search results.* Searches based on certain triggers, such as interest in a specific company.

- *Credit limit alerts.* Finding out that you are close to your credit limit before you make a transaction that is rejected.

Products/services and quote generation.   Being able to find out and investigate service offerings via a wireless device. This could result in a quote on that service.

Transaction.   The user should be able to conduct financial transactions through a mobile device.

Community.   Several community features will be available to the user, including discussion groups and FAQs.

We will also discuss localization features in the context of these applications.

Wireless Extensions to Existing Acme Mobile Brokerage Functionality

Let's start this section by driving into Figure 20.6 in more detail.

Figure 20.6 shows all the major components that constitute the Acme Mobile Brokerage. The AMB components are broken out according to the different layers specified in Figure 20.6. A few notes about this figure are in order:

- Localization components have been added in all of the formatting layers. These localization components are responsible for customizing the information and the presentation of information according to the pref-

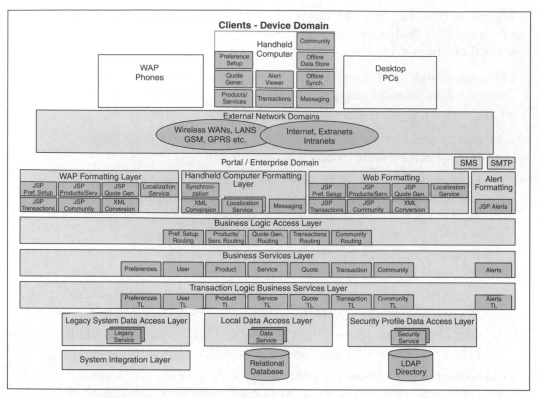

Figure 20.6 Detailed logical application architecture.

erences of the users. For instance, if an American salesman were working in Mexico, any information presented to him would be in English. In addition, he would only be presented with information relevant to his specific geographic location in the country were he to run a search for the nearest ATM, for example.

- Offline storage is required and included in the handheld computer client.

- Offline communication and replication is required and included in the handheld computer client. The reason for this is that all information entered in offline mode must be sent back to the back-end system.

- Every business function outside of messaging is structured in the same way. Each has

  - An offline component in the handheld computer. This component executes on the devices primarily when the device is offline.

- A Java Server Pages (JSP) component for each business function to generate the WAP and Web pages. JSP pages are a Java-based mechanism for implementing the presentation layer.

- A routing component in the business logic access layer to find the correct business function. This ensures that the presentation layer has no knowledge of the business component structure.

- A business logic component in the business services layer to execute the relevant business function.

- Transaction logic components in the transaction logic business services layer to find the correct data. These ensure that the business logic components don't know anything about the data and legacy system.

- The alerts component in the business services layer contains a trigger that can initiate a notification from the system. The trigger is responsible for running periodic checks to determine whether a notification is to be sent.

- A common service is used to access legacy systems, a relational database and an LDAP directory.

The rest of this section will give some examples of some of the business functions, including sample interactions between components.

Quote Generation

We will explain the quote generation functionality by describing the flow of a transaction launched through a handheld computer.

In the first scenario, let's assume that the handheld computer is online. The quote generation component in the handheld computer allows the user to enter the required quote parameters, as illustrated in Figure 20.7. These are communicated over HTTPS to the quote generation routing component in the general presentation layer. The quote generation component in the handheld computer passes it through localization and XML conversion components in the handheld computer formatting layer while sending it to the business logic access layer. The localization component is responsible for detecting information about the user: where the user is working, what device is being used, and so on. The XML conversion component converts the call into XML format.

The quote generation routing component executes the correct business functionality in the business services layer components. The logic is not necessarily contained in just one component—the quote generation routing com-

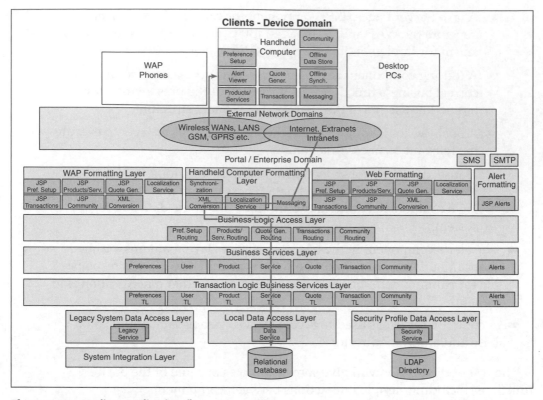

Figure 20.7 Online application flow.

ponent may call several functions in different components in the business services layer.

It may turn out that data is needed from back-end systems. If this is the case, the components in the business services layer send requests to the transaction logic business services layer. The corresponding component in this layer determines where the information is and how to get it. It sends the appropriate requests to the data access components, which then retrieve the information.

Once the quote generation component in the general presentation layer is finished calling the business logic, it returns the handheld computer formatting layer via the business logic access layer. Here, an XML stream is converted to a response the handheld computer can understand and send back to the handheld computer.

This kind of flow is standard for all the applications we are examining in this chapter. We will describe other aspects of the functionality using other types of flows that exist in the system.

Products and Services

The products and services functionality can be executed in exactly the same way as the quote generation application described in the previous section. We will not repeat this description. Instead, we will illustrate how this functionality would work offline using a handheld computer.

Figure 20.8 shows the application flow for running the products and services functionality offline. In order to peruse the information in an offline setting, we are assuming that the user has to run the functionality online once to obtain the data from the back end. Given that this synchronization has taken place, the user starts the application and runs the products/services component on the handheld computer. While the user works with the application, information is stored in the offline storage component on the handheld computer. When the user exits the application, the information remains there.

After using the feature several times during a day, the user decides to replicate with the back-end system. While connecting, the offline communication and replication component determines whether any offline data has been stored. If the answer is yes, it launches a transaction to the back end to

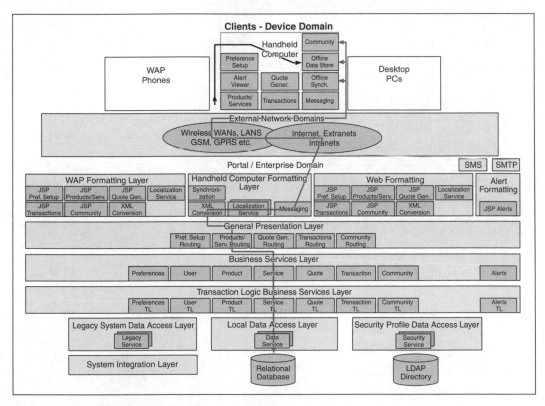

Figure 20.8   Offline application flow.

update this information there. This transaction will be the same as described in the quote generation flow, only using the products/services components in each layer.

Preference Setup

At this point, we have covered how different functionalities can be run using a handheld computer. The next feature will be demonstrated using a WAP phone.

Figure 20.9 shows how the preference setup process can be executed using a WAP browser. Let's say that one of the AMB users is giving a talk in London. Just before his presentation, he decides to use his WAP phone to change his preferences to vibration alerts so that the device doesn't interrupt his speech.

The only difference between this scenario and the proposal generation scenario is that the request from the WAP phone is sent to the JSP preference setup component in the WAP formatting layer. This component is used to translate the message and, using the localization component, to determine localization parameters for the user. The application now knows that the user is in London, using a Nokia phone model X.

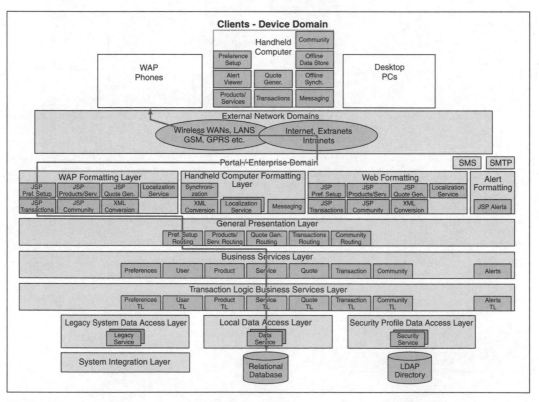

Figure 20.9 WAP browser application flow.

The JSP Preference Setup component in the WAP formatting layer reads and understands the request and forwards the information to the preference setup routing component in the business logic access layer. From here, the process is exactly the same as with the previous examples.

Alert

The alerts functionality is a little different from what we have previously described. The difference is twofold:

1. The alert that is being sent will not be sent back to the source of the request.

2. The system can initiate alerts on its own.

We will describe this scenario using a Web browser. Refer to Figure 20.10 for a detailed description.

An executive using AMB wants to find out when a certain security reaches a certain price. She has used the preference setup functionality to set this alert. The system periodically checks whether the stock price is at the specified

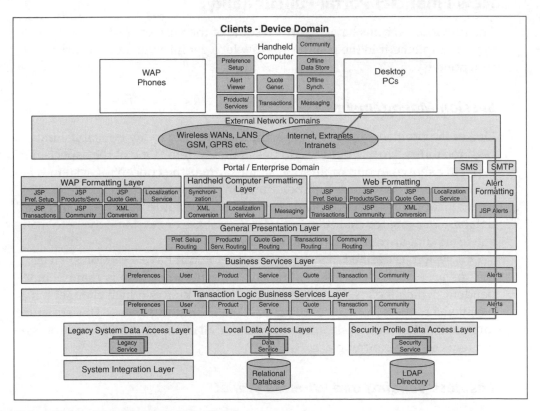

Figure 20.10   Web browser application flow.

price by periodically running the alerts component in the business services layer. This component executes the business logic to find out whether an alert should be sent. If an alert is sent, this component sends a message to the JSP alerts component. This component formats the request appropriately and sends it to the SMS server to be sent to the user.

The same process is used for email notifications. However, the requested message type will be email in this case, resulting in a formatted email message being sent via the SMTP server.

Transactions and Community

We have already gone through all the different possible scenarios in which users can interact with the system. Given that, we will not describe the transaction and community services.

However, going back to the point made in both Chapter 10 and Chapter 15, this is OK! In fact, we should be able to implement new business functionality in a predictable way. This is the reason we value architecture.

New Financial Portal Functionality

The previous sections have described some of the wireless functionality that we can implement in the near term. Now let's get into the more futuristic functionality.

Session Management

One of the problems many wireless applications face today is establishing high reliability. How would we do this in our system? One of the aspects not explicitly shown in our application architecture is the concept of system services. System services are common functions that are used to execute business functionality. An example of a system service could be drivers. You need database drivers, for example, to access a database or session drivers to manage users that are currently using your system.

Our system is defined in such a way that you can have a session such as the one previously described. All functionality in our system is channel independent, which means that the system no longer knows which channel it has received a request from once we get past the formatting layers. This structure implies that the same session can be used for all channels, as we are processing requests in a central place.

Cashless Vending and Wireless Wallet

From an application perspective, a wireless wallet requires that some application components execute on the client while others run on the back end. We have to have something that keeps track of our funds locally and a server component that can interact with the wireless wallet to authenticate itself and the wallet, to check the funds, and to deduct the appropriate amount.

The application architecture itself does not change—the same model as used with quote generation or any other functionality can be used. We can implement the back-end services in the application structure we have used so far and can implement a client application such as the ones that are shown on the handheld PC.

Summary

This chapter illustrates the benefits of a multichannel architecture. Several benefits have been described in some level of detail here:

New business functions and channels can be added in a predictable way since a clear architecture has been defined.

Adding new business functionality does not necessarily involve changing your existing applications and infrastructure. It may only mean that you are changing some components. For instance, if you want to rebrand a site, you can do this by changing the formatting layers only. The rest of the application layers will at least be very similar to what they were before, given that you are not adding any new business logic.

Wireless services and functionality are implementations of one of these channels, but the application architecture will support others as well.

Technical Infrastructure Architecture

Introduction

In Chapter 18 we saw how Acme Financial leveraged Internet technologies to achieve specific business goals including customer attraction and retention, cost reduction, and competitive positioning. We learned that adding a wireless dimension not only ported the applications' functionality to Acme's consumers on the go, but also, more important, added to the value received by the company's client base.

As a major mover in the financial markets, our client was by no means unfamiliar with the technical requirements necessary to provide for secure high availability and data protection that included linear scalability and consistent performance via a flexible infrastructure. However, the introduction of our wireless application did necessitate that we integrate the mobile components into the existing systems, ensuring that all functional, data, and informational requirements were still being met and were done so with the utmost measures related to security and system performance.

In this chapter we provide a brief overview of the entire system and the physical infrastructure related to the existing systems, data centers, and network architecture. We then look at the mobile elements that were integrated into the environment and the security systems and measures required for supporting them.

Technology Infrastructure

Figure 21.1 illustrates a high-level overview of the technology infrastructure, showing multiple databases with different business designs integrated and distributed on a global basis. This type of global integration is a central requirement to reach Acme's customer base. Although each geographic location has certain sets of business rules that are unique, many of the functional and information systems remain the same on a worldwide basis, often receiving the same data feeds related to market indices, company profiles, news feeds, analyst reports, and certain other third-party links. Similarly, account structures, profiles, and financial products and services are also the same at a high level, with specifics tailored to the unique geographical and political environments.

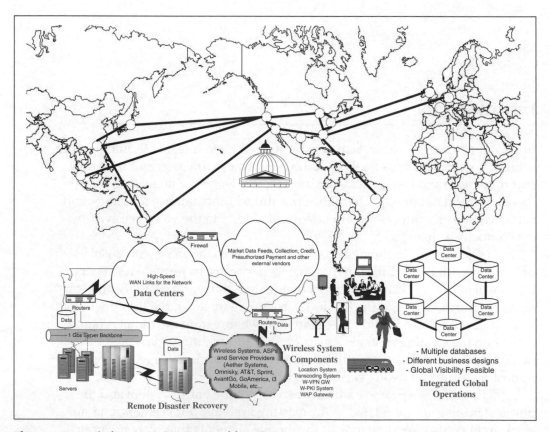

Figure 21.1   Wireless FSI infrastructure blueprint.

There were several mobile components involved in this solution. They include:

- *Location systems* for the delivery of timely customer information to both the account holder and the financial advisors, based upon their current geographic location.

- *Transcoding systems* for managing multiple devices and network protocols, including an onsite WAP server and the collocation of certain wireless application service providers' systems.

- A *W-VPN gateway* for cost-effective access into corporate systems by account managers.

- Multiple and redundant W-PKI and security systems.

In order to achieve a successful implementation of the wireless application, we needed an infrastructure that could deliver reliable information in real time. The distributed enterprise data centers had to function seamlessly. We were concerned with managing the flow of the time-critical information across a tethered and mobile-based, globally diverse community. As we mentioned earlier, our client had already done much of this work over many years and had done so in a way that provided high availability at the five-nines level (99.999) within a flexible, scalable, and secure infrastructure.

Our client's physical infrastructure consisted of the same three-tier architecture for enterprise storage, networking components, and data software supporting their application and data environments. Acme's move into the wireless space was predicated by and mandated a solid, bulletproof wireless security system. Our plan clearly addressed this mandate, but we cannot disclose details in this chapter. However, all of the components mentioned before in the area of security were present in our client's systems. The high-level view in Figure 21.1 shows the overall physical infrastructure that supports the applications. It is divided into three main areas: distributed and integrated data centers, disaster recovery centers, and the wireless systems and security components.

Our design efforts attended to the security requirements of our client's dynamic wireless business environments. Their highly available data center and network consisted of independent data centers connected to a distributed network of data and enterprise management systems. It anticipates inputs from a variety of wireless devices, carriers, data service providers, and, in some cases, different wireless application service providers. The designs were supplied a redundancy of hardware and software components that, quite simply, don't exist in most business environments due to cost considerations. In some cases our solution called for performing geographic load balancing (mirroring) across multiple data centers. However, even in this scenario certain portions were outsourced to collocated third parties, including a particular wireless application service provider with a global footprint. Our client's data environments provided for nondisruptive backup of their databases, using business continuance volumes (BCVs) and server-based I/O channel load balancing and failover.

Design Requirements

When we began looking at our client's technical architecture, once again there were many items we needed to assess. We needed to characterize and understand the current systems and how they interact with each other. We had to thoroughly understand the capabilities and requirements of the new solution and how these would best be integrated into the existing systems. Much of this type of work has been documented in the earlier TI chapters and we would refer the reader to those chapters for more details. What we want to emphasize here is that we conducted a very intensive and detailed analysis of current systems, sizing issues, and approval of software systems that would integrate into the existing environment while meeting specific performance requirements. Once this information is collected, it can be run through models to confirm or invalidate expectations and assumptions regarding the new structure. This modeling is very helpful and contributes valuable information into the development and testing phases of the project.

We were able to acquire a thorough understanding of the old and new data center and computing performance capabilities rather quickly because of previous work done with our client and ongoing maintenance and system documentation procedures they had implemented. We began compiling information about the physical systems components and configurations that existed and were required. We then introduced these into our modeling software. We needed to know the number of locations, the existing WAN and LAN links, the types of networks deployed, and the equipment and wiring supporting the infrastructure. We also needed to understand how existing applications were impacting the network in terms of protocols used, the QoS and types of class issues, and the times when applications were used. Fortunately, all of this information already existed and was provided to us by the client. We also conducted a similar survey of existing security implemented by our client and their service providers.

These activities enabled us to get an accurate view of the existing network and to document any existing gaps that needed to be addressed in upgrading the architecture. From here we mapped out our strategy for bridging these gaps and implementing the infrastructure upgrade to support the future state design.

Assumptions

As expected, when developing the recommendations we had to make certain assumptions. However, one assumption did assert that our analysis and models were accurate based upon the inputs from our client. These included network performance at various times of the day, week, and year; transaction volume estimates; topology designs provided to us; input from vendors and manufacturers; benchmarks for all of these systems; the number of reports; and types of information requests. Accurately modeling a large data environment such as our client's was no small task.

- We assumed that our physical component mapping to the application platform and network support was accurate based on information obtained from the client, product vendors, and the five nines (99.999) high availability mandated, already implemented, and measured by the client.

- ASPs were collocated with our client and might offer alternatives or replace some components but, in any case, would address the issues of high availability.

- High availability means 99.999 percent availability = 5 minutes of system downtime per year.

Let's now look at the various systems in one of the typical data centers. Figure 21.2 illustrates the major information system application servers and the data storage.

Data Center Architecture

In the SAN, network storage and switches are managed through the EMC ControlCenter. The ControlCenter provides centralized information management for the enterprise storage network systems. This comprises the monitoring, configuration, control, tuning, and planning capabilities of the information infrastructure, while keeping track of changing requirements. Status, configuration, and performance data, as well as key control features such as BCV control, are also available to system administrators through application program interfaces (APIs). Figure 21.2, which in fact is quite similar to the MMM data architecture, depicts the data center on a trunked gigabit Etherchannel backbone, supporting the LAN and WAN environments, and reflecting the SAN segmentation. Secure high availability was the major consideration over cost. Leveraging our due-diligence efforts from previous engagements, the SAN was installed with dual redundant systems, using the previously mentioned EMC, Cisco, and Oracle structure for the highest level of sophistication, flexibility, and scale.

In this environment, EMC software provides multipath access to disks and dynamic load balancing for enhanced performance. The software leverages the alternate path to increase performance by routing I/O requests on the fastest access path and avoiding delays that are a result of contention. It also protects against failure of a server I/O channel by failing over to a sec-

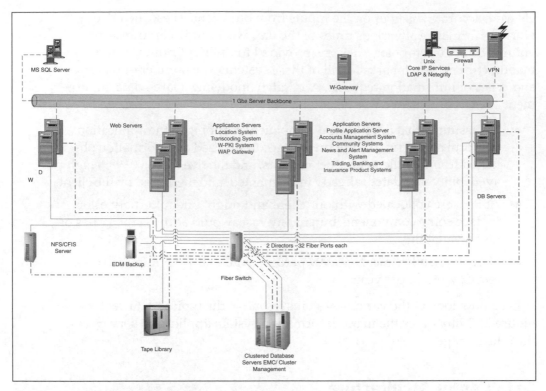

Figure 21.2 Acme Financial data center.

ondary channel and providing load balancing of the server I/O between channels.

It was important for the client to have point-in-time copies of the production database in order to off-load certain tasks. For example, here the copy is used to:

- Extract data for a data warehouse.

- Run reports.

- Test upgrades.

- Run database consistency checks.

EMC software provides for both a physical and a logical backup. Once the logical backup is complete, the entire contents of the database are exported and copied to tape. The backup host off-loads and isolates tasks from the production database host, resulting in greater system performance and higher availability of the data. Using this approach for maintenance tasks also enables greater frequency of backups and integrity checks to be run with less impact to the production database's availability.

NOTE ON SOFTWARE VENDORS

This chapter mentions several specific software applications and vendors, such as EMC and Oracle, relating to the technical architecture of the solution. We do not endorse or recommend a particular software vendor, but have listed these as examples.

The software also allows system and storage administrators to create independently addressable BCVs, in the background, for information storage without disturbing the production environment. This parallel processing capability offers workload compression and increases availability of the application via a resilient platform. Once BCVs have been created, they are split from their production mirrored volumes and used for backup. The BCVs again mirror to the previously paired production device once the backup is finished.

An Oracle Parallel Server (OPS) works in conjunction with a Cluster Manager (CM) to monitor the status of system resources and to provide redundancy via the ability to fail over host processes. There is another OPS entity called a Distributed Lock Manager (DLM) that tracks the status of a cluster node and informs the database about which node clusters are active. OPS protects against downtime should a node be lost. It provisions a node failover answer in case the primary host fails. The hot standby database instance takes over for the primary host without a lengthy start-up or data recovery. If an application is using a session with the Oracle Transparent Application Failover (TAF), almost no delay happens when reconnecting to the second host. In the data center we find primary and secondary database servers. The secondary, clustered host shares disk volumes with the primary host. The secondary host runs an OPS instance to provide node failover. The secondary database host, also running an OPS instance, will be idle most of the time. Therefore, it can be used to perform backups of BVCs and perform a variety of other simple tasks to offload the primary server. Also positioned in the data center is a server cluster for data warehousing routines.

Separate from the storage system are a data backup manager and tape library. This data manager is a centralized, high-performance backup and restore system. The system combines software, hardware, and support services to provide solutions to increase productivity and facilitate business continuity. The multiprocessor server handles the data transfer to an automated tape management system. A direct network connection is provided to keep this traffic off of the SAN. We also find a file server here for various nondata enterprise file storage. Our client has implemented it here to also serve as an NFS host.

Our client implements their cluster servers in groups of four to provide 75 percent production capacity in the event that a single node were to fail. This throughput is very important in situations characterized by high transaction volumes and in critical environments such as financial services. A great majority of our client's machines are on the Unix platform because of its robustness, resiliency, and open standards. Certain applications require NT machines, and these have been provided with their own VLANs in order to isolate their associated proprietary network protocols. This makes troubleshooting network problems much easier. A secondary firewall provides VPN services supporting wireless access to system applications. A WAP gateway is also provided. Security procedures required that the gateway and system servers be located on site behind the enterprise firewalls.

We did not use a single wireless network protocol due to the global nature of Acme Financial. As in previous applications, we had to be prepared to accommodate multiple different wireless devices, cell phones, PDAs, and notebook and notepad computers across multiple wireless networks and associated bandwidths and protocols. Thus, we again needed an extremely robust transcoding system. We brought in one of our partners, who, we believe, is the market leader in this space, to provide this service to our clients. It is important to note that we are really not just talking about Web scraping or XML transcoding. The vendor has designed an entire suite of software servers that are sensitive to the unique attributes of mobility and, as such, integrate with a host of middleware and enterprise application systems.

The ability to locate a customer or financial advisor, guided by alert notifications and specific user requests, in real time based on physical location, was paramount for our client. Again, a comprehensive, robust system was deployed and comprised several integrated tracking systems. This system also tied into other parts of the enterprise, customer profiles and business rules, and knowledge and event managers.

Security systems were implemented across a number of systems and global networks. A wireless private-key infrastructure (PKI) with the use of digital certificates was mandated by the business functional design. Let's look at some of the issues we had to consider in this arena.

Some of the specific issues associated with the mobile devices presently used by our client's customers and employees included slow interactions, limited processing power, and relatively low memory capacity. Yet another issue existed around the restrictions on exporting and employing cryptography. Round-trip times for transmitting data could become long and the connection might be closed, as a result requiring continuance to become part of the application design. A primary security objective was that the encryption protocol be lightweight and efficient with respect to bandwidth, memory, and processing power requirements, while at the same time remaining effective and robust.

As we have mentioned and defined elsewhere in this book, in order to provide secure transactions in any medium, there are four major components that must exist: confidentiality, authentication, integrity, and nonrepudiation.

One of the leading practices our client demanded to be implemented was the invisibility of security systems. It did not matter how the managed security services were delivered, or who delivered them, from the end user's perspective. Our client wanted us to see to it that their customers would never have to worry about security, wonder how it works or whether it is working, or even take extra steps to ensure that policies were being enforced. For the end user, security was to be automatic, invisible, reliable, always on, and up to date. We took several steps to make sure that this was the case. Implementations of various techniques and some modifications were necessary, depending upon the devices and network protocols being used. These included the following:

- Automatically ensuring all security requirements for the end user, in accordance with policies, types of devices, and network protocols. The end user does not knowingly download the latest antivirus software or scan his files or encrypt them each and every time; the system does that through and in accordance with certain defined policies and procedures.

- Transparently implementing security policies, scanning files, downloading the latest versions of antivirus software, and encrypting files. The customer doesn't know the system is working when using a portable device.

- Ensuring in a reliable way that hackers will not be able to pass off viruses and perform digital signature copies or corrupt files.

- Provisioning for an always-on $24 \times 7 \times 365$ status of protection. This is deployed extensively for all the actors within the enterprise's connected space.

- Server/gateway certificate services available in X.509 as well as WTLS mini-certificate, using both RSA and ECC technologies for certificate cryptography. Short-lived certificate service for WTLS server/gateway.

- Managed PKI services for clients, including digitally signed transactions from clients as well as client authentication. The wireless client must contain a private key and digital signing logic, but should not store or process a certificate for the corresponding public key. The certificate is stored in a repository directory within the wired infrastructure, and is picked up by application servers whenever a digital signature requires verification. With this approach, there is no concern about the size of the certificate. Therefore, standard X.509-format certificates, like those used in the wired Internet world, are also used for client key pairs in the wireless environment. This ensures that servers use the same functions to verify digital signatures from wireless clients as they do for wired Internet clients. The wireless client certificates are issued through a managed certificate issuance service provider and delivered to repositories for storage and use in the wired infrastructure. Certificates are also sometimes delivered to and stored on client devices or SIM cards.

- Wireless platform PKI enablement toolkits.
- Packaged services for wireless eCommerce (second phase).
- Certificate authorities (CAs) providing certificate services to both servers and mobile clients. The CA is hosted in a highly secure and reliable facility to ensure guaranteed service. It uses ECC and RSA encryption, providing X.509v3 and WTLS mini-certificates, using CRLs for revocation.

We undertook this mandated approach to security for our client, ensuring a mapped variety of duplicated and fault-redundant systems. In addition to these undertakings, we hosted many of the preexisting security measures already in place—some of which were discussed in the network architecture section of other chapters.

Network Architecture

Topology

Figure 21.3 illustrates the overall network infrastructure architecture that supports our client's environments. Once more, this topology is very similar to that described in previous chapters, aiming at providing the maximum throughput of data. The topology supports the flow of different data types that must be managed across this architecture. It provides minimum latency across the entire enterprise. It uses ATM in the distribution edge across many important WAN links to multiple data centers.

Multilayer Model

The network topology is the multilayer model that we have discussed in previous chapters and as illustrated in Figure 21.4. It supports the seamless transport across an environment where more than 80 percent of the traffic is not local. Layer 2/3 switching is provided with VLAN trunking. The distribution and core layers were implemented such that the network diameter is kept to a minimum of two router hops across the entire WAN in the majority

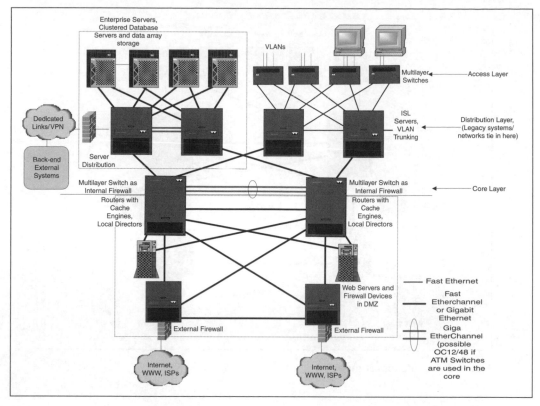

Figure 21.3   FSI network architecture.

of cases. This keeps data moving at near wired speed and reduces latency to within a few microseconds. This design model is inherently scalable. It is highly regulated in terms of performance and configuration. This supports troubleshooting and implementing upgrades to the network. It is highly structured and deterministic. It can support multilayered networks of differing protocols and provide throughput based upon content type and quality of service (QoS) needs. The network is designed to remain fault tolerant, robust, resilient, scalable, and high performing.

Security

Multilayered Approach

We have talked about installing security systems through a multilayered approach. The general principles expounded in the previous sections do hold true. However, we should not think of them as a catchall blanket for security. There are many facets to security. A good approach is to simply realize that

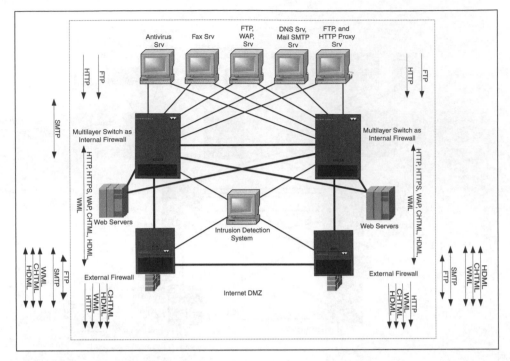

Figure 21.4 Multilayer model.

security holes do exist. Addressing those areas up front will not necessarily fully protect your systems, but will certainly provide a solid framework.

People have a tendency to think that they can avoid threats to security. For example, some individuals believe that a firewall stops unwanted traffic. It doesn't. A firewall is only as good as the person who manages it and sets up the access control Lists. Some people think that encryption algorithms are bulletproof, but we will see in the next section that they are not. Some believe that VPNs and IPSec offer the ultimate in security. However, VPNs are nothing but holes in our firewalls! The truth about security is that there is no way to avoid security threats. There are only ways to manage the risk.

One of the key lessons of this book is that security is an ongoing and never-ending process, and that a breach in security can happen at any time and almost anywhere in your organization and its business processes. The best that can be done is to educate people, especially IT professionals, to the many types of risk they need to be aware of and control. Fortunately, security risk is manageable; otherwise, we would all be out of business! So don't let us scare you completely, but do let us motivate you into action. Our financial clients know this and treat it as such. Let's examine some brief topics to keep in mind about mobility and security.

Where Is Your PDA or Mobile Phone?

People use their PDAs like little black books. They are full of personal treasures. Phone numbers such as 800 numbers to remote access points can be found on them. Personal identification numbers and a host of passwords are found in them. Bank accounts are on them, social security numbers, and addresses of loved ones. We use these devices to store time-critical information related to our most current business activities. They are our business calendars. So what else is new? Haven't we unwittingly been doing similar things like this with our notebook computers for a long time? The answer is yes but . . . PDAs and mobile phones are small. Do you know where your mobile phone is right now while you are reading this book? Imagine a system administrator who has accidentally misplaced a PDA containing an unencrypted list of passwords!

PDAs and mobile phones, for the majority of cases right now, are also unprotected. If anyone had your PDA, would they be able to look at your bank account, contact addresses, or personal calendars? These devices are generally not secure by default. Thus, wireless devices are potentially the largest new threat to enterprise and personal information security. These devices are small in size and relatively cheap, which makes them very easy to buy and equally very difficult to control. Analysts predict that by 2003, worldwide there will be more than 1 billion smart devices connected wirelessly.

During 2000 a Trojan horse called Liberty Crack was released that attacked devices working on the PalmOS, including Handspring's Visor. The code has the potential to wipe out all programs stored on the device. In addition, joke files have started to spread across PDAs and cell phones. Seemingly harmless activity at this time, joke files compromising security are only the tip of things to come. Because people are quick to share jokes, these files can easily provide a means for other, more hostile types to launch an attack.

How do you protect against critical data walking out the door with people who are carrying the PDAs in their coat pockets? Setting policies for the proper use of handheld devices will be a tricky proposition. This is even more difficult if the device belongs to the individual and not the organization. As you can see, cutting back on expenses for handheld devices might not be the most cost-effective policy in the long run. A disgruntled employee or a corporate spy could easily synchronize with a desktop computer and walk away with up to 2 GB of data in his or her PocketPC. Industry experts refer to this as *promiscuous synchronizing*.

PDAs are great devices for spies. A PocketPC with an 802.11b card can quickly become a great sniffer without having to gain interior access to a building. A hacker group called L0pht Heavy Industries has created a war dialer for the PalmPilot. This program allows one to scan a phone system or a range of phone numbers for any modem that answers its call.

Taking It All Seriously

The U.S. Department of Defense is very concerned about small smart devices. The department is conducting a top-down security review for the use of PDAs, two-way pagers, cell phones, and notebook and palmtop computers. This review is part of an initiative in the Pentagon to implement tougher security measures. We should all take wireless security very seriously.

Our financial client recognized the criticality and requested us to implement a multilayer approach to security across their entire enterprise and particularly as applied to the wireless devices used by their clients. Earlier in this chapter we mentioned some of the systems we helped them deploy. And although there are others we cannot divulge, we want to make you aware of the following:

- Change network security codes often. Default codes should be available only to a trusted and insured third party.

- Isolate the access path users will take to your network. This reduces the amount of activity that is visible.

- Monitor access logs because they point to source addresses and make it easier to identify attempts to penetrate network login security.

- Implement media access control (MAC) address tracking.

- Implement a multilayered approach across the entire enterprise in both technical and nontechnical systems and business processes.

- Manage your security risks in an ongoing and proactive manner. These risks change for every organization and depend on the unique attributes of the enterprise.

Again, our intent was not to scare you away from implementing a wireless network. Your organization will very likely need to do so in order to remain competitive. We simply want to stress the up-front and ongoing work that must be applied in the area of managing security risks.

Summary

In this chapter we looked at the entire data system and physical infrastructure for our financial services application. We then described mobile elements integrated into the environment and the security systems and measures required to support them.

As we had mentioned earlier, our client is a major mover in the financial markets. The company certainly was familiar with the technical requirements necessary to provide for secure high availability and data protection. Their systems were already scalable and flexible in an infrastructure that could deliver consistent performance. Many of these systems had already been deployed, maintained, and well documented. However, the introduction of our wireless application did necessitate that we integrate the mobile components into the existing systems, ensuring that all functional, data, and informational requirements were still being met with the utmost attention to security and system performance.

This concludes our discussion dealing with a financial service industry application and the methodology followed to build it. Part 5 of this book entails architecting a mobile office solution.

PART

Five

Architecting Wireless Mobile Office Applications

CHAPTER 22

Wireless Mobile Office Overview

Introduction

The wireless mobile office continues an evolution that has been under way for several years now. Unlike many of the wireless solutions described in other sections of this book, the wireless mobile office is not specific to any single business process and therefore is not a single *application* or type of application. The concept, rather, deals with enhancing office technology infrastructure to blur its limitations or to gain independence of physical location altogether. This chapter provides an introduction to how mobile and wireless technologies can be used to extend the reach of the traditional office environment and, in that process, add independence of location for business people constantly on the move. The chapter will give an overview of office infrastructure from the information technology perspective and show how wireless technologies can be applied to provide business benefits, given today's mobile business climate.

Traditional Office Infrastructure

Office automation tools have been available on most office workers' desks for over 10 years. Pioneering software including WordStar and WordPerfect led the way toward enhancements to office productivity when the task of producing documents went electronic in the 1980s. During the same period, products such as Microsoft Multiplan and Lotus 1-2-3 took spreadsheets to the next level and made the analysis of numbers available to those users who worked with PCs.

In the 1990s, word processors and spreadsheets were complemented by graphics packages that virtually turned ordinary office workers into graphic arts specialists, providing them with powerful yet simple-to-use features to create presentations and other graphics-intensive documents with near-professional quality. In the early 1990s, vendors such as Microsoft and Lotus jumped on the opportunity to create and market attractive software bundles, consisting of software suites that contained their popular word processing, spreadsheet, graphics, and personal database applications. These bundles of office automation tools provided users with a similar look and feel and good interoperability between the individual components.

To summarize, the software evolution has transformed the personal computer from a specialist tool to a ubiquitous productivity booster. Whereas back then it was used primarily by secretaries and professionals such as data analysts, the PC nowadays has become a basic office fixture similar to pens and paper.

Today's comprehensive office infrastructure generally provides a vast range of services including:

- Word processing
- Spreadsheet
- Presentation and graphics
- Email
- Videoconferencing/online meeting tools
- Shared calendar
- Document management
- Intranet
- Internet
- Knowledge management tools

To simplify further discussions and throughout the rest of this chapter, the first three applications will at times be referred to as *office automation tools*, the next two as *communication tools*, and the rest as *information management tools*.

Challenges in Today's Business Environment

During the same period that witnessed the evolution of office automation software, the general business environment became increasingly global, reflecting rapid advances in the information technology, telecommunications, and media sectors. Business managers need to be able to quickly react to changes in this type of environment. They need to constantly fine-tune their organization to avoid losing market share or shareholder value. In today's environment businesspeople seem to always be on the move. More time than ever before is spent in airports, waiting for delayed flights, in taxis, and at hotels. The frustration of being stuck in such a situation is easily exacerbated, as the individuals who are traveling are likely to be under significant pressure, trying to cope with ever increasing workloads. Since agility has become such an important aspect of successful business management, it is crucial to be able to respond rapidly to internal or external matters. This means that there is a significant demand from businesspeople all around the world to gain access to the services provided by traditional office infrastructures. Figure 22.1 shows a sample of environments where an extended reach of the office environment would be highly beneficial.

During the last 15 years mobile phones have also become rather ubiquitous. Although these phones have offered some relief for folks who spend a lot of time away from the office, the usefulness of these devices was fairly limited. In the next section of this chapter we will see how more recent advancements in wireless technologies have improved mobile communications.

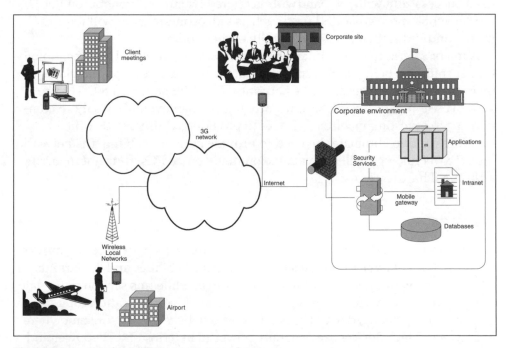

Figure 22.1 Wireless office usage scenarios.

What Is Needed to Create the Wireless Mobile Office?

This section will shed some light on the wireless technologies presented in Chapter 3 of this book, specifically in the context of wireless office applications. As much of the concept of the mobile office is about extending the reach of the office infrastructure, we will also provide an overview of what is needed at the back end to enable that extension.

Wireless Mobile Office Concepts

The core idea behind the wireless office is to extend the reach of the traditional office infrastructure to include virtually any location that is covered by a wireless network. At the time of this writing, major carriers around the globe are preparing for the rollout of third-generation (3G) cellular networks. While these will play an important role within a few years, existing digital cellular networks like GSM/GPRS can be used today to enable at least parts of 3G functionality. Although fairly limited in bandwidth, these technologies have good geographic coverage in most countries, and the phones to use with these networks are readily available. Some of the wireless mobile office applications, namely email and basic intranet and Internet access, can be used successfully with the bandwidths offered by current-generation networks, while more advanced uses such as videoconferencing will require more bandwidth than what is generally offered today.

A major alternative to cellular phone networks is wireless LAN technology that currently is being deployed by operators as a service in public places, including airports, hotels, and restaurants. The IEEE standard 802.11b provides 11 MB of bandwidth, significantly higher than even 3G. As more companies are investing in wireless LANs to provide mobility inside office buildings and to eliminate the need to provide network cabling, the network interface technology will be available on many office PCs, thus not incurring additional hardware costs.

Devices

Going hand in hand with the progression of wireless technologies, we are seeing a plethora of device types, each with its special abilities and constraints. Chapter 3 provides general guidance in this area, while this section will relate the qualities of various device types to our three office application categories.

Generally speaking, device types can be said to be *thick*, *thin*, or somewhere in between. Thick devices are generally powerful in terms of CPU power and storage capacity. Thus, they can host all sorts of office applications and be expanded to use many types of network technologies. The most obvious example of such a device is the portable PC, and anyone used to carrying such equipment knows that they are usually far from being as lightweight as one would desire. On the other hand, modern mobile phones tend to be highly portable, with form factors the size of a deck of cards, and weight accordingly. Of course, these phones lack most of the functional qualities of the laptop PC, but they can still provide some usefulness in conjunction with certain wireless office applications. Positioned somewhere in between these two extremes are the PDAs. Although limited in computing power and storage capacity compared to PCs, they still have enough muscle to run simple office applications while being able to allow for some local storage capacity. Several developments in the marketplace seem to indicate that the latter two types of devices are merging. A couple of the mobile phone vendors have models in their product lines that include quite advanced PDA functionality. Similarly, PDA vendors such as Palm have released versions of their models that include wireless connection capabilities. As we move forward, the lines between different device technologies and their capabilities are getting more and more blurred.

Generally speaking, though, we have to face a trade-off between mobility and functionality. These trade-offs have an impact on the investments needed at both the device and the back-end level. Figure 22.2 illustrates the relationship between portability on the one hand, and infrastructural investments on the other.

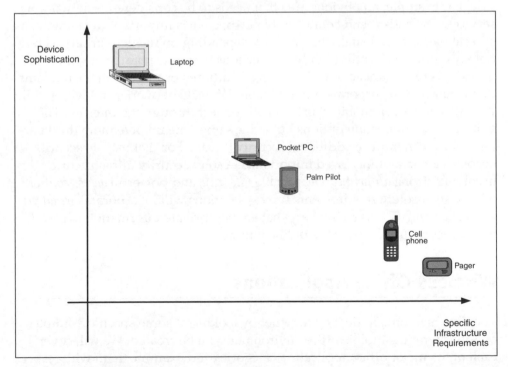

Figure 22.2   Portability versus functionality.

What this picture really shows is that powerful devices like laptops and advanced handheld devices generally require relatively small investments on the back end because of their inherent capacity to run the same applications as those designated for the PC. In addition, PDAs generally are delivered with out-of-the-box tools to synchronize locally stored with centrally stored content. On the other hand, because these devices are quite advanced, they are more expensive on a cost-per-unit basis. In addition, they are also more complex to use than most of the less advanced appliances. The more portable devices such as cell phones are usually cheaper per unit and may be expanded in functionality as technologies such as WAP gain a foothold in the market. Still, these types of devices may require more infrastructural investments due to their limited capacity in terms of screen real estate, memory, and computational power.

As a general rule of thumb, then, this means that the cost advantage of using mobile phones might be offset by the infrastructure investments required to support these devices. The next section will provide some insight into what kind of infrastructural support wireless office applications might require.

Back-End Requirements

To facilitate mobile access and use of the office infrastructure, depending on the device used, we might need expanded back-end support at the server side. For example, an obvious area that needs to be considered is networking services. The end-to-end connectivity perspective is important to maintain, as several parties will usually be involved depending on the networking technology employed. Until recently, connections to the corporate network from outside were usually achieved by using traditional or mobile phone modems that connected to corporate modem pools. With 3G networks and with public W-LANs, however, mobile workers will connect through the Internet. This will increase connection options but at the same time involve more third parties, which can impose additional security threats. For that reason, security becomes an area of increased importance. A solid security infrastructure involving strong *authentication*, reliable *integrity*, and powerful *access control* will be an absolute requirement. Typical solutions will incorporate *virtual private networking* (VPN) technology that enables private information to travel inside secure tunnels over the public Internet.

Wireless Office Applications

The time has come to tie it all together by looking at some specific examples of how a wireless mobile office environment can be created. We will cover each of the major office application categories and examine them with respect to wireless applicability.

Office Automation Tools

Timely access to information contained in documents is one of the most important drivers and requirements for mobile office implementations. Traditional office automation tools can be made available to mobile users quite easily. The simplest and most straightforward way of accomplishing this involves using standard laptop PCs accompanied by some sort of network access hardware, usually a mobile phone with built-in modem or a wireless LAN PC card. Some manufacturers also offer PC card-based cellular network modems that can be used with older phones or even as stand-alones.

A higher degree of portability can be achieved by using handheld PDAs. Users of these devices can, in most cases, gain access to scaled-down versions of popular office automation software for word processing, spreadsheets, databases, and presentation graphics. The Microsoft PocketPC-based PDAs generally include these office automation applications right out of the box, whereas their PalmOS-based brethren do so only in some instances. Most handheld computers are delivered with synchronization hardware and software that effectively enables PDA users to take parts of their documents with them on the road. Because it likely will take a few years before wireless networks deliver the reach and bandwidth required to gain completely reliable access to critical information, it will be necessary to locally store data including email and documents of various sorts to ensure information availability and to keep bandwidth usage at reasonable levels. Locally stored data will in turn need synchronization services to make sure that local updates are reflected centrally, and vice versa. Documents are synchronized with the user's desktop or laptop PC, using hardware usually referred to as *cradles* and synchronization software specific to the vendor of the device.

Communication Tools

One of the most frequently requested wireless applications is email. This should come as no surprise, since much of today's business correspondence relies on email as the prime channel. As there are a few important standards that are applicable to mobile use of email services, support for remote access to email is one of the easier services to implement. Devices ranging from advanced mobile phones to PDAs and laptops have generally built in support for access to email accounts using standard Internet email protocols like POP3 or IMAP4. In addition to that, many PDAs on the market today can access corporate email systems such as Microsoft Exchange—in some cases directly and in other cases through synchronization with a personal computer.

Another technique is to provide email/SMS gateways. These gateways are offered by many mobile operators around the globe. The way they work is that an email message sent to an account is automatically forwarded as an SMS message to the mobile phone.

Despite the communication and collaboration functionality provided via simple email, instant messaging, and chat applications, full access to video-conferencing or virtual meeting and other collaboration tools such as shared calendars can be much more difficult to attain on any type of wireless device. Obviously, access to the personal calendar is provided as a standard feature of any Personal Information Management (PIM)- oriented PDA, but the ability to access other people's calendars requires online access to a shared calendar system. Active use of such applications generally demands higher bandwidth than current 2G cellular networks, which means that in practice this can only be accomplished using a laptop/notebook PC or an advanced PDA to connect to the corporate network through a public W-LAN service or a high-speed cellular network. Even more bandwidth is generally required by virtual meeting tools, but once the more advanced network technologies become widely available, we will likely see widespread adoption of such applications in wireless forms.

Information Management

It's usually very difficult to provide office automation support to any device less advanced than a good PDA or a laptop/notebook PC, especially if editing capabilities are required. For example, shared calendaring and document and knowledge management are applications that by their nature require a device to have solid storage and processing capabilities. Other services such as access to intranet and Internet content are usually quite easy to enable at the network service level, but the utility of such access will ultimately depend on whether the content can be presented in a useful form. Because there are major variations between different device types in terms of how much content can be displayed on the screen, presentation layer services will need *clipping* capabilities. Clipping basically allows devices like WAP phones and PDAs to view HTML-based content by modifying the information to fit the small screens and limited graphics display capabilities of these devices. Yet another service included in the information management category is the executive dashboard application that provides its users with critical information regarding a company's business operations. We will explore the dashboard feature in more detail in the following chapters.

Other Applications

One of the positive side effects of implementing wireless mobile office services is the fact that once the infrastructure is established it can be used for many types of wireless services. Additional applications may include:

- Time reporting
- Expense reimbursement

- Enterprise applications such as enterprise resource planning and supply chain management packages

In some cases, this kind of mobile access will require significant changes to existing applications. This will be true especially if the applications are old and lack Web browser interfaces. If the applications are modern and provided by a major package vendor, chances are pretty good that some level of mobile access is supported out of the box or at least as an option. Major vendors like SAP and Siebel have already invested heavily in this area and the situation will naturally improve over time.

Summary

This chapter gave a brief overview of mobile office concepts, dividing them into three categories: office automation tools, communication tools, and information management tools. We looked at a couple of examples regarding how wireless may enhance these solutions by extending the infrastructure's reach. Given the various challenges that an increasingly mobile workforce faces, some of the benefits of wireless solutions include:

- Better use of waiting time
- Improved decision-making capabilities
- Timely access to accurate information
- Reduced response time to requests or exception reports
- More agile business management as a key performance indicator and other important information are available at will

The extent of a wireless mobile office application depends largely on the required services and functionality, driving the choice of device upon which to deploy the solution. The more sophisticated device types, the thick clients, will be less portable, and their acquisition cost will be higher while back-end support will be easier to provide. The less sophisticated devices, on the other hand, are less expensive, yet will likely require more attention to back-end requirements.

By and large, the reach of the traditional office infrastructure can be extended significantly by leveraging wireless technologies. More or less all office-related applications, ranging from the usual office automation via information access and information management to communications tools, can be provided on a handheld device. Chapter 23 will introduce a real-life case study to illustrate wireless business, systems, technical, and information systems architectures.

Conceptual Business Architecture

Chapter 22 provided us with an introduction to the wireless mobile office, reviewed traditional office applications, and spent time addressing some of the challenges inherent in today's increasingly complex and global business environment. We looked at some wireless concepts, briefly discussed devices, and offered a categorization scheme that separated wireless office applications into three major categories: office automation, information management, and communication and collaboration tools (see Figure 23.1).

One of the applications within the information management category is the *executive dashboard*. This and the following chapters will use the executive dashboard to illustrate a wireless office application. Whereas the ability to work on spreadsheets or to fill out and submit a time and expense report while on the go are valid applications whose functionality is significantly enhanced by adding the wireless dimension, our executive dashboard case study nicely illustrates a few of the more traditional mobile office features while allowing us at the same time to explore critical functionality that adds tremendous value to the application's user and organization.

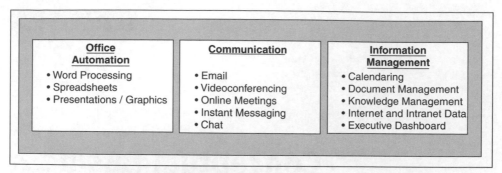

Figure 23.1 Wireless application categories.

Executive Dashboard Overview

An executive dashboard is a management tool that provides an executive with key corporate performance measures. By having critical performance indicators at their fingertips, executive decision makers can quickly gauge the state of a particular business process and initiate action. Although capable of showing metrics for the company as a whole, the dashboard usually focuses on providing information specific to the company's individual business units. The dashboard monitors the performance of such business units, and displays the data in aggregate form, sparing the dashboard user from getting lost in the data.

Presentation Modes

The preferred information presentation mode is highly visual; key performance metrics are frequently displayed via graphical representations, including bar, column, line, and pie charts. If the executive desires a more detailed view, he or she can drill down into the supporting data structure and access the raw data that provided the input for the graphic. Figure 23.2 illustrates a few examples of dashboard visuals.

Because the dashboard is a tool whose purpose is to provide meaningful insights into critical business processes, it must do more than simply relay performance data. To become a truly useful application, a dashboard is configured prior to usage with a set of business rules. These business rules provide the intelligence that triggers alerts or offers other visual clues as to the state of a particular measure. For example, it surely would be nice for an executive to be able to use a wireless dashboard application to view this week's inventory turnover statistics. However, without any business rules the data is not formatted yet for executive usage. For the executive to be able

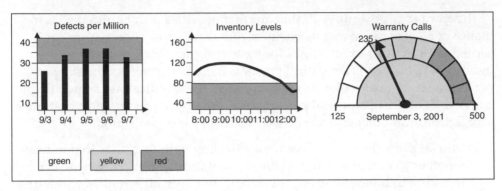

Figure 23.2   Executive dashboard graphics.

to immediately gauge a situation and determine whether he or she needs to take action, the data must be massaged to allow for the rapid extraction of meaning.

To be able to zoom in on what implications are hidden in the raw data, the dashboard must be preconditioned with a set of business rules or business targets. These business rules specify what is considered *standard* and what constitutes *out of norm.* Any deviation from the standard can then be highlighted in the presentation of the data. In other words, the business rules determine how data is displayed. Usually, green color denotes data that is within specifications, whereas red is reserved for data that is out of the normal range. In addition, the business rules frequently determine when data is presented. For example, if a process is out of spec, an alert might be triggered that pops up on the user's screen.

Key Performance Indicators

When explaining the functionality of an executive dashboard, we like to fall back on an analogy often used in textbooks. Think of an executive dashboard along the lines of an instrumentation panel of a 747 jumbo jet. The pilot at the helm relies on his or her dials and gauges to indicate altitude, speed, fuel level, direction, and so on. Yet there are way too many gauges in the cockpit to constantly monitor. If needed, all the data is in reach, but to make the panel truly useful, the pilot relies on its exception-reporting capability. When something goes wrong—say, a sudden drop in cabin pressure occurs—a warning light will light up. Similarly, when fuel runs low, a warning sound is emitted. That's the true value behind the dashboard. Alerting the user—in this case, the pilot—about an event that deviates from the norm and allowing him or her to take immediate action is what separates critical information from supplementary data.

To return to the executive dashboard that provides insight into the performance of a corporation and its business units, the application we are concerned with in this section allows for measuring specific key performance indicators (KPIs). KPIs are critical business metrics that reflect the state or performance of a particular business process. These indicators include financial performance, operational performance, employee productivity, customer satisfaction, and other critical success factors of the organization.

Financial performance.   Here we are dealing with measures that indicate whether a company's strategy implementation and execution are contributing to bottom-line improvements. In effect, all activities within the organization will ultimately impact financial performance, which means our financial metrics give the executive a high-level, bird's-eye view of the health of the company. The financial metrics our dashboard might present include cash flow, sales growth, market share, burn rate, and accounts receivable (A/R) collections.

Operational performance.   This, on the other hand, captures more specifically how we are doing in terms of executing our strategy to deliver value to our customers. Metrics being measured may relate to production (equipment utilization, up-/downtime, scrap, rework, defects per million, etc.), warehousing (inventory turnover, spoilage, picking accuracy, inventory accuracy, order cycle time, store time, returns, etc.), or distribution efficiencies (routing, scheduling, load optimization, on-time deliveries, etc.).

Metrics.   Metrics presenting employee productivity allow the executive to quickly gauge the efficiency of the workforce. A human resources (HR) professional may be especially interested in keeping tabs on the organization's staff levels, retention, attrition, sick days, on-time arrivals, time and expense reporting, vacation schedules, and so on.

Customer satisfaction.   Customer satisfaction measures may include returns, call center contacts, time to resolve, satisfaction ratings, number of product inquiries, account growth, and account attrition.

After this brief introduction to executive dashboards and how they differ from generic reports or data, we are ready to introduce our case study, to be used in subsequent chapters.

Acme Toys: A Case Study

Our client—let's call the company Acme Toys—manufactures action figures for children, young adults, and avid collectors. Acme does not operate any retail stores, but distributes its wares via large retail outlets and toy stores across the nation. Acme Toys has office locations and warehouses across the country, whereas manufacturing takes place primarily in Asia. The Acme organization in the United States is involved mainly with the design and marketing of the action figures, often in conjunction with major movie studios from which many of Acme's best-selling characters are licensed. The company provides its manufacturing facilities with blueprints for the plastic injection molding process, costume style, and fabrics, as well as other accessories that come with the figure. Finished products are boxed and crated predominantly in Asia, although some of Acme's toys require final assembly at a few Acme locations in the United States. Coming from Asia the product is shipped to six Acme-owned and -operated warehouses in strategic locations throughout the United States, from which individual retailers are serviced on a geographic basis.

Business Requirements

Acme's headquarters are located in Los Angeles, yet the entire organization consists of various locations that have been added over the years, usually via acquisition. The organization is growing at a healthy pace and expects to incorporate additional distribution centers within the United States and Canada over the coming years. Whereas Acme's top executives used to be able to manage their organization mostly from headquarters, the recent and upcoming expansions as well as a growing customer base will require increasing travel. Existing facilities, acquisition candidates, and customer organizations, spread throughout the country, require Acme executives to be on the road more frequently, away from their desktops at corporate HQ. With the increased travel requirements, however, come those challenges associated with time spent at airports, waiting for flights, living in hotels, and riding in taxicabs.

In anticipation of an even larger network of facilities and customer location, Acme realized that its traveling top executives would have to be provided with means to stay in touch with the organization, and vice versa. Making occasional telephone calls or checking emails via dial-up connections would not suffice to deal with critical operational issues in a time-effective manner. The decision was made to equip the executives as well as other mobile key employees with an executive dashboard, run from a mobile device. The application would allow Acme executives to do the following:

- Stay in touch and communicate with the office in a real-time manner.
- Work on office applications such as email and spreadsheets.
- Gain access to Internet information sources.
- Review corporate key performance indicators on a regular or exception basis.
- Take corrective action if required.

We will return to these business requirements and how the wireless dashboard addressed each in the *Acme Dashboard Benefits* section of this chapter.

Project Challenges

This project, unlike the others described in this book, did not start with the development of a Web-only application. Instead, CGE&Y was called upon to take an existing dashboard application that had been deployed at our client for years and extend it to the wireless realm. There were several implications that the project team had to deal with. First, because the team had to work with a legacy application, it would be severely constrained by the way the application was architected in the first place. Having been built in pre-Internet times, the executive dashboard was not configured with the notion of ubiquitous access in mind. The application was a proprietary solution that was well integrated into our client's back-end legacy system, yet it lacked adaptors that would provide for future functionality to be integrated easily. Lacking provision that would render the application flexible and scalable, the legacy dashboard had to be significantly modified to allow for a wireless dimension. Although our client considered replacing the homegrown solution with an off-the-shelf package, it was decided that the amount of time and energy that would be required to integrate such a new package with Acme's legacy application would eclipse the resource requirements to modify the old solution.

The second challenge faced by the team involved the extent to which the application was to be made available to our client's employees. Whereas the traditional application was available only to C-level executives, our client wanted the mobile version also to be available to traveling sales directors, sales representatives, and other key representatives at the senior and middle level of management on the road. Extending access itself was not so much a problem, as the dashboard could easily serve data and information to multiple concurrent users. What was tricky, however, was to design and build a security scheme that granted access to content based upon the user's clearance. All data and information accessible had to be filtered according to a user's authorization level. For example, highly confidential data, such as employee compensation records and performance evaluations, had to be effectively shielded from sales representatives, who really only needed access to sales-related data and customer information.

Third, because our client operated from several locations in the United States as well as overseas, the global integration piece added another dimension of complexity the team had to carefully address. Whereas the tethered version of the dashboard allowed access only via corporate terminals located in a few executives' offices at Acme's headquarters, the mobile application had to serve a large number of individuals across time zones, states, even continents. The global aspect required the team to coordinate with client offices around the world regarding which features would be offered, which information would be critical to various stakeholders, and how such information would be relayed.

At first glance, this activity might seem somewhat trivial. After all, an alert announcing a delayed shipment of raw materials could very easily be sent to the vice president of manufacturing, who would then pick up the phone and take corrective measures. But, why stop there? Adding the wireless capability and opening the dashboard to a larger community of Acme employees could be used to really drive value from the application. Because critical data could be relayed to all parties involved, the dashboard provided a means to significantly shorten the reaction time to out-of-spec events. To use our delayed raw materials shipment example again, truly leveraging worldwide capabilities meant not only alerting the VP of manufacturing, but also notifying the general manager of each affected production facility so that they could adjust production levels, notifying the raw materials procurement manager to look for comparable materials elsewhere, to notifying the VP of customer support to prepare a statement announcing a possible delay in finished goods shipment to the company's VIP accounts, and alerting the VP of sales to put a break on those go-getters who were out in the field selling the finished goods whose production might be interrupted. Coordinating this type of information flow, routing the alerts, allowing each recipient to respond with an appropriate action, and then routing these actions to all affected parties required a massive effort from a work flow design perspective.

A Day in the Life of an Acme Executive

Let's look at what the application can do. We'll again start with an overview of the wireless dashboard application by presenting a figure that captures the enterprise-internal data back end and dashboard front end. Figure 23.3 illustrates the various features provided by the solution.

From the diagram you can see that the primary components of the dashboard include access to both the Internet and Acme's corporate intranet. In addition, there is a feature that supports traditional personal information management, such as a shared calendaring tool. Of course, we also find a communication feature that entails cell phone connectivity, there is a routing and approval mechanism, and we find reporting features, including regularly produced periodic reports, ad hoc reports, and exception alerts. We'll describe the benefits of these application components in a minute. But first, let's have a little fun.

If you remember, the case studies in the previous parts of this book presented the content in such a way that we first looked at the Web-centric version of an application, its functionality, and its benefits. After we had

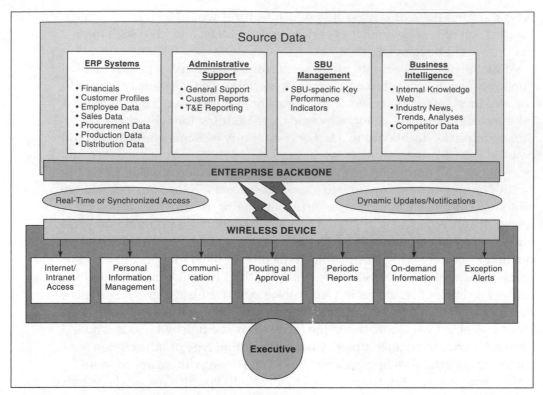

Figure 23.3 Wireless executive dashboard application.

explored the tethered version, we would then take a peek at the wireless solution, exploring how application components translated to the mobile dimensions and outlining the benefits associated with this adaptation. We also discussed entirely new functionality that was enabled via the wireless medium and that did not previously exist in the wireline version. However, in this part we are going to do things a bit differently.

For this case study, we are going to use a new presentation approach. Instead of describing the application along the lines of its functionality in a stand-alone fashion, we will portray a day in the life of a busy executive on the road. We will follow this executive throughout his day and explore how the wireless dashboard application enables and facilitates the various work activities that make up his busy schedule. The purpose here is twofold: Not only do we have the opportunity to discuss what the dashboard can do, but we also witness the application in a real-life setting, which allows us to relate functionality to critical business requirements. Let's get started.

Mr. Smith is a senior VP with Acme Toys, a company in the business of designing, manufacturing, and selling action figures and other toys. It is Thursday at 7:00 A.M., and Mr. Smith is already at the Los Angeles airport to catch a flight to Dallas, Texas, where he will spend the day delivering a speech about past-quarter results to the local office's senior executives, inspecting the Dallas warehouse, and later meeting one of the company's VIP accounts for dinner. A full day, no doubt. Let's see how Mr. Smith uses the executive dashboard application as he progresses through the day.

First, we should establish the hardware that Mr. Smith is using. Because he likes to travel light, Mr. Smith carries a duffle bag with some overnight accessories and a change of clothes, but he left his laptop computer in the office. Instead, he brought with him a PocketPC, a wireless modem, and a spare battery. The PocketPC, as we will see later, is Mr. Smith's PDA of choice because it not only allows him to work with traditional PDA applications such as calendaring and email, but also provides him with access to various full-color Web sites; allows him to work with word documents, spreadsheets, and other email attachments; and permits him to receive information from his company's back-end applications, including Acme's ERP system and other packages.

7:10 A.M. At the gate, Mr. Smith checks into his flight to Dallas and is delighted to learn that there would be no delays this morning. Still, he has about 30 minutes before the boarding process would begin, and he decides to quickly check his email. Because of the wireless modem attached to his PocketPC, Mr. Smith is able to log into the corporate email system and retrieve a batch of messages that had accumulated from the day before. In addition to these emails, Mr. Smith retrieves the latest toy industry news and a few articles that cover news about Acme's competitors. Because Mr. Smith previously had set up a personal news profile that automatically searches multiple Web sites for news containing certain keywords specified by Mr. Smith, he doesn't need to spend much time surfing the Web for such intelligence, but instead just downloads whatever articles the system found on his behalf.

While the system performs the news download in the background, Mr. Smith diligently answers several of the emails. Just before he has to board the plane, he remembers to download the latest monthly performance report for the Dallas location, containing mostly financial data pulled from Acme's ERP system. Mr. Smith disconnects after the financial data is completely downloaded, although not all news articles have made it across yet. Mr. Smith doesn't worry about this, as the news download will pick up from where it left off the next time he connects to the Internet.

8:00 A.M. The flight has taken off without delays and is now well under way to Dallas. In midair, Mr. Smith continues to answer his emails, using his PocketPC in disconnected mode. In addition, he reviews the latest performance report for Dallas, which he downloaded while waiting for the plane to take off. Finding a few items that would warrant in-depth discussion with the Dallas executive team, he writes himself a few short notes to be included in the presentation he will give.

9:00 A.M. Ahhhh, breakfast.

9:30 A.M. One of the emails Mr. Smith received this morning came from his administrative assistant. The assistant stayed in the office late last night to incorporate some last-minute changes Mr. Smith asked for. After that was done, she attached the latest version to her email message to Mr. Smith, knowing that he would check his email first thing in the morning.

Mr. Smith opens the attachment and familiarizes himself with the presentation's new format. "Looks much better than mine," he said to himself, and made a note in his to-do list to send her some flowers for working late the night before. He next decides to relax a little and read, after having worked off most of his email received thus far. He knew there would be many more to come throughout the day. Mr. Smith opens a few of the articles on industry trends he received this morning, and incorporates some noteworthy facts into his speech.

1:00 P.M. Mr. Smith lands at the Dallas/Fort Worth airport and adjusts his watch. After exiting the plane, his PDA detects that the airport is wired with a local area network. The PDA logs onto the network to synchronize with Acme's home office. The emails that Mr. Smith worked on are being sent, new emails are retrieved, and the industry news download continues. All this takes place without requiring Mr. Smith to intervene. Within 15 minutes, Mr. Smith has exited the airport and jumped into a taxi to take him to Acme Dallas.

1:30 P.M. While en route to the Dallas office, Mr. Smith receives an alert on his PDA. The GM of the Dallas location is frantic; against all expectations, the new regional VP of sales will attend the meeting. The GM hasn't prepared for this and urgently needs an analysis of the office's performance over the past eight quarters. Mr. Smith uses his PDA to create a custom query that shows the confidential metrics for the Dallas location for the past two years. Within a minute, Mr. Smith receives the complete report from Acme's financial application and forwards it to the GM. In addition, Mr. Smith pulls background information about the new VP from Acme's employee database to refresh his memory. While online, Mr. Smith also uses Acme's calendaring system to schedule lunch with the VP tomorrow, thinking that this would be a perfect opportunity to better get to know this recent addition to the Acme executive team.

2:15 P.M. Mr. Smith arrives at Acme's offices, and, after briefly saying hello to the GM and his staff and grabbing a leftover sandwich, he heads to the conference room to make sure everything is ready for his presentation at 3:30. The conference room is wired with a local area network that is immediately detected by Mr. Smith's PDA. "Great," he says to himself, and prints out the modified presentation and the notes he took on the plane. Then, he quickly orders a bouquet of flowers and has it delivered to his assistant. By now, the PDA's battery is running low. Mr. Smith replaces it with the fresh one he brought and has the other charged.

Next, he connects his PDA to the conference room projector, has a few copies of his presentation prepared as handouts, and awaits the Dallas executive team.

3:30 P.M. The presentation of Acme's quarterly results goes by without major incident except for the sales manager's question regarding his sales team's performance in comparison to the national average. Mr. Smith answers this question by switching from the presentation contained on the PDA to a live connection. Still projecting against the conference room's white board, he pulls up the income statement, and then drills down into the numbers by repeatedly tapping on the statement's sales section. Each tap produces the supporting numbers that collectively constitute the income statement. Thus, Mr. Smith shows exactly where the numbers are coming from and how they compare to each other, first at the national level, then at the regional level, and finally at the Dallas location.

4:45 P.M. Mr. Smith and the Dallas GM head over to the warehouse for a brief inspection tour. The warehouse manager welcomes Mr. Smith and proceeds on a walk around the facilities. Suddenly, Mr. Smith receives an urgent alert from his administrative assistant that Acme's VIP client, whom he was supposed to meet for dinner at 7:00 P.M., has to push back that appointment by two hours. Mr. Smith engages in brief real-time chat with his assistant to confirm the details and to change the dinner reservation. While online, Mr. Smith also quickly pulls the latest periodic (monthly) report covering performance statistics for the Dallas warehouse, including inventory level, turnover, delivery statistics, returns, and level of obsolete inventory. After confirming these statistics with the warehouse manager, he then runs an on-demand query for the same data from Acme's flagship location in Los Angeles. A heated discussion ensues.

5:30 P.M. The warehouse inspection behind him, Mr. Smith is settling in Acme's office to capture his notes for the day and answer urgent emails, when he receives an alert from Acme's Hawaiian office that the new client the company has been trying to sign for months is finally ready, but has now demanded an additional 5 percent discount. Mr. Smith issues a conditional approval on the spot and sends an alert back to the sales director in Honolulu who is presently at the client site. In addition, he forwards his communication to Acme's Customer Acceptance department to verify the prospective client's credit rating, and he also sends a copy to Acme's legal department.

6:30 P.M. Mr. Smith is still engrossed with answering voice mails and emails, when he receives another alert. The local VIP client, whom Mr. Smith was supposed to meet at 9:00 P.M., now is able to keep the original dinner appointment at 7:00. Mr. Smith jumps into a taxi.

6:45 P.M. Mr. Smith arrives at the restaurant just in time to access the Acme network for some last-minute information. He pulls the client profile that contains the latest data on client issues as captured by Acme's customer relationship management (CRM) application. Mr. Smith knows that it is usually during dinner that his customers open up in regard to any critical issues between Acme and its clients. In addition to a review of customer support summary information, he pulls the client's latest financial data. Mr. Smith is especially interested in an account aging that would show any major unpaid invoices. Luckily, it seems that he will not have to ask for any checks this evening. Instead, he will offer to pay for the meal and have it charged to the wireless wallet he calls his PDA.

9:30 P.M. It is getting late when Mr. Smith returns from dinner and checks into the hotel. To add insult to injury, the hotel's guest reception is hopelessly understaffed. A long line of guests who want to check in has already formed, but Mr. Smith is not worried. He uses his PDA at the hotel's check-in kiosk to obtain his room key and quickly is on his way. Once settled in his room, Mr. Smith continues to finish his day's tasks and prepare for the day tomorrow.

11:30 P.M. Just before going to bed, Mr. Smith uses his PDA to establish a video chat with his wife, Shannon, and nine-year-old daughter, Michelle. Both have just returned from Michelle's high-school theater performance and can't wait to show Daddy the pictures they took with Mom's digital camera. Mr. Smith congratulates his daughter and sends her to bed, after which he briefly shares the day's events with his wife. At 12:15, Mr. Smith makes sure that both his batteries are charging, sets his PDA to wake him up at 6:00 A.M., and falls dead asleep.

Although this day-in-the-life example may seem a bit dramatic, it serves our purpose of illustrating the various applications and features provided by the wireless executive dashboard. Let's look at some of the benefits the solution provides to Mr. Smith and those that happened to cross his path, whether in person or electronically, during the day we've described here.

Acme Dashboard Benefits

The business drivers we outlined earlier in this chapter, including the need for instant availability of key corporate performance metrics, provided the basis for the development and deployment of the wireless dashboard. These business requirements built the case for and served to justify the financial expenditures and deployment of nonfinancial resources to bring the project to fruition. As an outgrowth of these business requirements and the resulting wireless solution, there are several specific benefits that accrued to the stakeholders who were directly or indirectly touched by the application. Direct beneficiaries include executives on the go. Indirect beneficiaries include those individuals the executive interacts with. Ultimately, it is the entire organization that receives the dashboard's payoff.

The following discussion summarizes the value received by the executive dashboard user—in our case study, Mr. Smith. In general, the dashboard provides for better decision making based on timely access to key operations performance measures. Specific benefits include enhanced exception reporting, ubiquitous information access, improved communication, and higher productivity away from the home office.

- Exception reports communicate that something within the organization deviates (or is about to deviate) from the standard. These deviations can be positive or negative. However, if the recipient of such a report does not learn of the potential problem until hours or days after the fact, these reports become rather worthless, as the exception is allowed to exist until action is taken. In the case of a particularly large sale, for example, the negative impact resulting from an executive not learning about the incident may affect the sales rep who brought in the deal and whose morale would surely be boosted by a timely acknowledgment from the top. In the case of faulty production equipment, however, the impact of not knowing and being able to take immediate action can be much more severe and damaging to the organization.

- The wireless dashboard enhances, if not outright enables, effective exception reporting. The dashboard allows for the sending and receiving of instant, actionable alerts. These alert messages pop up on the recipient's PDA screen immediately, provided he or she is connected to the network. In case the recipient is disconnected, the alert is the first information that is displayed after the unit is logged on again. Actionable alerts are a vehicle to instantly relay time-sensitive information to the dashboard user and allow him or her to immediately take action—including preventive measures or error correction. An alert can be sent by another employee of the organization, such as was the case when Mr. Smith's admin had to cancel the original dinner appointment. Similarly, other originators of alerts could include enterprise-external users, such as your brokerage's financial advisor, your suppliers, customers, or business alliance partners. In contrast to these alerts that are composed by people, there are system-generated alerts. System-generated alerts include urgent notifications that are triggered by a system-monitored event, such as when a major sale is logged in the organization's accounting system or a production line malfunctions.

- The wireless dashboard provides access to information contained within or outside of the corporate network. Today's executives require increasingly accurate and timely insight into the performance of the organization to be able to prevent derailments or fix problems as they occur. Aside from previously discussed exception reports, executives require access to the internal email system, as well as periodic reports or supplementary data pulled from the company's ERP, CRM, or other legacy systems. Access to enterprise-external information includes access to the Internet for industry news or competitive intelligence, financial analyses, or information pertaining to or originating from customers, suppliers, or alliance partners.

- The wireless dashboard can be used as an effective communication tool. Effective communication is critical for the executive to ensure that a company's employees receive and understand the direction provided by the organization's leadership. To be able to steer the company, such as in the case of taking action after receiving an exception alert, the executive must be able to contact in real time those lieutenants who are responsible for the affected areas of the business. In addition to reaching these individuals, effective communication entails an element of speed. For example, during his visit to Dallas, Mr. Smith received an urgent alert that an important target account of Acme required an additional discount before signing the bottom line. Mr. Smith's ability to instantly approve this discount resulted in signing the client while the sales director was still at the client's site. Who knows whether Acme would have received the business had Mr. Smith learned about the request using traditional communication channels such as phone, fax, or email. Because Mr. Smith could be reached at the very time the critical decision needed to be made, Acme won the business.

- The wireless dashboard significantly enhances its users' efficiency and productivity while away from the office. Mr. Smith was able to perform critical management tasks on his PDA—tasks that Acme executives engaged in from their PCs or laptops before the wireless application had been deployed. The portable dashboard allowed Mr. Smith to make better use of his time. Whether he was waiting for the boarding call at the airport, traveling for hours on the plane, or giving a presentation, Mr. Smith had the information he needed at his fingertips at all times. Especially for executives who are frequently on the go, the wireless dashboard provides essential business management functionality in both connected and disconnected modes, eliminating nonproductive use of time.

Now that we have explored some of the most visible benefits—that is, the benefits that accrue to the wireless dashboard user—let's think about the value received by those who are not immediately exposed to the dashboard's features. These secondary parties include primarily other company employees as well as third parties such as customers, suppliers, and partners that the executive interacts with. For example, if you recall, Mr. Smith was able to provide the Dallas GM with a critical report, he scheduled a getting-to-know-you meeting with the regional VP, and he helped the sales director to close a deal. So, while the imminent user of the dashboard seems to be the primary receiver of its benefits, the individuals the user interacts with also stand to gain from the application, yet in a roundabout way. Value emanating from the user cascades through the organization. The end result, of course, is an improvement in corporate performance, providing the company's entire group of stakeholders, including owners, directors, managers, employees, partners, suppliers, and customers, with a return on their investment.

Summary

This chapter introduced us to the case study that will be our real-life example to be discussed in the coming chapters. As a representative of a mobile office application, a wireless executive dashboard allows us to explore some traditional mobile office applications while also reviewing several new types of office-related uses that the mobile technology affords. At its core, the application provides its users with information about a company's key performance indicators generally relating to financial, operational, employee, and customer satisfaction performance.

The particular client we described needed a wireless executive dashboard application to better manage a growing operation that required an ever increasing contingency of key employees to be on the road—and thus out of immediate reach. We learned about the functionality of a dashboard, how it works, what it measures, and the value the application can provide to the organization as a whole. Various features were explored by following a senior executive throughout his busy day, making extensive use of the application while on the go. The application allowed this executive to stay in touch and communicate with the office in a real-time manner, work on office applications such as email and spreadsheets, gain access to Internet information sources, review corporate key performance indicators on a regular or exception basis, and take corrective action if required.

Having an understanding of what a dashboard application is and the benefits it provides to its users, we can now move on to the next chapters that will explain the information, systems, and technical architecture behind the solution.

Mobile Office Information Architecture

We have constructed many information architectures throughout this book, including a wireless B2B commodities exchange, a B2C financial application, and a specialized B2E structural design to automate a mobile sales force. Our final architecture also applies to the B2E realm, but it serves a much broader purpose: the wireless mobile office. The mobile office was demonstrated in Chapter 23, in which we introduced Acme Toys. Acme took an application that has existed for quite a while, the remote mobile office, and spiced it up by adding a wireless dimension.

Overview

Some of the information requirements of the wireless mobile office have been touched upon in the previous parts of this book, under the auspices of the connected enterprise. Essentially, the wireless mobile office comprises software and network connectivity that allows workers to perform roughly the same set of tasks when mobile as they can when physically located in the office. Whenever possible, the steps taken to complete these tasks are the same whether local or remotely connected to the enterprise. As we will see, both local and remote connectivity can benefit from the addition of wireless access.

The mobile office was one of the key areas in which businesses are concentrating their wireless efforts, as stated in Chapter 5. That's one of the reasons the application developed for Acme Toys was selected as our case study. Acme's executives are a highly mobile group that needs to interact with other Acme employees and enterprise data on a continuous basis. The overall efficiency gains are felt throughout the organization: How many times has an employee or an entire group been blocked because they need five minutes of an executive's time for direction or an approval? With wireless access, chances are that the wait state will be much, much shorter.

And that is the ultimate goal of the connected enterprise: facilitating the connections between those in the organization who depend on input from other people and/or systems. The streamlined connections can take frustrating hours and days of waiting and compress them to mere minutes, if not less. As we will see, it can also mean the difference between winning and losing a time-critical battle. We will now look at the data architecture required to make that enhanced corporate infrastructure possible.

A Day in the Life of Mr. Smith: His Data Needs

In Chapter 23, we followed Mr. Smith through a fairly typical day as a senior VP for Acme Toys. This day gave us insight into a very mobile employee, who needed dynamic, specialized, and sensitive data in real time. This need creates some interesting pressures on the Acme information systems department, which has to work hard to connect Mr. Smith with the rest of the enterprise as often as possible. It also creates some interesting balancing acts between providing rich functionality and ensuring ease of use, something vital to a user group such as senior VPs.

To reiterate, Mr. Smith is using a PocketPC with both a wireless modem and a wireless LAN card. The latter gives him high-speed access to public LANs and Acme's internal network when on Acme sites. The wireless modem connects him when he's outside these pockets of high-speed connectivity. Together, they provide a high degree of probability that Mr. Smith is either connected or has the ability to connect at any time. As we shall see, having connectivity at the drop of a hat facilitates every aspect of Mr. Smith's workday. Let's look at each step in this day and see the data requirements they hold.

7:10 A.M., Los Angeles Airport. Mr. Smith has some time before his flight boards, so he uses the high-speed LAN connection provided in the airport to get some work done.

1. Checks email.

Checking email is quite straightforward, assuming firewall and proxy issues do not block access to the mail servers. Sending mail is handled via Simple Mail Transfer Protocol (SMTP), and all mail clients have built-in mechanisms for storing unsent messages locally. As for receiving mail, both the POP3 and IMAP mail protocols allow for copies of emails to be stored locally on the client, so all data transmission issues can be handled at the application level.

2. Receives toy industry and competitors' news.

This functionality requires a series of steps. First, a service has to exist for the aggregation of toy industry news. The aggregator will have *bots*, or programs, searching the net for specific content. The service can be either part of Acme Toys' IT infrastructure or a third party contracted to provide toy industry–specific content. Mr. Smith has already taken the next step—a one-time client configuration to select the areas in which he is interested. Typically, these interests are stored as value-attribute pairs on the client and are known as *cookies*. Cookies are a standard means by which a Web site can generate customized content.

In the airport, Mr. Smith notifies the system of his desire to retrieve the latest content matching his parameters. The bots set to work, creating the list of content to be downloaded. In an ordinary Web model, this set lasts only as long as the session—if the client drops off, the set goes away and the whole process has to be restarted from the beginning. However, this is not acceptable behavior for the executive dashboard.

Because of today's issues with connection reliability, the set needs to be created on a server and persist until successfully pushed to the requesting client, or until user action on the client cancels the request. Figure 24.1 shows a conceptual view of how the dashboard client accomplishes this communication goal. Requests coming from a dashboard client are routed through the dashboard request manager in the Acme network.

The request manager assumes the identity of the client for the purposes of talking to the targeted server, behaving much like a proxy server. The request manager receives the request from the dashboard client, opens a connection to the desired server, makes the request of the server as though it were coming from the dashboard client itself, then retrieves the content in the server's response. The request manager stores that content locally, then begins pushing it to the requesting client. Periodically, the client will send a message to the request manager acknowledging the receipt of a set of data. That acknowledged data is then removed from the persistent storage of the request manager, having fulfilled its mission of guaranteed transfer.

Should the client lose connectivity, the request manager retains the content not acknowledged as received by the client. When network con-

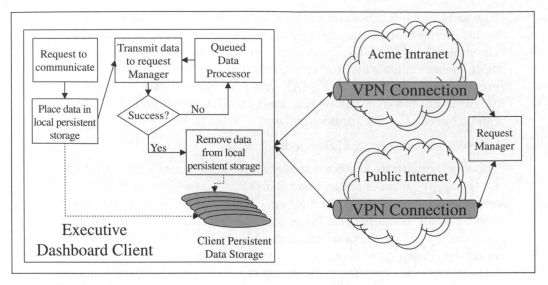

Figure 24.1 Request manager persistent storage mechanism.

nectivity resumes, the client will notify the request manager, pumping any remaining data to the client until the queue on the server side is empty.

So Mr. Smith can begin his download of industry news and get on the plane with whatever percentage of the transmission that was successful before his connection was lost. Sound impressive? Well, later on we shall see just how smart the request manager is.

3. Pulls the latest monthly performance report on the Dallas office from Acme's ERP system.

On the surface, pulling a report from an enterprise resource planning (ERP) system seems very different from downloading news reports from the Web. Upon closer inspection, however, the data requirements of the two are quite similar. The ERP system has its roots back in the days when dumb terminals populated the business landscape—a model unfortunately duplicated by the Web browser. Both models use synchronous communication, and a client request will lock up the application until the server's response is received.

In our case study, Mr. Smith makes the request for Dallas's monthly report, and it travels through the request manager to the ERP system. As far as the ERP system is concerned, the client is the request manager (though Mr. Smith's logon profile is used to determine whether he is authorized to access that report). His probabilistic connection with the network does not affect the ERP, since the request manager masks it. This is a powerful feature—Acme's legacy systems do not have to be altered in any way, yet they are completely integrated with state-of-the-art mobile technology.

8:00 A.M.—1:00 P.M., flying from Los Angeles to Dallas, Texas. Due to U.S. FCC regulations, Mr. Smith cannot use either his wireless modem or his wireless LAN card at any point during the flight. He could use the onboard Ethernet LAN that the state-of-the-art aircraft provides to flyers (at a premium), which has its own wireless connection to the Internet—but Mr. Smith does not have an Ethernet card to plug in. He grumbles, writing a note in his PocketPC to pack the card in his travel bag on his next trip (how could he have forgotten it?). He makes the best of it, being able to get just about everything done in disconnected mode.

1. Continues to answer emails.

    The executive dashboard does not have to enhance or alter standard mail client behavior. Those clients are already adept at maintaining their own local client storage of mail and have built-in capacity to queue pending outbound messages. The dashboard simply provides an easy mechanism for launching the mail client.

2. Reviews the Dallas performance report.

    The report pulled from the legacy ERP system is stored locally on the PocketPC, available for offline use.

3. Reviews the updated version of his presentation.

    The mail client in Mr. Smith's PocketPC is capable of storing more than email messages—it can handle attachments, too. This powerful feature gives Mr. Smith freedom from his laptop, which would otherwise be required to receive such rich content. Instead, he can use the updated presentation provided by his administrative assistant as a baseline for his own changes. His assistant deserves those flowers, indeed.

4. Reviews the articles on industry trends and the competition.

    Just as the performance report pulled from Acme's ERP system is locally available for offline use, so too are the articles located by our Web news search. They are incorporated in Mr. Smith's presentation, all being done by a disconnected palm-sized device. Mr. Smith briefly marvels at this and then prepares for landing.

1:00 P.M., arrival at the Dallas airport. Mr. Smith is now able to reconnect to the online world, and the wireless LAN available in the Dallas airport ensures a broad pipe into that world.

1. Email is synchronized.

    Even though the functionality of the mail client has been virtually untouched, it has to be tweaked a bit in this case. The executive dashboard detected the network in the airport, then had to activate the mail client to force queued mail transmission and download. All of this occurs without Mr. Smith having to interact, however, so the intrusion is minimal.

2. Download of toy industry and competition news continues.

Here's where the executive dashboard shows off a little. Remember that the download began in the Los Angeles airport lounge, but was interrupted by Mr. Smith's boarding his aircraft. Now that he has arrived in Dallas, the dashboard client detects the network connection and a message is sent to the dashboard connection manager that Mr. Smith is back. The downloads resume immediately, so Mr. Smith will be able to finish reading an article on Malaysian toy factory construction (it was cut off in midsentence, and Mr. Smith is interested to see the article's conclusion).

The original request from the dashboard indicated that the content of the response is highly dynamic. The connection manager respects this and holds onto the original request until the response is complete. In the case of Mr. Smith's reconnection message, the connection manager not only begins pushing data to Mr. Smith, but also re-asks the news service provider for any updates that have been posted since the time the original query was issued. Sadly, Mr. Smith learns of the scandalous resignation of his longtime friend and rival, Mr. Jones of Molten Lava Toys, which occurred while he was in the air.

1:30–2:15 P.M., travel from the Dallas airport to the Dallas Acme site. One of the traditional connectivity dead spots, Mr. Smith's connection is alive and well in the car ride from the airport to his meeting. He is able to arrive at the Acme site without delay, yet fully prepared.

1. Alert of regional VP of sales meeting attendance.

Here is a clear differentiator between the Acme executive dashboard and a standard Web application: a server-initiated message. We have discussed some of the characteristics of the traditional Web, including the reinvention of the dumb terminal. Another characteristic is the request-response synchronous communication model, where every interaction between client and server is initiated by the client and responded to by the server.

In the Web model, pushing real-time alerts to mobile clients is not practical, so something has to be done in the dashboard to give it this functionality. The ability to reach out to mobile personnel is one of the greatest advantages of wireless enablement, and it must be leveraged to realize the full potential of the dashboard.

There are many ways in which alerts could be incorporated into the dashboard, but every strategy has to perform similar tasks. They must register the client with a server, indicating how that client can be reached if needed. The server registration mechanism is already in place with the dashboard, since it sends "I am alive, and here is where I can be reached" notification to the connection manager. (This kind of message is known as a *heartbeat*.)

The connection manager can use the heartbeat to show who is actively connected to the system and provide the means to deliver alerts. Just as the manager queues all other traffic, alerts will be queued for delivery should the intended recipient not be available when the alert is sent. Fortunately for our Dallas GM, Mr. Smith is available when he sends the alert about the plans of the regional VP of sales to attend that afternoon's meeting.

2. Query for last eight quarters of performance metrics for the Dallas office.

This Dallas office performance query is similar to, and yet quite different from, the performance report Mr. Smith obtained while waiting in the Los Angeles airport. That request was for a standard report, which is essentially a query run once, the results formatted and stored to prevent the same query from being repeated endlessly by Acme personnel in need of that commonly used data. This new request for the last eight quarters will produce essentially the same result; the numbers and their presentation will not be the same, but both will be reports generated from the same data source.

On the other hand, the two requests are quite different. The L.A. request for the standard report probably involved little more than dropping down a couple of menus in the dashboard, until the proper report was found. The request for two years worth of Dallas data, however, is a custom query and has power and potential complexity far beyond the earlier request. To fulfill this request, we will make the following two assumptions (both quite reasonable):

- The system from which the report is generated is a relational database, and therefore accessible via SQL queries.

- Mr. Smith is *not* knowledgeable about that database's schema, nor is he adept with SQL.

These two conditions pose a considerable challenge to the creators of the dashboard's user interface. On the one hand, it has to provide sufficient functionality to allow Mr. Smith the power to fully customize his query, whether he wants eight quarters of financial data or some other request not possibly foreseen by the system's designers. On the other hand, Mr. Smith cannot be exposed to the semantics and complexities inherent in SQL, but valid SQL must be generated for the request to be processed successfully. We will assume that the application designers overcame this challenge with a keyword-driven query engine or other similar approach, and Mr. Smith is able to get the reports he wants.

3. Forward performance metrics to the Dallas GM.

Once the report is generated, it is sent to Mr. Smith via the dashboard's communication engine, including the connection manager. Now Mr. Smith wants to forward that report to the Dallas GM, who caused this whole uproar in the first place. This is the first nonmail data that Mr. Smith is attempting to send, and it introduces a whole new client issue: What if the client loses its network connection before the transmission is complete?

Never fear, the executive dashboard has that covered as well, as depicted in Figure 24.2. Just as the server has a persistent data queue for transmission to the client, the dashboard client has a persistent data queue for transmission to the connection manager. The client sends data until the server acknowledges receipt of the entire message, then it is deleted from the client queue. The server now has the message in its persistent queue, where it will remain until the recipient (in our case, the Dallas GM) acknowledges successful download of the report on his client. Bear in mind that these acknowledgments are at the network level, not the application level, so no users are even aware they are being exchanged.

4. Obtains background information on the new regional VP of sales.

This activity is very much like Web browsing Acme's HR intranet site. The only difference is the connection manager's presence, guaranteeing the transactions from end to end.

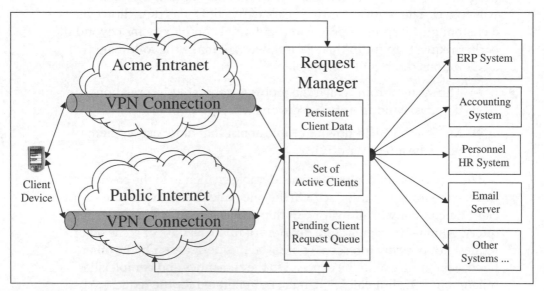

Figure 24.2   Client-side persistent storage mechanism.

5. Schedule lunch with the new regional VP of sales.

   Common among business systems are shared calendars, useful for scheduling just such an event as lunch with the new regional VP of sales. There are two factors affecting how useful this kind of shared calendar can be: the frequency of synchronizing one's local copy of the calendar with the server's version, and the diligence with which one enters events into the calendar. The former is easily dealt with—the connection manager will keep the local copy of the calendar in sync with the publicly shared server version as often as the client is within the realm of the network. The latter, however, will not benefit from any technological breakthrough. If people do not conscientiously enter their engagements into their calendars, it will be useless. The Acme executives, however, live by those calendars, so it is in their own best interest to keep them accurate.

2:15 P.M., arrival at the Dallas Acme office. Though very productive during his travels, Mr. Smith still has some tasks that need to be performed on-site. Upon his arrival, he does *not* need to do anything special to get his PocketPC to recognize the availability of a wireless LAN connection to Acme's intranet. All Mr. Smith knows is that he shows up at any Acme site and his connection works.

1. Prints out the presentation he modified on the plane.

   Mr. Smith finds a printer next to the conference room in which he will soon be presenting. The printer's network address is clearly labeled, and Mr. Smith sets out to create a connection to it from his PocketPC. Since he is at an Acme site, the dashboard identifies network resources on that site, so Mr. Smith has a small, closed set of network resources to comb through before finding the right printer. After getting a local IS person to help him plug his PocketPC into the projector, he is ready to make his presentation.

2. Orders flowers for his administrative assistant.

   Mr. Smith has a few minutes of downtime, so he checks his to-do list. Seeing the entry to send his assistant some flowers, he opens a Web browser and goes to his favorite online florist. Having visited the site before, his PocketPC has cookies storing his login and account information. This makes the transaction little more than entering his password (for security purposes, this is *not* stored), selecting the flowers, selecting the recipient, and confirming the credit card data.

   This transaction looks exactly the same as online transactions conducted from wired workstations. And, if nothing went wrong in the transaction, it would appear to be exactly the same. But Mr. Smith's battery was dying, so after he made his request, he was forced to change it.

The request went from his executive dashboard to the connection manager, then off to his online florist. The response came back from the florist, but Mr. Smith was not available to receive it. Ordinarily this throws a monkey wrench into the Web model, but the connection manager is designed to handle just such a situation. It stores all of the information pertinent to both the transaction and the session. The former is passed back to Mr. Smith's client when he reconnects; the latter is used for any subsequent communication with the florist. In this way, the server is able to initiate communication with the client, even though the Web model does not.

3:30 P.M., presentation is given. Mr. Smith takes the up-to-the-minute data he had available on the plane to make his presentation as timely as possible. Not fully aware of this capability or Mr. Smith's savvy at using it, the Dallas sales manager questions the validity of the asserted lack of performance by his sales team.

1. Defense of the Dallas sales team assessment.

    One of the most important, stressful, and sometimes frustrating tasks for the sales team is to determine the veracity of the data upon which their very future depends. How often have we heard the question, "Where did you get those numbers from?" answered with something inane like, "Same way I always do—I pushed the button on my spreadsheet, and the numbers appeared." Mr. Smith is far more sophisticated than that.

    He not only knows where the data that supports his position comes from, but also how to find and present that data. The live connection Mr. Smith uses to obtain that source data is a session on the legacy ERP system, but handled at the network level by the connection manager. In this case, its use is invisible, as the high-speed wireless LAN connection is rock-solid, so queues are emptied just as soon as the requests come in.

4:45 P.M., arrive at Acme's Dallas warehouse for inspection. This scene demonstrates both the ability to handle unforeseen events (dinner plans changing) and on-demand access to enterprise data and services.

1. Receives change in dinner plans alert, and engages in real-time chat to confirm changes.

    Mr. Smith's administrative assistant uses the real-time chat feature of the dashboard, one of the possible means for delivering alerts (introduced earlier in this chapter), to send the change in dinner plans. This flows naturally into the two-way chat required to quickly confirm the update to his day. (Yes, the dinner reservations have been changed. Yes, the updated reservations have been confirmed.)

2. Pulls the latest monthly report on the performance statistics of the Dallas and Los Angeles warehouses.

    These are the same types of requests handled by the connection manager as previously described. These are standard reports, so they are obtained by going through a few menus in the dashboard to find the report needed. The effect on Mr. Smith's companion, the Dallas warehouse manager, is quite predictable, but not as easy to manage as the report downloading mechanism.

5:30 P.M., back at the Dallas Acme office. When most people are thinking about wrapping up their workday, Mr. Smith is still going strong, still ready to handle unforeseen events as efficiently as possible.

1. Answering email.

    As has been stated, the mail client handles the dubious network connection easily on its own. The dashboard does not need to get involved in mail, making it one of the easiest pieces of functionality to enable wirelessly.

2. Receives update alert about Hawaiian VIP client demands.

    Though the data needs for this type of alert have been covered already, the benefit it provides in this kind of example is worth mentioning again. This is the heart and soul of the connected enterprise, giving people access to the key decision makers in just about real time. It is so vital to the success of any organization, and Acme Toys could well have lost the Hawaiian bid without it. The reduced turnaround time gives Acme a competitive advantage over the competition, allowing them to close a deal they have been pursuing for months.

    The numbers look appealing off the top of Mr. Smith's head, so he uses his session with the Hawaiian sales manager to give a conditional approval. Since Mr. Smith does not have the time to analyze the merits of the business proposal in detail, he delegates that task to the appropriate team of analysts, in this case Acme's Customer Acceptance team. He also forwards the proposal to Acme's legal team, to ensure they are doing everything properly.

    These requests are no longer real-time alerts; they are sent via standard corporate communication, presumably email. This shows one of the limitations of the connected enterprise: Not even wireless access can speed up the turnaround time of lawyers. Mr. Smith has, however, minimized the delay between client request and Acme response by forwarding the request to the proper desks less than a minute after the client asked for it. This has traditionally been one of the challenges faced by Acme (and every other large enterprise), and wireless connectivity tightens the response loop.

3. Receives change in dinner plans alert—yet again.

No new data capabilities here, just another example of how his wireless connection makes Mr. Smith capable of handling the pace at which his day changes. Mr. Smith heads off to dinner with his clients.

7:00–9:00 P.M., dinner with VIP client. Watching his VIP client with utter amazement, Mr. Smith is certain he has never seen so much filet mignon consumed by one human being in a single meal.

1. Pulls general client and financial profiles of his dinner companions.

Before his client arrives, Mr. Smith leverages his wireless connectivity to Acme's network for the final time today. He browses the Acme Intranet for pertinent data on his dinner companions. The content is automatically made available offline by the client communication piece of the executive dashboard, should he need it again at a future time in which the network is not present. Other than the persistence of the data on the client and the involvement of the connection manager on the server, the interaction is just like any other Web browsing, though performed on the Acme intranet. That illusion is strengthened by the fact that the different underlying data transmission model works without the knowledge of Mr. Smith.

2. Pays for the meal with the wireless wallet feature of his PocketPC.

Most of us have gone out to a restaurant and paid for the meal with a credit card. Typically, such a transaction consists of the following sequence of events:

- The bill payer is presented with the bill.

- The bill payer hands it back to the waiter along with a credit card.

- The waiter swipes the credit card into a machine, which calls the credit card company for authorization of the transaction.

- If approved, the credit card company deducts the appropriate funds from the bill payer's account and transfers funds to the restaurant.

- The waiter obtains a printout of the credit card company's authorization, which is presented to the bill payer for signature.

Stated another way, a financial transaction occurs in which the payee authenticates himself or herself (currently done by signing the receipt). The transaction occurs in real time via a connection between the restaurant and the credit card company, by which funds are transferred from a customer's account to the restaurant.

When looked at from that perspective, wireless bill paying becomes a logical evolution for such transactions. Let's look at the changes required to move from the traditional credit card model to its wireless successor. The restaurant needs a compatible data connection with Mr. Smith's PocketPC, by which he can initiate the funds transfer. The network bandwidth can be trivially small, since the amount of data sent is negligible.

From the credit card company's perspective, the transaction is the same whether wireless or not. The company simply takes a small set of transactional data, which in the current model is the data encoded in the magnetic strip of the credit card. Together with the ID of the restaurant, the data is processed against Mr. Smith's account information on file at the credit card company. This process remains unchanged, assuming the data made the jump from the wireless to the wired world outside the credit card company's domain.

From Mr. Smith's perspective, he needs the account information stored in a secure way on his PocketPC. It is important that his credit card information not be compromised should he lose his PDA or should it be stolen. He needs to feel secure that the transactions he initiates will be billed exactly once, and only those transactions will occur. Once those security concerns are allayed, he is enthusiastic about the ability to leave his wallet at home with his laptop.

> **NOTE**
>
> This example simplifies somewhat the actual detail interaction between the transaction acquirer, the credit card issuing institution, and the settlement process that takes place between them. But nonetheless, it is intended to illustrate the functions of a wireless wallet.

9:30 P.M., hotel check-in. Mr. Smith is not sure which is more amazing—the long check-in line at 9:30 at night or the hotel's inability to handle it. No matter, he doesn't need to stand in line.

1. Checks in via wireless self-serve kiosk.

   The hotel check-in kiosk performs wireless financial transactions, exactly as described. However, the kiosk is capable of much more. Since Mr. Smith frequents that particular hotel chain, he took the time to enter his hotel frequent traveler membership information into his PocketPC. The kiosk receives this data as soon as the transaction begins, so the credit card information and room type preferences are already known, eliminating the need for just about all data entry. Mr. Smith's bonus program points are automatically tallied, and he is informed that he qualifies for a free breakfast tomorrow morning.

His key is dispensed, and before a single person can get off the traditional check-in line, he is on his way to his room.

11:30 P.M., talks with his family. Thank goodness Dallas is two hours ahead of Los Angeles—he can see his daughter without violating the bedtime he and his wife have worked so hard to enforce.

1. Participates in a video chat with wife and daughter.

   Finally, some family time! The videoconferencing feature is important to analyze from a data perspective, for a couple of reasons. First, it contains huge amounts of data, all of which must be sent in real time. Second, since it's such a time-critical application, the store-and-forward model of the connection manager does nothing but add overhead.

   These are very special data needs unique to this class of application. It's real time, and only synchronous communication makes sense. If data is lost, or one side or the other drops off, then the conference is dead. Thus, the main requirement of the videoconference is a wide pipe—there are a lot of bits flowing and not a lot of time to deliver them.

Mobile Office Information Principles

We have reviewed a great number of uses Mr. Smith has for the mobile office and the data needs accompanying them. We now need to take a step back from these specific features and data requirements to look at the overall information principles. Our architectural framework mandates the application of these principles so that we may create an information architecture that not only solves specific use cases, but can be applied to the general solution as well.

Mr. Smith was an excellent choice for our test case, because his data requirements exceed those of just about every other Acme employee. The typical Acme sales representative is concerned with a specific product line, a specific client account, or a local geographic area. Mr. Smith is concerned with all three, and scaled to an international level. Mr. Smith is also as mobile as an employee gets, traveling far and wide to oversee Acme Toys' operations. Deriving the information principles from his activities will yield a very credible set. The principles can be summarized as follows:

1. The mobile executive workforce needs to be connected as frequently as physically possible.

2. The mechanics for connecting to the network need to be hidden from the mobile user.

3. The network must have the ability to identify wireless clients and push requests to them (the traditional Web model only allows clients to make requests of servers, with synchronous responses).

4. Absence of network connectivity should not prevent users from performing any requests, and those requests should be carried out without user action once the network connectivity returns.

5. A request in progress should be able to be resumed at any point, without the loss or duplicate transmission of data.

6. The network connection handler needs to be smart enough to know whether the connection is high-bandwidth or low-bandwidth, and filter the tasks available given the connection pipe size. The connection handler also has to recognize whether the connection is to the public Internet or to Acme's intranet, and handle proxies, firewalls, and authentication accordingly.

7. The network connection handler should be aware of the high-speed and low-speed wireless connection vehicles available. It should only use the low-speed connection as a fallback if high-speed access is not available.

Now that we have an enumeration of the Acme Toys' information principles, we should look at the class of client machines that can meet those needs. We will compare the capabilities of Mr. Smith's PocketPC against the data needs of the dashboard, to see whether it is overkill (too thick), overmatched (too lightweight), or right on. (For a detailed description of the advantages and disadvantages of the thick- as opposed to the thin-client models, please see Chapter 14.)

What are the characteristics of Mr. Smith's usage? He interacts with the Acme infrastructure, regardless of whether the network is available to complete those requests in real time. If the network is not there, the request must be queued on his client until such time as the request can be forwarded. This clearly means a thick client, one of whose key characteristics is the ability to persistently store application data locally. So we can safely conclude that the PocketPC is not overkill, because its storage capacity will be well needed to perform all of Mr. Smith's tasks. Since this is a custom B2E application, a thick client is both reasonable and appropriate.

Mr. Smith needs at least a PocketPC, but will it be enough? The data and computational needs are not excessive—some spreadsheets and presentation software are about the most severe challenges of Mr. Smith's usage. The user interface and data storage capacities of the PocketPC are up to the job. So, too, is the network infrastructure capability: It has the ability to connect both a wireless LAN card and a wireless modem card, needed for ubiquitous access. The next two chapters will go into more detail about the application-level and network-level requirements, but for now we will take the view that a PocketPC will not be overmatched by Mr. Smith's usage.

Having made a preliminary determination that the client hardware is right for the job, we now have to look at the characteristics of the thick client. As we have seen elsewhere in this book, there are pros and cons to using a thick client. We need to investigate how we can mitigate the negative aspects associated with the thick-client model.

First of all, we are dealing with an executive dashboard. The word *executive* appears in the title for good reason: It is neither cost effective nor necessary for a sales representative, for example, to have full access to all information, including enterprise data. As a result, the dashboard will not be widely distributed. A small user base is much easier to support, drastically reducing complexity over an enterprisewide deployment.

Second, the distributed data issue is minimized, because the users of the dashboard are data consumers, not data producers. As a result, dashboard users do not run the risk of conflicting with or corrupting enterprise data (they're just reading it).

Third, the term *heavyweight client* is relative—we're not talking about lugging around a Cray Supercomputer here. The PocketPC is aptly named, since it can fit in many pockets, and it has the capacity for keeping data to be transmitted across the network queued between uses (even when it's turned off). That is the primary requirement of the thick client and is not very prohibitive.

Finally, the requirement that all messages be authenticated is not as cumbersome as it may first appear. In reality, every *session* must be authenticated, with a session being defined as a connection between the client and the server for the purpose of exchanging one or more messages. Since the messages are queued, typical usage would have several messages making up one session. Authentication strategies usually include something along the lines of a PIN number with a time-sensitive pass code. Details of security strategies are beyond the scope of this data architecture, but a brief mention was warranted here. If you would like to know more, please refer to Chapter 6.

We can conclude that, although the PocketPC is not a requirement per se, a client with that order of computing power and data storage capacity is needed for the executive dashboard.

Summary

This chapter provided an overview of the data requirements generated by the wireless executive dashboard, typical use of which was outlined in Chapter 24. From those use cases, we derived a set of information principles with which to create a data architecture. We demonstrated how the productivity of the mobile executive could be greatly enhanced by using the dashboard.

The wireless executive dashboard is a B2E application, meaning it was developed by a business to connect its employees. One characteristic of B2E applications is forced usage—a business can place systems burdens on its employees that are not acceptable in a B2C model. On the other hand, since the user base is known and reachable, lowest-common-denominator technologies do not necessarily have to be applied, but may give way to more sophisticated solutions.

One manifestation of the B2E nature of the dashboard is the thick-client model. The advantages of that model are stronger client capabilities and usability despite unpredictable network connectivity. The disadvantages of the thick client are mitigated by the small distribution of the application and the fact that all users are data consumers, not data producers, in the ERP systems.

The data requirements and information principles discussed in this chapter will be key inputs to the next step of our process, creating the information systems architecture.

Mobile Office Information Systems Architecture

Introduction

Once again, it's time to discuss the information systems architecture behind our application. In this section, we've been discussing the mobile office and the various applications of which it is composed. It might be a good time to quickly review the business concepts and functionality of the mobile office before we continue with a more technical discussion around the logical information systems architecture. However, before we review the conceptual business, it's important to note the approach we'll take in this chapter.

The approach of this chapter is slightly different from that taken in the rest of the information systems chapters in this book. As before, we'll start by covering the current state and future state architecture, but this time focusing on key principles regarding the mobile office infrastructure. We won't dive into a great amount of detail around the applications or services, mostly because mobile office applications are generally out-of-the-box solutions. For instance, word processing, spreadsheets, and even email are generally not custom applications. Nor is the infrastructure to support these applications very customizable, at least not in the architectural sense. Consider a mobile office example such as Internet access. Generally, it's fairly well understood that Internet access is obtained through some Internet service provider (ISP) that in turn is leveraged to allow your organization's users to access the Web. The primary services here are the Web client, such as Mosaic, and the Web server, such as Apache. Depending upon the application, an application layer would be required as well as a data layer. Most of what architects need to know is how that infrastructure should look to support a mobile office design. With that in mind, we can briefly review the conceptual business design of the mobile office application, followed by a current state analysis.

If you recall from Chapter 22, there are three main categories of mobile office application functionality:

1. *Office automation.* Generally the Microsoft type of applications like Word, Excel, PowerPoint, and Access, which cover word processing, spreadsheets, graphical editors, and so on.

2. *Communication.* This category covers traditional killer apps such as email, chat, videoconferencing, and online meetings.

3. *Information management.* Personal information management (PIM) applications have long been a foundation for the office environment. Functions including calendar management, document management, Internet/intranet/extranet access, and knowledge management are staple applications in this environment.

Certainly, almost anyone who works in a white-collar environment can relate to the mobile office. How many businesses are you aware of that do not use some sort of word processing package or email system? Generally, the mobile office infrastructure allows for all of these applications to be integrated into one dynamic system with seamless interactivity between systems.

In Chapter 23 we demonstrated how the executive could take advantage of the mobile office in a typical day-in-the-life scenario. You might remember that throughout the day, the executive used applications such as email, word processing, and Internet/intranet. The true mobile office functionality was seen through the collaboration and automation of various tools that an executive might use on a daily basis. The mobile office enables the executive to dynamically access information through such processes as receiving email, reacting to that email by writing a note or creating a document, and

> **NOTE**
>
> In this chapter, we use the term *mobile office* interchangeably in the current state and in the future state. The main difference is that the current state resides in a nonwireless environment. The future state resides in a time when all aspects of mobility have been applied to the current state scenario. In other words, the future state is the mobile office nirvana.

sending/printing/faxing that document to the addressee. Of course, executives aren't the only users of the mobile office—almost anyone with a job that requires travel or who needs collaborative access to their business application is a potential user of the mobile office. Let's talk a bit more about how the components of the mobile office look in a current state analysis.

Current State Analysis

We mentioned earlier that this chapter takes a slightly nontraditional approach, at least when compared to the other information systems chapters in this book. Instead of covering detailed application services and concepts, we'll focus a bit more on how current mobile office applications can be integrated as a holistic solution. Let's start by taking a look at Figure 25.1, which illustrates how a mobile office architecture might look.

You'll see that there are two primary touch points, or subsections, in the design with separate integration points and functions for the mobile office. These two subsections are:

- Common office platform.
- Common services.

The mobile office infrastructure comprises all of these components, and potentially many more. Let's cover each of these touch points in a bit more detail.

Common Office Platform

These applications generally consist of standard office applications and traditional front-end tools such as Web browsers, word processors, utilities, graphics viewers, and even enterprise resource planning (ERP) facing tools. The common office also includes advanced office services such as personal databases, graphical editors, and possibly custom internal applications for time and expense, personal information managers, and groupware. Finally, the critical piece that integrates the standard and advanced clients is some sort of local area network or local office equipment. This component gives our mobile office users a connection to the back-end services that are so critical to the collaborative design of the mobile office. Services include:

File sharing services.   In our executive example, file sharing integrates with any standard or advanced office application to seamlessly share files or data to accomplish tasks.

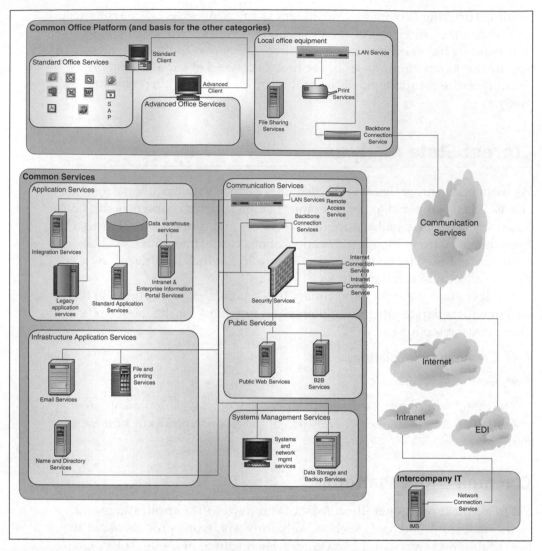

Figure 25.1 Mobile office current state architecture framework.

Print services Standard printing and, potentially, faxing capabilities.

LAN services. Connects the clients to the network services.

Backbone connection services Serves as the liaison between the LAN and the communication services components.

Common Services

The true backbone of the mobile office infrastructure resides in the common services components. Each of the services within the common services component plays an important role in the mobile office architecture. The reason that all of these components together make up common services is that they are common to all the users. Whether the application calls for an ERP tool or an email application, all of the tools are common to everyone in the mobile office enterprise The five services that are important to our discussion are:

1. Application services.
2. Infrastructure application services.
3. Communication services.
4. Public services.
5. Systems management service.

Let's get right into a discussion on each of the services in greater detail.

Application Services

Application services are the heart of the mobile office infrastructure. In Figure 25.1, we've outlined several possibilities that may be served by the application services. Most of the applications that the user is in touch with on a daily basis reside here. The five main application services in our mobile office architecture are:

Integration services.   Enterprise application integration (EAI) tools and middleware applications that integrate unlike systems with each other.

Data warehouse services.   Business intelligence services supply the executives and analysts with much-needed reports to assist in the strategic business decision-making process. Online analytical processing (OLAP) tools that dynamically generate reports also fit into this category.

Standard application services.   The standard application services are those applications that are the lifeblood of the common user. Those applications may vary depending upon the user, but they may include time and expense applications, project management or costing tools that are specific to your organization, or sales force automation and CRM tools—any intra-company application that is common to your user.

Intranet and enterprise information portal services.   These are the services that provide the knowledge management and intellectual capital possible for the enterprise. Besides the intranet, Web-enabled news, financial analysis, human resources, community home spaces, and general company knowledge bases fit into this category.

Legacy application services.   This is the system service that nobody wants to talk about, but is a must-have for any mobile office architecture. Terabits of data can reside in mainframes, old ERP systems, archived financial data, and countless other databases. The legacy systems can quite often be considered antiques, like that old mainframe that looks like it's been sitting in your data center longer than your company has been around. Regardless of their age, these are components that store data necessary to company operations and will always play a role in the mobile office.

Infrastructure Application Services

Certainly, the mobile office infrastructure would not be complete without the services that are used the most on a day-to-day basis. These applications generally are not associated with any specific business process, like accounting or human resources, but are commonly employed by all the users of the organization.

Generally, the infrastructure application services can include, but are not limited to, the following:

Electronic mail/groupware services. The components that make up the email services can certainly vary based upon how complex the infrastructure is and how many clients it must serve. In general, email/groupware applications are usually package-based on such programs as Lotus Notes/Domino, Microsoft Exchange, and Novell Groupwise.

As Figure 25.1 illustrates, the email/groupware services require connections to the backbone communication services that allow the users in the common office a platform to leverage their services. You may be thinking, *How does the email/groupware services architecture differ from what I am using today?* The answer is simply, *collaboration.* It's rare for an organization to exist in today's Internet world without email or some sort of group communication toolset. But the mobile office is much more than simply giving email access to your users; it's a common forum where business users can access all of their critical applications with ease. For example, instead of simply using Lotus Notes for email in the nonmobile office environment, why not allow them to access a data warehouse report and react to information in that report through an automated email message containing relevant portions of

> **NOTE**
>
> This book and its authors do not necessarily recommend one specific email/groupware package over another. The intent of the mobile office architecture and this chapter is to show how the components within the infrastructure can be integrated as one collaborative solution. There may also be components that are not discussed in this chapter that could fit in the design.

that report sent to the decision makers who can immediately react to that information? Let's continue with the other areas of the infrastructure application services, which will help this to make more sense.

File and printing services.   Generally, file and print services are a necessity in any enterprise architecture. These important services provide the storage and accessibility for sharing files and printing documents throughout the mobile office. File and printing services and email/ groupware leverage each other when necessary, as in email attachments or printing documents.

Intranet services.   One of the foundation tools that almost every large enterprise has implemented over the years is intranet services. The company intranet provides for Web-enabled accessibility to company information, knowledge bases, human resources data, and a multitude of other applications.

Name and directory services.   One of the areas that probably doesn't get enough attention in any mobile office design is the name and directory services. These services are important to the overall architecture because they provide the authentication of users. Components such as Lightweight Directory Access Protocol (LDAP) are used here to store information about the corporate users to allow them access to certain applications within the mobile office. You can see again in Figure 25.1 that the name and directory services are directly connected to the mobile office backbone to provide data about the users when called upon.

Communication Services

The brains of the mobile office reside in the communication services segment. Again, as Figure 25.1 depicts, the communication services act as the primary backbone connections for LAN connections, remote user connections, and Internet firewalls. These data architecture components allow the entire mobile office to communicate and collaborate as one. The primary segments served by the communication services are as follows:

LAN services.   The LAN services are a local area network connectivity mechanism that allows all the users in the network to communicate.

Remote access services.   Since not all of your users are stationary, remote access services are required to allow the mobile (not necessarily wireless) user access to the mobile office network.

Security services.   Firewalls and antihacking security tools make up the security services components. Security is a primary concern for every architect, as well it should be. (Chapter 6 covers security in greater detail.)

Backbone connection services. As described in the common office platform section, backbone services serve as the primary mechanism to route data and bind networks together.

Internet connection services. Serves as the router to the Internet. This service can be as simple or as complex as you'd like. But usually it serves a similar purpose of connecting your users to the Internet backbone.

Public Services

Access to the outside world is obtained through the public services, as outlined in Figure 25.1. In our mobile office architecture, these services include the following:

B2B services. As Part 2 describes, exchanges are fast becoming a mainstay enterprise application. B2B services in the mobile office architecture allow users accessibility to the exchange functionality.

Public Web services. What would the mobile office be without Web services? The public Web services house the Web server components that allow the corporate Web site to coexist on the Internet.

Systems Management Services

The systems management services are those applications that are generally used only by the administrators of the network and rarely even seen by the user community. Those applications are:

Systems and network management services. These are mostly package-based network administration tools that may or may not come with your network operating system. Usually Windows NT, Novell Netware, Unix, Linux, and a host of other systems come with some sort of system administration tools. It's not unheard of for network administrators to use custom network management tools or shareware, for that matter, as well.

Data storage and backup services. Backups are usually the part of the network administrator's job that is easily forgotten, until a fire needs to be put out. Suddenly, the backup service is their best friend. Again, these are components of the mobile office that are never seen by the user and sometimes rarely seen by the network administrators, depending upon how well the service is automated.

To quickly summarize, the current state architecture of the mobile office is probably not too unlike what many IT architects already use today. Many of the key services, such as email, Internet access, and ERP, are fairly common in the IT environment. It's important to reiterate, however, the collaborative nature of these applications and what's behind all of them that makes them tick. Let's take a look at the future state architecture and see how wireless technology enables the mobile executive through our principle-based approach.

Future State Analysis

We will take a different approach to describing the future state of mobile office applications than we did in previous information systems chapters. In the previous chapters, we described the structure of a specific system built for a specific purpose. These systems consisted of integrated products and custom components to make sure they worked as designed. The mobile office consists of several different applications that are either bought off the shelf or are extensions of existing systems.

If we took the same approach as in the previous chapters, we would describe how to build an email system or how to build an extension to an ERP system. Instead of doing this, we will focus on giving you a high-level idea of how different business problems can be solved using different applications. At the end of the chapter, we will summarize by offering some guiding principles to show you how implementing the mobile executive dashboard will change your office environment.

Before we get started, let's review the mobile executive dashboard functionality. Since you have already read Chapter 23 describing Mr. Smith's day, we will just list the different types of applications that he uses:

Email.   Mr. Smith uses email extensively to communicate with his coworkers.

Web browsing.   Mr. Smith browses the Web to get the latest news and to check the credit ratings of clients using an online service.

Reporting.   Mr. Smith receives several reports about Acme's performance. All of the reports are based on specific queries he makes to the back-end systems.

Alerts.   Acme sends alerts to their employees based on the preferences of the specific employees.

Hotel check-in.   Mr. Smith checks into his hotel using his PDA.

Video chat.   Mr. Smith uses video chat to call his family.

We will now describe how the mobile office can provide this functionality. While reading these sections, please reference Figure 25.2.

Email

Acme already has an email system. Providing the email system for wireless devices basically means that the device has to have a functioning email client

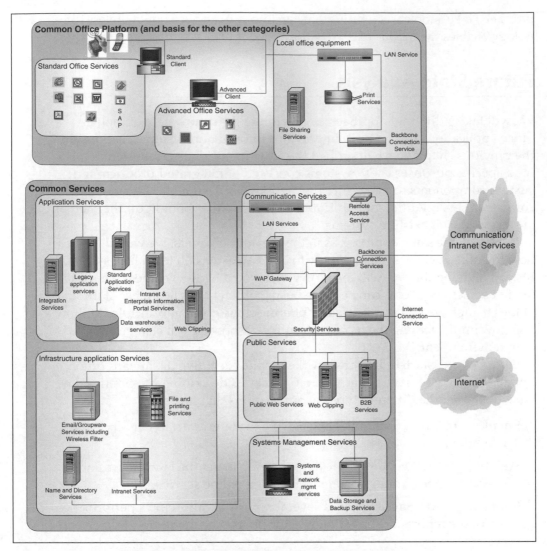

Figure 25.2 Entire state of the mobile office architecture.

and that the email server is modified to adjust messages to wireless devices. This doesn't mean that the email server is rewritten, but that there may be some filter that adjusts the messages when they are transferred to the wireless device. This filter resides within Acme. You can find it in the infrastructure application services on the same server as email/groupware services.

To illustrate, when Mr. Smith picks up his email using his laptop, he has a Microsoft Outlook client that goes to his mail server and gets his email. He gets the whole message with all attachments, viewing them as he goes along. If he now does the same with his WAP-enabled device using an email client written for this environment, he gets only the text in the message because his WAP device is not powerful enough to get the whole thing. The attachments are stripped off on the server side by a filter that allows Mr. Smith to access his email. Finally, let's say that Mr. Smith is using his PocketPC to get his email. He is using a reduced version of Microsoft Outlook here, but he gets all attachments up to a certain size. If they are too large, the server strips them off.

Based on which PDA or wireless device you choose, different email clients will be available. In general, these clients should work with the email systems that are already in place or at least access email in a standards-based way. By *standards-based*, we mean using something like IMAP or POP3.

Web Browsing

The wireless device that is used will also determine the Web browser. In Mr. Smith's case, a PocketPC is the primary platform. Given that, a reduced version of Microsoft's Internet Explorer will be the likely browser choice. If a Palm Pilot is being used, other browsers may be available.

Three approaches to Web browsing are possible. First, there is WAP, which requires a gateway and specifically written pages to access the desired information. The second alternative is Web clipping, which converts existing HTML pages to pages that can be used on smaller screens using small devices. Finally, there is standard Web browsing, which would likely be best served by the addition of a transcoding engine. Strictly speaking, this is no different from what we see on the PC. However, most wireless devices will have smaller screens and some adjustments may be necessary to existing content and systems to ensure that the information can be viewed on the wireless device.

Acme has no control over external content, but they may have to make some adjustments to their internal systems in the application services and public services areas to support the wireless devices of their choice. The changes refer to adding application logic to create pages for a WAP browser or implementing a Web clipping mechanism. In the wireless dashboard, we saw this in the context of the reporting requirements. In order to access information in the Acme ERP system, someone—whether it's the ERP vendor, Acme, or a third-party product that can handle device conversions—has to support a wireless interface.

You can also see that we have added some servers in Figure 25.2. Web clipping servers and transcoding servers have been added in the application services and public services, while a WAP gateway has been added in the communication services.

Reporting

Reporting is a little more interesting. Generally, Mr. Smith needs information from Acme's ERP and supply chain management systems located in the application services area; for instance, during his presentation in Dallas, he needed to know how his sales team performed against the national average. These systems will be packaged solutions in most cases and will have some kind of reporting functionality. The real question is whether this functionality can be accessed in a wireless way. The easy answer is that many of the large vendors are already providing this kind of functionality in a limited fashion. However, many companies are probably interested in having a more short-term solution. Several approaches are possible:

- Acme can buy a reporting system and extract data from their existing systems. The major effort here will be to make sure that the data can be extracted in a cost-effective way and that the reporting tool has the ability to show reports effectively on wireless devices.

- Acme can custom-build this functionality. This also means extracting data in a cost-effective way. The difference between this and the previous option is that the reporting functionality must be custom built. This is not a preferable solution, as this kind of functionality will be available from several vendors in the very near future.

We have not added anything to Figure 25.2, as the long-term solution is preferred here. However, you would have to add a reporting server in the application services section to implement the first short-term solution.

Alerts

In the near term, all alerts at Acme can be sent as email messages or SMS messages. Acme also needs an application that can generate the alerts. If we look at Mr. Smith's day closely, he receives alerts either from his assistant or from the ERP and supply chain systems. An assistant sending an alert is a manual process and does not need to be solved here. The interesting issue to solve from an application perspective is how to send alerts from a system.

In previous chapters, we have sent these alerts through a custom application. This is not the preferred solution here, since Acme already has systems that have all the relevant data available to them. Ideally, the ERP and supply chain systems should support the configuration and sending of the alerts.

In the short term, this may not be supported by many of the vendors. If the functionality is urgently needed, it can be custom built by writing an application that allows users to configure alerts and run all the appropriate business logic to detect when an alert needs to be sent. In addition, it needs to format the message as an email or SMS message and send it. We have already described this kind of solution in the other chapters of this book (see Chapter 20, for example) and we will not describe it here since it's not the preferred solution.

Hotel Check-in

Many of us, on occasion, have had to wait in line for some time to check into a hotel. Checking in via wireless technology would certainly speed up the process. Let's think about what has to happen to check into a hotel. First, the hotel has to identify you. Once they know who you are, they have to have payment security from you, assign you a room, and give you a key.

Your wireless device plays the largest role in identifying you for the hotel. In an ideal world, we would have some standard way to identify ourselves to any other device. Perhaps this can be accomplished by an application run in the hotel that asks your wireless device who you are. A corresponding application on the wireless device would answer, using some secure way to do so to avoid fake identification. The hotel application could then ask for your credit information, which would then be transferred in a secure way.

What would these applications look like using today's technology? Most likely, the hotel application would initiate a request using the Secure Socket Layer (SSL) protocol. Your PDA would ask you whether you wanted to respond and would send the response using SSL and your personal digital certificate. This certificate would identify you to the hotel. Once this transaction is complete, a similar transaction would take place to provide your credit information.

As already mentioned, it would be ideal for this to be standardized and we hope to see this in the future. Until then, it may be the case that the hotel chain you are staying with will provide the wireless application for you.

Wireless Wallet

Wireless wallets allow users to pay for services online in a secure and traceable way. In general, wireless wallet functionality is implemented using a package solution, as is used in conjunction with secure transfer protocols. Future wireless wallet functionality may include being able to use your fingerprint or your smart card to authenticate yourself to the wireless wallet and to launch a transaction.

Video Chat

Mr. Smith has most likely bought a product to implement his wireless chat capability. This product should allow him to make a video call to his family from his hotel room. Ideally, no back-end server is needed to implement this, though there will be some variation in the product architectures.

We will likely see many different products with this capability once bandwidth to wireless devices is adequate to support such transactions and once wireless devices are powerful enough to run the software.

Summary

Implementing the executive dashboard does not mean that you have to implement all new technology, but it does mean that you will have to deploy some additional services within your existing applications and environment. Here are the basic principles to guide these changes:

- We may see increased integration between existing applications. If we go back to the reporting functionality for Acme, we can see that there may be a need for Acme to implement a custom solution that integrates a reporting tool with their ERP and supply chain systems.

- One of the primary benefits of the executive dashboard is access to critical data. This may require some companies to build additional data warehousing services.

- Intranet and enterprise information portal services will also support wireless access.

- Additional products focusing on collaboration may be deployed to implement new ways to communicate.

- The security around wireless services will be absolutely critical. At this point, there are many different and inconsistent solutions that are being implemented on different devices.

The primary changes lie in the application services and public services areas, where many of the applications that will be accessible through the wireless devices reside. In addition, Acme would have to deploy some gateways to enable certain functionality.

CHAPTER 26

Technical Infrastructure Architecture

Introduction

Our executive dashboard application provides Mr. Smith, the Acme Toys Senior VP, with time-critical information as he manages a very hectic schedule. In the earlier chapters we saw him downloading large office files and conducting a videoconference. Typically, we think of such wireless scenarios as futuristic. As technologists, we dream of them as doable, ultimately when the carriers deploy their 3G infrastructures. What may come as a big surprise, though, is that these applications are feasible today! The executive dashboard, as described in our case study, can be supported through the use of open standards and technologies such as the Institute of Electrical and Electronics Engineers (IEEE) 802.11b, IPSec, Mobile IP standards, and virtual private networks.

In this chapter we look at the overall technologies that support our executive while he travels inside and outside the Acme organization. We'll take a look at the different topologies of 802.11b and their technical considerations. We will also examine the issues associated with roaming and how to address them through the use of Mobile IP.

Technology Infrastructure

Figure 26.1 shows our executive traveling across the country to various meetings in different geographic locations. Some of these are facilities owned by Acme and are thus within their network control. Other locations lie outside of the corporation and their jurisdiction. The top left corner of the figure illustrates an off-premise, early-morning client meeting in Seattle. Later that same morning, we find our VP in the executive lounge at the airport, waiting to make a connecting flight to Dallas. This flight takes him to an executive meeting to review performance data and industry outlook statistics for a board meeting scheduled late that evening in Chicago. The next day, he will travel from Chicago to New York to make a public statement to the markets.

Throughout this trip our VP must have access to information in real time and be able to execute his duties. Whether on site or off, this executive needs access via a simple tool that can store and transmit time-critical information. It must be a device that is robust, easy to use, and capable of using a broadband wireless network. This technology is available today, through a fixed wireless solution known as IEEE 802.11b and a PDA based on the PocketPC platform.

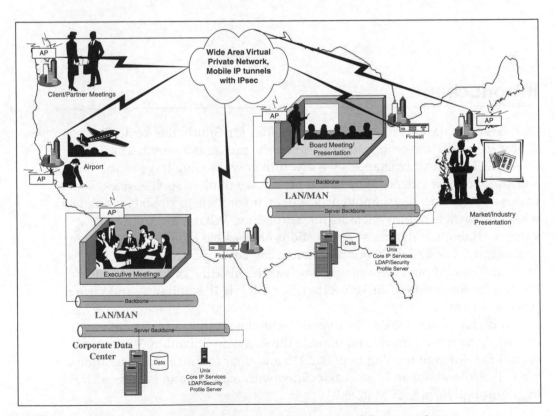

Figure 26.1 Mobile office application utilizing 802.11b and Mobile IP roaming.

The 802.11b standard, also known as the high data rate standard, has received backing from several global regulatory bodies and industry vendor alliances. The market potential of 802.11b is very promising, as can be witnessed by the entrance and promotion of companies such as Cisco, Nortel, 3Com, and other giants into the space. Many wireless industry leaders have united to form the Wireless Ethernet Compatibility Alliance (WECA). WECA certifies vendor equipment related to interoperability and compatibility with IEEE 802.11b. WECA's WiFi (Wireless Fidelity) label is your warranty that products will interoperate with each other. Still, there are many arguments that favor standardizing with a single vendor when possible.

The true value of 802.11b is its data rate that supports wireless throughput comparable to wire-speed Ethernet. An additional benefit is the fact that 802.11b is compatible at the physical and data link layers with IEEE 802.3 Ethernet technologies widely deployed around the globe. The primary features and services of 802.11b were defined by the original architecture of the 802.11 standard. The new specifications only affect the physical layer in the area of robust connectivity and higher data rates. The 802.11 standard defines two pieces of equipment: (1) a wireless station, that is, a computing device with a network interface card, and (2) an *access point* (AP) that acts as a bridge between the wired 802.3 and wireless 802.11 networks. The AP performs much the same way as a base station in cellular networks. It aggregates the access of multiple wireless stations to the wired network and transmits information to these stations.

Three physical layers were originally defined in 802.11: two spread-spectrum (SS) radio techniques—one using *frequency hopping* (FH) and the other using *direct sequence* (DS)—and a diffuse infrared specification. The two spread-spectrum techniques have caused some confusion and competition in the industry. However, this was addressed in the 802.11b specification that supports only DS because of the technology used to achieve higher data rates. This results from the FCC regulation restricting subchannel bandwidth to 1 MHz. This makes FH systems spread their usage across the entire 2.4-GHz band. The result is that FH systems must hop often, which leads to an overhead limiting the higher rates. It also means more users—something that can't be achieved with DS. DS offers superior range and more robust performance than FH. However, it does not scale as well in terms of its number of users. The 802.11 standards operate in the 2.4-GHz spectrums of the Industrial, Scientific, and Medical (ISM) band. These are globally recognized by international regulatory agencies as dedicated for unlicensed radio operations. This means that 802.11 products do not require licensing by the user or other industry players. As such, though, 802.11b is subject to interference caused by other devices (like microwave ovens) operating in this spectrum.

It's important to keep in mind that 802.11b is a local area network (LAN) technology. It was designed to cover a set of floors in a building or even several buildings. It functions at the data link and physical layers of the Open Systems Interconnect (OSI) model. This includes the *logical link control* (LLC) for data, IEEE802.2, as well the *medium access control* (MAC) for the link. It is not a metropolitan area network (MAN) technology network (128 Kbps). It is not a wide area network (WAN) technology like Cellular Digital Packet Data (CDPD, 14.4 Kbps). As a LAN technology 802.11b can operate at 11 Mbps, 5.5 Mbps, 2Mbps, and 1 Mbps. What is important is its ability to carry higher network, transport, and session layer protocols such as TCP/IP, IPSec, and SSL128-bit encryption, among the many other network protocols in use today. This also means it can utilize the wired Internet and its available throughput rates for MAN and WAN connections. We will look at this infrastructure shortly, after we have covered some of the network topologies described by 802.11. First, though, we should take note of emerging technologies in the ISM bands as well as personal area network (PAN) technologies such as Bluetooth and IEEE 802.15.

Additional Wireless Standards

IEEE802.11a is designed to operate in the 5-GHz ISM bands to deliver up to 54 Mbps. This spectrum is not recognized around the world by the regulatory agencies as is the 2.4-GHz spectrum, but effort is under way to open the bands globally. Also, we should keep in mind that, while existing access points may have upgradeable radios, they would also need the ability to support the 54-Mbps data rate at the network interface to the wired LAN in order for such an upgrade to be effective. This could require new APs and possibly a newly wired infrastructure if the current architecture is not sufficient. The latter is unlikely these days. Still, it warrants consideration. Another technology known as HiperLAN2 is also being developed to work in the same spectrum. However, it uses a different MAC, which is still in development. 802.11a is already written.

Another working group, HomeRF, is developing a wireless home networking protocol for many different types of home-based cordless appliances. HomeRF uses the ISM band and can achieve rates of up to 1.6 Mbps at about 150 feet. While this technology might become a gateway in some residential homes, it will not compete against 802.11 protocols designed for enterprise WLANs and other fixed wireless Internet services. The Home RF group is petitioning the FCC for rules modifications that will permit high-speed FH using the 5-MHz channels.

Bluetooth (BT) technology and standards provide a means for replacing cable that connects one device to another via a radio link. Even so, it is now being viewed as a technique for connecting several devices to create small radio LANs and/or PANs (called piconets). The big promise of BT is that its chips will be made small, consume little power, and remain very inexpensive.

A BT Special Interest Group (SIG) was formed in February 1998. The group announced itself globally in May with the hopes of forming a de facto standard for the BT interface and software. The SIG's mantra is that BT products are about to sweep the manufacturing world and take the consumer market by storm. The reality is that many questions need to be answered by manufacturers on how best to realize BT in order to drive critical mass in its adoption.

Bluetooth technology offers some rather unique opportunities and applications. However, time-to-market delays could rush chip set development and overlook cost, functionality, and ease of use. This challenge could create another technology fragment adopted only in certain niches and not with the universal application envisioned by its founding members. Competing technology and standards adaptation issues still need to be decided upon. Without a doubt, we are already seeing expensive BT products on the market, with nothing to talk to.

BT is still in a development stage, leaving several architectural and technical considerations to be answered. We are likely to see an evolution to universal market penetration by 2003 if the SIG traverses the dangerous terrain of compatibility, functionality, ease of use, and cost.

Another challenge is the fact that the IEEE 802.11b standard and Bluetooth are known to cause interference problems with each other. The size of the problem often depends on the platform the company supports. Viewed from the user perspective, the interference may not be noticeable except in large data file transfers. Recent studies seem to support this viewpoint, showing that interference is not an issue within 20 feet of an access point. However, at high power, the range of a Bluetooth device is about 30 feet. BT proponents often consider 802.11b as an enterprise solution due to its power requirements, cost, and range. They see 802.11b not addressing the interconnectivity of multiple devices in PANs. Such a view ignores 802.15.1, which is basically BT expressed as an IEEE standard below the BT Logical Link Control and Adaptation Protocol (L2CAP) layer.

Version 1.1 of the BT standard is close to being approved. To date, there are no basic or fundamental changes in BT v1.1, aside from edits, clarifications, and errata. However, the version does cover issues of interoperability between the ever evolving and growing numbers of profiles created by the BT specification. On the protocol front, the IEEE 802.15 working group, chartered by Lucent Technologies and others, is an interesting solution. It will eventually scale to higher bandwidths and provides a more robust specification and methodology. Following are the task groups for 802.15:

802.15.1 Task Group 1 on 1-Mbps Wireless PAN (WPAN) based on the Bluetooth Radio 1 Specification, authorized March 1999.

802.15.2. Task Group 2 on Coexistence for Wireless Applications operating in the Unlicensed Bands, authorized January 2000.

    802.15.3.   Task Group 3 on 20+ Mbps WPANs for Multimedia and High Performance Digital Imaging, authorized March 2000.

    802.15.4.   Task Group 4 on Low Rate WPANs for less than 100 Kbps and very long battery life applications, authorized November 2000.

Because of its price and lower power consumption, Bluetooth technology offers some enticing and imaginative applications in the wireless space. However, its time-to-market delays could rush chip set development and overlook cost, functionality, and ease of use. The SIG, developers, and manufacturers need to be especially active and judicious in the area of BT embedded systems, drivers, and system integration. Bluetooth's play in the application space will increase at the time when its industry has successfully addressed interoperability/functionality and ease of use concerns.

802.11b Topologies

There are two basic topologies, or *modes*, that are defined by the 802.11 standard. One is often called an *ad hoc* mode, where access to a wired network is not needed. This is a peer-to-peer mode usually set up where a connection to wired network service is not required or where a wireless network may not exist. This mode is referred to as an independent *basic service set* (BSS). Examples include a single room or an entire floor such as in a hospital, or two people meeting to exchange information at a dinner meeting. The second mode is often called an *infrastructure mode*. Here wireless stations need access to network resources such as file servers, printers and fax machines, and other wireless and wired computers. However, such terminology can be misleading. What the specification defines is something called an *extended service set* (ESS).

A BSS consists of a single cell created by two stations with a definite propagation boundary. An ESS consists of multiple cells interconnected by access points and a distribution system where communication needs exceed the range of a single BSS. We should note that for the majority of our application we are concerned with the infrastructure mode or ESS topology. There are situations where a BSS can function for an executive or anybody needing to exchange data with another wireless or wired entity. The setup for the BSS is fairly simple and straightforward. Also, the use of IR 802.11b for secure wireless links between buildings should be mentioned here but is beyond the scope of our application. We will instead focus our attention on the issues concerning connections to enterprise systems and mobility.

Figure 26.2 shows a collection of BSSs in a corporate setting, grouped together as an ESS with Ethernet as the distribution environment. The ESS supports cell roaming and handoff via the APs and other networking technologies. The standard presently acknowledges three types of mobility: (1) A

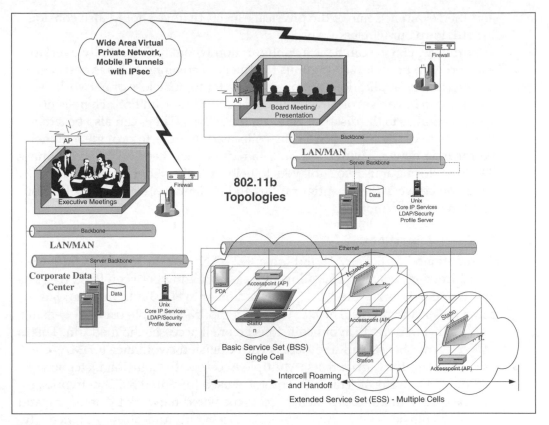

Figure 26.2   802.11b Topologies.

situation where stations do not move or at least move within a local BSS is referred to as a *no-transition mobility state.* (2) When a station or stations move from one BSS to another BSS within the same ESS we have a *BSS transition mobility state.* (3) When station(s) move from one BSS to another BSS within a different ESS we have condition called an *ESS transition.* The 802.11 standard clearly supports the first two states. However, it doesn't ensure that a connection will continue when making an ESS transition.

The standard identifies the distribution system as an element that connects the BSSs within an ESS through the use of access points. An AP is an address-able station that handles the mapping of an address to destination function and integrates multiple BSSs. The BSSs and ESS network are invisible to the LLC layer. The distribution component of 802.11 provides the logical services, but the standard does not limit the makeup of the system. It can be 802 compliant such as Ethernet or Token Ring or some other network that provides a logical integration point known as a *portal.* In the case where the wireless network complies with the 802 standards, the access point becomes the portal. Let's look at the BSS configurations that 802.11 will support within an ESS and then examine some of the physical aspects that we should also consider in setting up our wireless LAN.

Within an ESS we can have a configuration where BSSs partially overlap each other to provide a contiguous treatment of coverage. This is best where a synchronous application cannot tolerate a temporary loss of network use. We can also have a set of physically disjointed BSSs in an ESS. There is no specification as to the distance between these BSSs. There can also be situations where two BSSs are physically collocated in order to provide a higher-performing network with some built-in redundancy. Let's take a look at how 802.11 performs at its two sublayers to help us make better decisions for choosing between these configurations and understanding their implications in relationship to mobility.

MAC Layer Functions

As mentioned earlier, 802.11 and other 802 LANs use the same LLC and 48-bit addressing schemes. This allows for easy bridging between IEEE wireless and wired networks. The 802.11 MAC is similar to 802.3 Ethernet, but has some major differences. The 802.11 MAC supports multiple users by sensing the channel in order to avoid collisions before it accesses the medium. This is called Carrier Sense Multiple Access with Collision Avoidance (CSMA/CA) as opposed to the Carrier Sense Multiple Access with Collision Detection (CSMA/CD) protocol. This is the protocol used for wired 802.3 networks. In this case the wireless components access the wired network by detecting and handling collisions when two or more devices communicate simultaneously. In an 802.11 WLAN, collision detection is not performed because a radio transmission overpowers the station's ability to hear, or detect, a collision. Therefore, the 802.11 standard uses CSMA/CA or what is sometimes called the *Distributed Coordination Function* (DCF). CSMA/CA attempts to avoid collisions by using unambiguous packet acknowledgment (ACK). This means that an ACK packet is sent by the receiving station to confirm that the data packet was received in its entirety.

CSMA/CA works in this manner. A station senses the channel for activity and, if nothing is detected, it will wait a random period of time and then transmit its packet if the channel is still clear. When the packet is received intact, the receiving station sends an ACK frame that completes the process when the sender receives it. If the ACK is not detected, a collision is assumed and the packet is retransmitted in the same manner. Thus, CSMA/CA provides for a shared medium the same as CSMA/CD. The ACK function also enables the protocol to effectively handle interference and other radio-related problems.

Sometimes two stations can hear an AP from opposite ends but not each other due to distance or an obstruction. The 802.11 standard defines an optional Request to Send/Clear to Send (RTS/CTS) protocol. With this feature, a sending station sends a request to send and waits for the AP to send a

CTS. Every station in a BSS is supposed to hear the access point. Hearing CTS causes them to delay a transmission in order for the sending station to send its packet and receive a packet acknowledgment. RTS/CTS adds overhead by temporarily reserving the medium. As such, it is used only for large packets of data where retransmission would be an even greater overhead.

The 802.11 MAC layer provides two other features: *cyclic redundancy checking* (CRC) and *packet fragmentation.* Each packet has a CRC checksum to ensure that the data was not corrupted in transit. In 802.3 Ethernet, the higher-layer protocols handle such error checking. Packet fragmentation breaks large packets into smaller units before sending them. The MAC layer reassembles the fragments and is not seen by the higher protocol layers. This can be useful in congested areas or when interference is an important issue. This is because big packets are more likely to become corrupted. Packet fragmentation, like RTS/CTS and CSMA/CA, improves the overall performance by reducing the need for retransmissions.

Another interesting and proactive aspect of the 802.11 MAC specification is the *point coordination function* (PCF) which can support time-sensitive application data such as voice and video. In PCF mode the AP controls access to the channel. A BSS set up in this mode has its time spliced. When a BSS has switched to this mode, the access point will poll each station for data and, after a set time, move on to the next station. The stations are not allowed to send data until it's polled. Similarly, stations receive data only when they are polled. This setup guarantees a maximum latency. However, this scheme becomes impractical in large networks, and so it carries with it some issues of scale.

Additionally, the DS signaling for 802.11b divides the 2.4-GHz ISM band into 14 channels. Under U.S. FCC rules there are 11 distinct channels. This factors out such that we can use no more than 3 channels in any one physical location or BSS cell. In addition to such performance improvements, DS provides robustness by using a technique called *chipping.* DS must send data across one of these channels without hopping to other channels. Chipping is used to compensate for noise on a channel. Each bit of data is converted into a sequence of bit patterns called *chips.* The inherent redundancy of each chip mixed with the spreading of the signal across a single channel provides for a unique form of robustness. Data can still be recovered from the chip even when part of the signal is damaged. Chipping reduces the need for retransmission.

To support noisy environments and extend its range, 802.11b uses dynamic rate shifting. This feature can automatically adjust data rates and counteract for changes in the radio channel. When devices move beyond the optimal range for operation at the full rate or if there is a lot of noise, the device transmits at lower speeds, decreasing to 5.5 Mbps, 2 Mbps, or 1 Mbps, for example. If a device moves to within an AP range for a higher transmission rate or the signal-to-noise ratios improve, the transmission rate adjusts automatically to the higher rate. Rate shifting is a physical layer mechanism. It is invisible to the higher layers in the protocol stack.

Installing multiple access points on the same frequency will increase the span and fault tolerance of the WLAN. When one access point in a segment fails, the stations can roam to the other access points without an interruption of service. However, this won't increase the bandwidth. Then, if we place three stations on three different frequencies, we can improve the bandwidth performance. Doing this is similar to divide-and-conquer strategies used in wired networks today, whereby a switch provides the necessary segmentation that a repeater cannot give. Typically, considering the signal-to-noise and protocol overheads, WLANs get about 5+ Mbps of actual throughput for each wireless channel used. Our three channels will add up to a throughput of around 15 Mbps. As we mentioned earlier, under U.S. FCC rules there are 11 distinct channels available. In practice we can have only three operational frequencies in any given area. This is similar to cellular networks, where we must deploy our cells in such a way that adjoining cells do not utilize the same frequencies and thus do not overlap. If cells do share the same or over-lapping frequencies, then there is a chance for some interference.

Cell Roaming

The 802.11 standard specifies services for providing the functions required by the LLC to send *MAC service data units* (MSDUs). These are broken out into two categories: station services and the distribution system services.

Station services were defined for providing functions between stations such as our executive's wireless PDA or a wirelessly enabled fax machine. The nature of the wireless medium opens it up to eavesdropping by those inside a building and on the outside of a structure. Because of such integrity issues, WLANs use authentication, deauthentication, and privacy services in their communications with each other.

Distribution system services are required in order to provide functionality across the distribution system in delivering the MSDUs. Each station must associate with only one AP. However, an AP will associate with many stations. This maps a station to the distribution system via an AP. It also sets the basis for station mobility between BSSs. Station and APs may disassociate from each other. This is a notification that neither device can refuse. Stations will disassociate when leaving a BSS. A station may do this in order to reassociate with another BSS or a stronger signal. This informs the distribution system of the mapping between access point and station as the station moves about from BSS to BSS within an ESS but not between ESSs! We'll discuss more of this latter aspect a bit later in this chapter. For now it will suffice to say that, in this situation, integration services enable the delivery of MAC frames through a portal between the WLAN and other non-802.11 networks.

The 802.11 standard also does not define an AP handoff for a station roaming from one BSS to another. Some vendors use proprietary handoff protocols to coordinate station handoffs within an ESS. What we typically find is a mechanism whereby the mobile station will monitor the signal strength or, rather, the quality of communication with a current AP. If the quality drops below a certain threshold, then the station will begin searching for another cell with better quality and switch to that cell or an alternate AP within the same cell. Some APs encode information for quality checks into a beacon message that can also contain other WLAN domain and data information. For example, an 802.11 MAC supports power conservation for portable devices. There are two power modes: *continuously aware* and *power save* polling modes. The continuous mode is obvious: the station is always on and drawing power. In the power save mode, the station is half asleep and the AP is queuing data destined for this station. The station will wake up at set intervals and receive its data. There are many other beacon and station parameters that can be watched and set up, such as responsiveness and time-outs based upon the nature of the application. If it's an interactive application, the responsiveness or the beacon frequency may need to be set high. The sensitivity or frequency with which a station performs a cell search can also be set.

Many large corporations have worked together to develop an Inter-Access Point Protocol (IAPP) in order to provide a common roaming protocol to support roaming between multiple vendor products. IAPP develops upon 802.11 by using the distribution system interfaces of the APs. It functions between the APs, using the User Data Protocol (UDP) and Internet Protocol (IP) for communication. The specification defines an announce protocol and a handover protocol. The announce protocol informs the network of when a new AP has been added and maintains configuration and station information across the WLAN. The handover protocol tells an AP when a station has reassociated with another AP. When this happens, the old AP will forward its buffered frames for the station to the new AP and the new AP will update the MAC filtering bridge tables. The presence of multiple-vendor roaming protocols, such as the de facto IAPP, should be placed high in consideration when designing and implementing a WLAN.

What about when a station roams beyond the range of an extended service set? When a station roams to a new ESS, it might require a new IP address appropriate for the new subnet. *Dynamic host configuration protocols* (DHCPs) can be used to renew the station's IP address. However, in wireless networks, security becomes an issue with this approach. We don't want to be handing out IP addresses unexpectedly to a bunch of black hats sitting in cars in our corporate parking lot! There are alternatives that can work at the MAC and network layers. We will examine these shortly. One solution for ESS transitions is Mobile IP. With this protocol, a user keeps the same IP address while roaming. This protocol supports certain challenges that we find in the area of virtual private networks (VPNs). First, though, let's take a look at the issues of security for our WLANs.

Security

As we mentioned, we do not want to be assigning IP addresses or granting access permission to unfriendly agents attempting to infiltrate our network from an undisclosed but wirelessly receptive location. However, without certain security measures in place, we are simply broadcasting information for free to anyone within hearing range. With this consideration, the 802.11 standard specified an access and encryption mechanism known as the *Wired Equivalent Privacy* (WEP) protocol.

With WEP, the WLAN service area ID is coded into every AP and is required knowledge in order for a station to gain access and associate with an AP. There is also a provision for maintaining a list of unique MAC addresses as a controlling filter table. The standard uses a shared key system; administration for distributing data representing the correct wireless-encryption key is the responsibility of the network owner. If that key is leaked, the network is compromised. The difficulty of changing keys varies from vendor to vendor. Shared keys can be difficult to distribute in a secure fashion. A unique public/private 128-bit key exchange algorithm, where every session is encrypted, might be more attractive.

In the first quarter of 2001, a group at the University of California at Berkeley was able to intercept and compromise data transmissions by means of holes they had found in the WEP algorithm. This doesn't mean one shouldn't activate the optional WEP to prevent the occasional listener from eavesdropping. Remember, in other sections of this book we stressed a multiple-layer approach to security. The Remote Authentication Dial-In User Service (RADIUS) protocol can be used to add another level of protection that authenticates remote clients to a centralized server. A RADIUS server, as defined in the IETF's RFCs 2138 and 2139, can be used to authenticate the media access control addresses of an organization's 802.11b network interface cards (NICs). This provides complete control of access to the WLAN.

Even though the IEEE has developed a firmware upgrade patch that should soon be available, the security hole still makes the case for stronger security measures being deployed at multiple levels. WLANs can support the same security standards that other 802 LANs implement for access control and encryption such as IPSec. A corporate VPN can also be deployed to secure connections between devices and the corporate network by installing a personal firewall on clients accessing the LAN. These technologies and others can create end-to-end secure networks for both wired LAN and WLAN components.

ESS Transitions via Mobile IP

In smaller wireless LANs where mobility across different ESSs is not a critical issue, one can likely get by with the deployment of static IP addresses. However, this quickly becomes an administration nightmare in larger networks. DHCP (Dynamic Host Configuration Protocol) is one solution often deployed where an IP address is automatically assigned to a device when it powers up on the network. The device leases the IP addressed for a specified amount of time. However, there are some issues we need to consider with the use of TCP/IP over a WLAN. TCP carries overhead that often doesn't have any payload data. Also, the protocol is not very flexible when it comes to wireless LANs. A marginal connection between a station and its AP can cause the protocol to terminate the session, creating a need for the user and/or application to reconnect to the network. Finally, problems arise when a mobile device associated with an AP on one network domain roams to another on a different domain. IP addressing was designed with an assumption that devices connected to a network domain remained stationary.

In situations where 10 or more devices may access a WLAN or where mobility between different domains is critical, one should contemplate using some type of middleware. These products can provide communication by using lightweight protocols and other techniques to move data between TCP/IP devices and mobile stations. Chapter 3 mentioned some of the transcription services that are handled by such middleware. It is important to note that a few middleware products provide home and foreign agent functions to support Mobile IP. These products can be of tremendous benefit to organizations and are used in our executive dashboard application.

Figure 26.3 illustrates a notebook computer labeled *mobile station A* moving across two different network domains, indicated by the presence of a router. This particular scenario involves an executive about to leave her office for an important meeting with our executive, Mr. Smith. This is the same meeting Mr. Smith has just flown to from L.A., and it takes place in the same building in Dallas where she works. One of her direct reports and another coworker developed an important spreadsheet needed for this meeting while together on the same flight into Dallas. The direct report has just arrived in the executive's office, fresh from the flight, and wirelessly transfers the file. Meanwhile the other coworker receives important information that substantially alters the figures. This person quickly incorporates the changes and sends the revisions in an email message with an embedded audible alert to the executive. The executive, however, is now in the meeting located on a different floor and on a different network domain. She has with her the wirelessly enabled notebook, mobile station A. It is still in the same powered-on state as it was when she left her office.

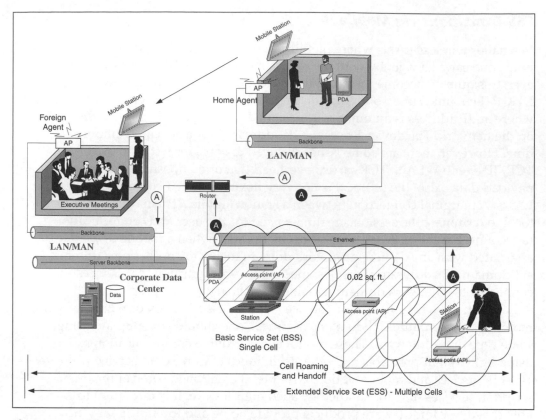

Figure 26.3 ESS transitions LAN/MAN roaming with Mobile IP.

Unfortunately, in this situation mobile station A and its home IP address make it difficult to route mail to the station unless there would be a physical change of the IP address. However, a physical change would require a reboot of the device by the hurried executive and the deployment of a DHCP server. A reboot is very unlikely. This situation is a good example of where Mobile IP saves the day.

Mobile IP uses an address-forwarding function to provide continuous delivery of packets to a mobile station as it moves from one network domain to another. The solution requires no changes to routers or Domain Name Service (DNS) but only a few software components. The components are a mobile node on the station that communicates with the other mobile IP components. This node is built into the TCP/IP stack or can exist as a shim to the stack. The other components are a home agent and a foreign agent. The home agent resides on the home network for mobile station A and forwards packets for the station to its new care-of address, which is the appropriate foreign agent. The foreign agent receives the packets and forwards them to the mobile station.

Mobile IP delivers data payload between the foreign and home agents via a network tunnel. The tunnel has two endpoints, one at the care-of address, the foreign agent, and the other at the home agent. The home agent encapsulates and forwards the data to the foreign agent, which decapsulates it and forwards it to the mobile station. Data sent by the mobile station is usually handled with normal IP routing, without using the home station. While this approach works well, the home agent is a single point of failure. The IPv6 mobility standard enables the mobile node to advertise its care-of address, enabling the correspondent to send its messages directly rather than through a home agent. If an external network doesn't support a foreign agent, then the network will have to assign a temporary IP using DHCP. This can be made transparent to our mobile executive. Finally, in some cases we must properly configure our firewalls for both the home and foreign networks to support Mobile IP traffic.

Figure 26.4 shows Mobile IP being applied to a wireless device across a wide area network and potentially hostile territory. In a situation like this we must ask ourselves, What is the address of the foreign agent?, How can we assure its integrity?, and How do we authenticate the address and its user? The mobile node must have a way to securely and unquestionably be identi-

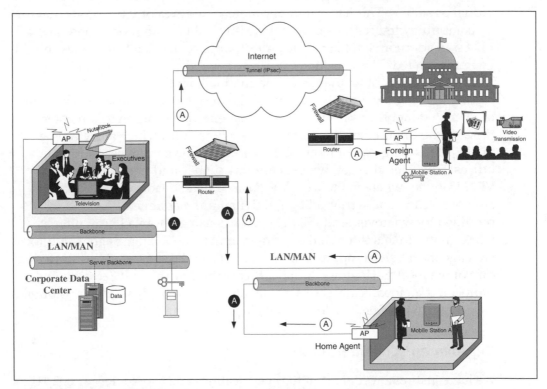

Figure 26.4 ESS transitions WAN roaming with Mobile IP.

fied by its home agent. In addition, we must be able to accomplish this while the node travels from one foreign domain to the next.

Mobile IP can utilize the strongest authentication scheme, whereby a station proves its identity by meeting certain cryptographic challenges and/or secrets. IPv6 Mobility uses IPSec. IPSec should be used with 128-bit encryption and asymmetric public/private-key arrangements that offer the strongest authentication available today. In a message thus signed, the mobile station tells the home agent whenever its care-of address changes. In IPv6 the station tells this to the corresponding node. When it moves from one foreign agent's domain to another, for a while the station is reachable by both. This occurs because the foreign agent the station is leaving agrees to forward information to the new care-of address for a short time. The mobile station— or, better, the mobile node—executes the handoff with the old and new foreign agents and updates the records with its home agent.

We should note that the IPSec private key is the critical information necessary in order to commandeer a network session and impersonate the mobile node and its digital signatures. This attack can originate from within the user's own mobile device. If the key is ever revealed to a system other than the node itself, then our security provision breaks down. In a situation like this, someone else could sign for the mobile node and appear to be that individual or station. This has real implications in the area of virus protection for all computing devices, especially wirelessly enabled and mobile ones like PDAs. In addition to robust virus protection, several other techniques should be implemented. These include expiring or evolving private keys, bio-authentication, or other unique and individual identifiers.

Mobile IP is now being used in VPNs because it overcomes the limitation of restricted mobility. VPNs are a very cost-effective solution for an enterprise, but are limited by the number of locations from which entry can be made. VPNs are great for nonmobile work-at-home employees or remote offices. However, they fail when people are mobile and need to enter the VPN from anywhere in the world. VPNs use static IP addresses. Thus, they are not able to handle temporary and dynamically assigned IP addresses provided by whatever local ISP the person is using. Mobile IP can offer a workaround to this through the use of its mobile node, home agent, and foreign agent. In a sense, our executive is still using the same static IP address in care of the local ISP address. Mobile IP, combined with the IPSec tunneling protocol, is an ideal feature for VPN manufacturers. Because of this solution, we are starting to see it included in their product offerings.

Deployment Considerations

We have just reviewed most of the specifications and topologies for 802.11b networks and discussed the various configurations of its mobile components.

We've also taken a look at roaming within a BSS cell and between cells, and how roaming across subnets and wide area networks can be accomplished. We should now briefly mention a few considerations that should be incorporated into the design and deployment of a WLAN.

The most important activity of deploying a WLAN is the up-front work necessary for the proper placement of access points. Many environments include areas out of range within a wireless LAN. It would be easy to deploy a WLAN if the environment were open and unobstructed. However, the reality is that multiple hurdles, including walls, desks, window blinds, microwave ovens, and other potential sources of interference, must be overcome or at least accommodated. A site survey is always required. We should think in terms of the applications using the WLAN, the potential number of stations, and the maximum/minimum required throughput. We'll need to think in terms of how best to arrange our AP and establish our BSS and ESS cells: where we will need to extend the same frequency bands and where we will multiplex them to increase throughput without additional interference. We should also understand the types of internal and external networks where the mobile stations will potentially reside and how we will provide for transitions to a new BSS or one in a different ESS and network domain. We also need to consider how WLANs are managed. This is fairly straightforward because the 802.11 specifications use the same Simple Network Management Protocol (SNMP) that was developed for other 802 networks. Finally, as always, we will need to carefully think out our security and addressing schemes across multiple levels in order to insulate the organization's network while allowing it to merge with other foreign networks and form new relationships of information exchange.

While this is a rather cursory treatment of the considerations necessary for deploying WLAN, there are references at the end of this book where one can obtain additional information. The intent here was to demonstrate the feasibility for deploying our executive dashboard application through the use of 802.11b, Mobile IP, IPSec, and VPNs. To do this, we attempted to provide a view of the mechanics of this architecture and how the various wireless components work together to support the application.

Summary

This concludes Part Five, which considered the executive dashboard application. We looked at the underlying technical infrastructure supporting the Senior VP, Mr. Smith. The transmission of large data files and the use of time-bounded (synchronous) applications like videoconferencing made us look for a technical solution that works today. In the United States and most of the world, carriers have just begun implementing their 3G networks. These networks could have supported Mr. Smith. However, 802.11b WLANs can offer even more bandwidth than 3G networks and provide for seamless connections to distribution systems using any 802 protocols.

We saw how 802.11b supports Mr. Smith in a personal, ad hoc, one-on-one breakfast meeting. We looked at the technology supporting corporate users and the executive in a WLAN environment. We saw how a WLAN solution could be extended across a wired WAN, providing the mobile executive with a virtually global presence. In essence, this can be accomplished today by using open, nonproprietary, standards and technology such as IEEE 802.11b, IPSec, Mobile IP, and virtual private networks.

Bringing It All Together

The Future of Wireless Technology

After all that has been discussed so far, you might wonder what's next. Trying to predict the future is always a risk, so we'll confine our comments here to trends and directions that are straightforward, and then disclaim any ability to define the detailed timing.

The discussion is broken into four main areas:

1. *Bandwidth and infrastructure.* This section focuses on the implications of raw bandwidth becoming available to mobile users, based on technology in evidence today. The short summary is that we will get more bandwidth and our utilization of it will improve, much as has been evident in the wireline space for the past few decades.

2. *Devices.* This section discusses what is likely to exist in new devices, and some of the problems that need to be overcome. The trend is toward more devices rather than fewer, with the opportunity for devices to satisfy the increasing prevalence of niche markets.

3. *System services.* In this category we include the set of services that are needed to make a wireless application function, but are not necessarily part of the application itself. The trend here is a continuation of the fragmentation of services as new opportunities are identified. The trend to incorporate these into the base operating systems and application

servers will also continue, rendering many of these services as commodities a few years after they are introduced.

4. *Applications*. This is the category that actually gets the work done. While all of the aforementioned are required, the applications themselves provide the value to the user. There are many examples floating around of possible functionality, including many that are of interest to end consumers. We believe the trend will be to focus heavily on the intracompany applications in the near term, and provide real business value to the companies implementing them. Longer term, the consumer applications will begin to take off. Identifying the key applications is the problem here, and the safest prediction is that the next-generation killer application in the wireless space is probably one that has yet to be envisioned!

Bandwidth and Infrastructure

Opening up the bandwidth for higher speeds to become available to more people is the thrust of the carrier industry. Behind all the discussions and concern about the spectrum auction and the amounts that have been bid is the realization that the base ability of a person to connect with the applications to enhance this wireless and mobile experience starts with bandwidth and the controls that have been exercised by the governments around the world. Without the bandwidth, the rest is moot.

The planned build-out of the new spectrum with higher bandwidth is proceeding slowly, with delays due to handset difficulties, capital outlay slowdown, and the normal challenges that scramble even the best plans. These obstacles will be overcome, and 2.5G and 3G will eventually be deployed. The timing will be more extended, and we expect some winnowing of the technology options along the way. However, the infrastructure, which really has to precede the applications, will be delivered over the next 5 to 10 years. The comment that it is too little, too late will also continue to be heard.

On the other hand, there are emerging connectivity options in the unregulated portions of the spectrum that cannot be ignored. The successes of 802.11b are, in some cases, dramatic. There are grassroots efforts to expand the limited distance that 802.11b allows by linking access points to cover much greater areas, at higher bandwidths than the carriers foresee. The uptake of this connectivity option is proceeding faster than anticipated.

It is too early to predict whether one of the aforementioned strategies will be the winner, or whether both will prove successful. Our general inclination is to believe that both approaches will be pursued, potentially outstripping what has been anticipated for 802.11b, yet falling short of the full 3G direction.

There is a third player in this game: Bluetooth. This technology is not expected to reach significant portions of its potential user groups until 2003, even though it has been hyped greatly for some time now. The hype and predictions for Bluetooth to become a general-purpose technology with widespread acceptance beyond its initial goals are probably misplaced. While the technology is satisfactory for the cable replacement marketplace, there are many barriers to its achieving the penetration that the optimists have forecast.

Devices

Discussing the future of devices in today's environment is an interesting challenge, as this area is changing so rapidly. The most certain prognosis is that device proliferation will be unpredictable in terms of specifics, yet predictable in the variety that will be tried. Device variety will come in several flavors. First, the bundling of various capabilities in specific devices will continue to occur. The interesting part will be the inventiveness of options that will be tried. The current pattern of nontraditional combinations will continue and accelerate over the coming years. Today we are seeing MP3 players in cameras, calendars in phones, PDA and phone combinations, and quite a number of developments that are proving interesting and unexpected to the historically technology-minded people. Even more exotic combinations should be expected in the future.

Among the factors propelling variety in wireless devices is the transformation of interest from being technological to consumer driven. This transformation brings fashion and designer sensibilities into the mix—a much-needed augmentation of the more linear thinking that has been the case for the past few years. The influx of creativity has opened up new possibilities for applications that are being envisioned and sought, rather than being logical extensions of what was done before. This combination of technological capability with designer opportunity will provide us with many extremely interesting devices to try out, accept, or reject as things progress.

Another driving factor for device proliferation is the continuing modularity of handheld devices—that is, the option to add on a variety of hardware modules. The best instantiation of that trend today is the Handspring Visor line of devices and the modules that are being built mostly by third parties. This device-as-a-platform theme has been adopted by Palm and Compaq as well. We expect to see more of this variability as the basic hardware and software engines become more powerful and option friendly.

With all the creativity in evidence today, one question that is being asked revolves around limiting factors. Today, there are at least two technology limitations that are key to what might be possible.

First, there is a roadblock concerning the power capabilities of batteries of given weights. This may be the most significant limitation of all, as it determines the weight and duty cycle of the various devices that can be manufactured. To be reasonable, battery life needs to extend for at least a full day of typical usage. This allows for an overnight charging cycle—a process to which it seems reasonable to expect people to become adjusted. There is some interesting work going on with zinc air batteries and tiny fuel cells, but it is not clear that either of these will provide reasonably priced improvements to existing alternatives. True breakthroughs here would be welcome, but are not evident in research on technologies today.

The second limitation is the effective, ergonomic design of the device interface as relating to data input. There are lots of options here, but few are practical for widespread usage yet. Examples include voice input (with recognition, not just recording), RFID tags (to potentially replace bar coding), and input anticipation software.

Voice recognition and input are often considered the ultimate in input effectiveness, providing the optimal user experience. However, there are a lot of arguments both for and against this technology. Currently, the limitations of portable power to drive the powerful recognition engines render this option useful for only specialized areas. As chips with higher processing capability and lower power requirements come to market, voice recognition will become a more viable option. Only then will we see how effective the technology can become for the enterprise user, after which it may begin to enjoy widespread consumer acceptance. We expect voice recognition and input to be a major trend in the upcoming years, and one that merits significant examination and tracking to see at which point the technology becomes mature and cost effective.

Bar coding has become a major way to transfer information into handheld (and other) devices. However, to provide that input, an unobstructed visual path is needed between the bar code itself and the input device. In a number of areas, the emerging need/desire is to scan the information without the visual contact, at a reasonable price. This, coupled with the declining price of RFID devices, is leading to a predicted major uptake in usage as the RFID devices become feasible for inventory control, luggage tracking, shelf space management, and other opportunities that will emerge.

We use the term *input anticipation* to characterize a number of tools that are available already today, with more being produced as we move forward. These tools attempt to guess your input based on context, history, abbreviated keystrokes, and so on. And while these tools have become ubiquitous in word processors and Web browsers, they are currently making their way to phones, PDAs, and almost any other device that needs input to guide its operations. As various algorithms are being tested in the marketplace, we expect some of them to become quite robust in their abilities, allowing more accurate and speedy data entry. When coupled with some of the attached pointing devices, effective usage of these tools could dramatically improve the usability and effectiveness of many mobile devices.

The one option that we have not included in this list is the keyboard. This device has worked well historically, and we do not expect it to disappear anytime soon. However, as the size of the devices is shrinking, the keyboard is becoming too bulky to carry along. The most popular PDAs today have keyboards available only as accessories, and that does not seem to have diminished the popularity of these devices. We can expect improved folding keyboards, perhaps even ones that will roll or fold up, to be developed and introduced into the marketplace in the coming years.

The opportunities for new devices that we've not yet been able to imagine (except in *Star Trek*-like flights of fantasy) are immense. As indicated earlier, the most certainty comes in the variability that we should expect. We will continue to witness many experiments, with some failing to find sufficient following and others emerging to fill the needs, both real and imagined, that the designers and technologists envision. These successes and failures will, in turn, lead to even new generations of devices. The exciting part of this is the speed with which the ideas can be conceived, developed into products, and delivered to the marketplace, where they can either succeed, or fail, or evolve, as the market renders its judgment.

System Services

We expect that the evolution of system services will follow the trends witnessed in software packaging of the past several decades. That trend has been for companies to introduce new products based on specific visions of new capabilities and how they should or could be packaged. Then, as competitors emerge to validate the product and its utility, the capability becomes genericized, and eventually it gets integrated into the operating system.

As an example in the mobile arena, we expect the current generation of transcoding products to become a commodity-like service to be integrated into all common application servers over the next few years. This will change the market for products and services in ways that are difficult to predict today, but it will then make way for improved approaches to solving the problems and spur yet another round of innovation and creativity.

Other capabilities that are starting to come to market and that we would expect to see evolve into the core of the mobile services include location-based services, text-to-speech offerings, additional security services, and personalization services. While some of these may retain their independence as product offerings for quite some time, due to patent protection or limited applicability, we would not be surprised to see them integrated into future operating system or application server platforms as ubiquitous, commodity-type capabilities.

Applications

Using the tools and capabilities discussed previously, we expect the biggest explosion in the area of applications themselves. As bandwidth increases and coverage becomes more ubiquitous, the number of operations that can be performed reliably with mobile and wireless devices becomes unlimited. There are three major trends that factor into what will be achieved.

First, there is the straightforward trend to move everything possible onto mobile and/or wireless platforms. The first major thrust of this effort was the introduction of laptop computers. Laptops allowed most everything that could be done as a stand-alone task at one's desk to be accomplished anywhere there was power. (The early batteries were so short-lived that to consider those laptops capable of operation independent of a power connection was really wide-eyed optimism.) This trend will continue, augmented by the efforts expended on synchronization of databases and other information. Eventually, all the common activities done today at our desktop computers will be possible on laptop and handheld devices. This trend toward making office applications accessible and robust independent of location will be the direction taken by many companies and device evolutions, particularly in the next generations of PCs and PC work-alikes.

The second trend is more specialized and concerns the continuing mobilization of current enterprise applications for mobile and wireless access. Into this category we place the ERP, CRM, and supply chain systems that are evolving by making their information accessible to untethered users. Based on development histories, this is not an easy task, as it requires capabilities beyond the relatively simple translation of screens to fit onto mobile devices. The reason is that back-end systems were engineered with the presumption of full screens and robust, real-time access to the nonreplicated data. The option of running an operation in environments that are sometimes connected, with nonreliable connections being the norm, was not built in. Adding this functionality after the fact is proving more difficult than was anticipated. Nevertheless, these problems will be solved, and we expect to see increasingly robust, mobile, and wireless utilization of enterprise information in the foreseeable future. Increasing bandwidth will allow more and more to be possible in a fairly linear extension of what we've seen in the past. Again, the evolution will frequently follow the progression we've experienced previously with the coming of remote terminals, then intelligent terminals, and finally individual PCs on every desktop.

The third trend is the one of most interest and variety. It encompasses the incorporation of capabilities that are important primarily to the mobile user. The key areas that we have extracted are location-based processing, personalization, and device-level integration.

One of the key capabilities that mobile applications can draw on is our ability to detect their location and take action based on that knowledge. This capability is exploited in even the early telematics applications. For example, location awareness is a critical component of the OnStar emergency service, illustrating how mobile applications can make use of that information in novel and important ways. There are lots of scenarios, many bordering on fiction, for what could happen with this capability available to applications at all times. We'll not speculate on these scenarios other than to note that government regulation and personal privacy considerations may allow this to evolve in ways we've never dreamed of (both positive and negative). The indication we have is that regulation and privacy issues will become central, as application designers learn effective ways to exploit the location capabilities that are becoming easily and cheaply available.

We included personalization here, even though the technology is common on existing Web sites and background processing today. However, personalization takes on a special meaning for mobile users. Here, personalization must become much more effective and be targeted to be acceptable on devices on which it is not as easy (or as cheap) to just ignore and click past personalized information that really isn't relevant to the individual. One of the complicating factors is that in some countries (such as the United States), the receiver of personalized messages usually ends up paying for them. These messages are not free, as they are often viewed in the Web environment. The other challenge to personalization for mobile and wireless devices is paying attention to location (see preceding paragraph), device type, and, potentially, other factors such as time of day that have not been explored in tried and tested applications today.

The third area is the device-level integration. What we mean by this is the effective handling of the various options available on a device for input/output and processing. Think of the section on device trends and how these trends will be addressed in the applications. One option (that will undoubtedly be used initially) is for the application to be aware of the specific device options, such as scanners and voice recognition, and have separate programming to handle the situations. As standards and the sophistication of the devices and software infrastructure evolve, some of this distinction will disappear. Even as that is happening, though, we expect new device options to emerge that will continue to require application awareness of the specific capabilities of the device and take appropriate action depending on the available options at the time. This has been done for years, but will become more extensive as the range of device options increases.

The result that we expect to see is a period of innovation and evolution from where we are today to a future of devices and capabilities that has only been foreseen by some of the visionary science fiction writers. Many of the scenarios that are being played out in the press and other publications will come to fruition, but almost certainly not as anticipated today. The experience in the marketplace, with real consumers evaluating the offerings, will largely determine what will survive. Government regulation in this space will also play a major role, as debates on privacy, security, and similar issues are resolved. Overall, the future of mobile and wireless applications will present exciting challenges for us all.

CHAPTER 28

Putting It All Together

Introduction

The final chapter of this book is intended simply to provide a review of some of the book's key concepts. We want to summarize the major take-aways and then leave you with a few additional thoughts. As we mentioned in the introduction, there are actually a couple of ways to read this book. Most people will probably pick a few sections that are of interest to them. Nevertheless, if you read the book straight through, from cover to cover, then we commend you! Either way, we hope that the book provided everything that you were looking for. The approach of this final chapter will be to review what you learned in each section, following an executive summary type of format.

If you were one of the brave souls who read this book from beginning to end, you would have noticed that each book section covers how a unique wireless solution is applied to a very real business case. In each of these business cases we used the CGE&Y approach to enterprise architecture, an approach that has been developed and refined over many years of successful system implementation and integration. In each scenario we showed how to build a wireless application while taking into consideration that the solution should solve some business need and/or is derived from some tangible value proposition. In client situations we oftentimes witness cool technology being the driver for business decisions. Many organizations are quick to make substantial technology investments simply to try and get ahead of their competition—only to find that the solution did not solve their business needs. Applying CGE&Y's proven methodology to wireless enterprise architecture, we look at all the aspects of a solid technical design. Just as a building architect needs to understand what the structure will be used for, so should a wireless enterprise architect understand all the aspects of the application before embarking on a wireless solution design and implementation project. Let's go through each of the sections of the book and summarize the key lessons we learned.

The Foundation of Wireless Architecture and Beyond

Part 1 of this book is probably best referred to as the foundation of, or the framework for, the subsequent parts. You might remember that the chapters in Part 1 gave you a firsthand look at the wireless landscape. One of the key lessons learned in this section was an understanding of the timeline spanning from the early public usage of the Internet around 1993 to the wireless Web of today. It's important to note here that mobility and going wireless are much more than simply building your Web site for a Palm PDA or a WAP phone. This objective will likely be a *component* of an enterprise wireless architecture, but should not be the sole reason to go wireless. Mobility brings new and exciting opportunities for organizations that never would have existed before wireless technology came into the picture. New business processes and business models are popping up in response to the anytime, anywhere personality of mobility. Exciting and cool technologies are taking users by storm. It's important to put all those technologies in perspective and learn how mobility is truly an enabler at both the individual and the corporate levels.

Going forward, you also learned about wireless infrastructures and some of the behind-the-scenes aspects of wireless architecture. From wireless service providers to wireless application service providers, Internet service providers, application services, and security, there is certainly a great deal of complexity that needs to be considered when building the wireless network infrastructure. One of the many confusing aspects for any organization planning a wireless infrastructure is the plethora of devices available in the marketplace. Figure 28.1 makes only a small dent in illustrating the possibilities that anyone might go through when deciding what devices to leverage for their design. These devices are constantly evolving and transforming, giving users more and more possibilities when designing their wireless business.

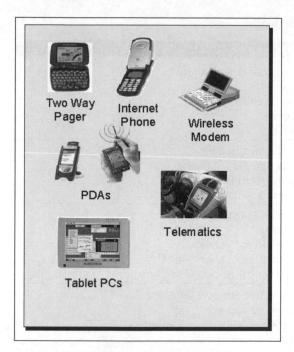

Figure 28.1   A multitude of devices.

Other very important aspects of wireless enterprise architecture include methodology and structure. Too many times we have seen organizations frustrated with system implementations as a result of inadequate planning or design. Who can forget the early days of the Web when every company seemed to be in a big rush to put up their very own Web site? Frequently, we witness large enterprise systems or Internet infrastructures being implemented in a rush, without being given adequate time for planning or design. Comparatively, the excitement of a new technology and a significant investment can drain expectations once a system is implemented that doesn't meet the buyer's expectations. One of the most critical steps any organization should take is to architect their system properly. As we discussed in Chapter 4, architecture design is much more than simply applying a methodology to a project. From CGE&Y's perspective, the purpose of architecture is this: *Architecture describes overarching designs of individual components so that their assembly results in a complete and working product. This design is needed to guide the construction and assembly of components.*

There are several very important aspects to architecture that are defined in Table 28.1.

These aspect areas also serve as the framework for the rest of this book, as we delve into the various case studies of wireless application design.

Table 28.1　The Aspect Areas of Architecture

ASPECT AREA	CONTENT
Business	Commercial organization Personnel Administration Processes Finance
Information	Information structure Knowledge management Data warehousing Processing structure
Information systems	Automated IS support IS services Integration of IS services Collaboration between services and components Critical use cases describing functional requirements
Technical infrastructure	Infrastructure support required IT services required Processing platforms and volumes Network diagramming Hardware/software standards and requirements Governance and security services per platform 　(failover, load balancing, encryption, etc.) Communication methods and standards
Governance	Ongoing support of the different business processes Areas of concern related to IS systems: 　Availability 　Controllability 　Performance
Security	Business security needs at the different levels of 　technology platforms, applications, and network 　infrastructures Security prevention services and elements, such as 　integrity and confidentiality

Electronic Exchanges of the Future

One of the ever growing manifestations of the Internet world is the exchange. Electronic marketplaces of today are mirroring their offline predecessors in that they serve as a forum in which sellers and buyers of products, services, and information gather to conduct trade. The efficiencies of aggregating sellers, buyers, or both in a central location have been the foundation of physical marketplaces of the past, and will be the drivers behind the growth of present and future electronic trading exchanges. These exchanges are serving various user groups: consumers, businesses, and governments. Each of these user groups has developed a range of specific exchange business models, coming in all sorts of flavors, including B2B, B2C, B2G, B2E, B2G, C2C, G2C, and G2B.

As supply chain relationships evolve from individual, straight-line processes via strategically linked networks to comprehensive, collaborative *value Webs* we are witnessing a similar evolution in trading exchanges. Whereas the electronic exchanges of the early days served primarily as consolidators of content and community features within specific industry sectors, today's e-marketplaces have matured into electronic trading floors, complete with commerce capabilities. In the future, both information aggregation and commerce functionality will be augmented with true value-added, premium-priced features, including data mining and tools for extensive member collaboration services.

The exchange model is constantly growing and transforming into new and more efficient ways for customers to access products, services, and information from anywhere and at any time. Using our MetX metals exchange case study, we saw that building a wireless exchange model entailed three compelling areas of functionality:

1. *Source data collection.* Supply chain monitoring and fulfillment is taken to a different level through the ability of accessing critical supply chain tracking data through wireless devices.

2. *Real-time monitoring and event-driven feedback.* Supply chain monitors receive real-time alerts and can react to those alerts at anytime.

3. *Dynamic optimization.* Finally, the optimization of supply chain business processes is transformed with wireless technology by giving real-time supply tracking functionality, allowing shipments to be redirected quickly.

The MetX Exchange provided its constituents with some very tangible benefits, including enterprise resource planning connectivity, catalog management, contract management, order management, and transportation management, as well as logistics and supply chain optimization. Using wireless devices, sellers were able to better penetrate a marketplace of buyers via increased reach and streamlined distribution processes as exemplified by on-the-fly reconfiguration of supply chain fulfillment activities at any time and from anywhere. Buyers were able to use wireless devices to better manage inventories by moving closer to a just-in-time model and leverage the exchange for one-stop-shopping convenience, while enjoying the environment's price competitiveness.

Figure 28.2 illustrates the key areas of the wireless supply chain solution for the MetX Exchange. The chapters that dealt with this exchange leveraged

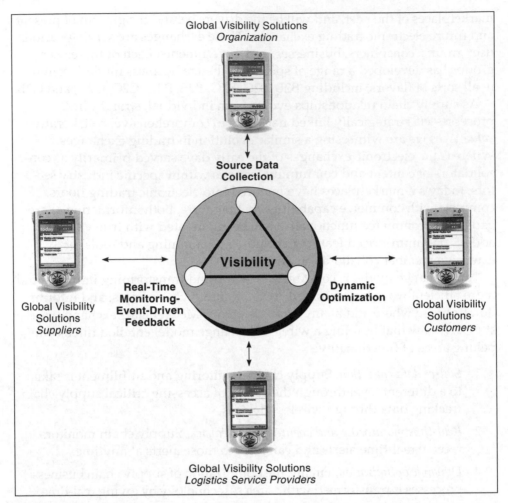

Figure 28.2 Wireless global visibility clients.

the architecture framework by defining the conceptual, logical, and physical layers that were defined earlier in the book. We defined the critical data needs of the wireless exchange supply chain tracking tool and showed how the various architectural layers are built through a logical current state and future state definition. Finally, we discussed some of the requirements to actually build the physical wireless infrastructure for our wireless exchange.

Building a Mobile Sales Force

Part Three covered a wireless sales force automation application we called MobileMediaMaven. Sales force automation tools generally fall into the category of customer relationship management (CRM) applications. We saw that enabling the sales force through mobile applications provided numerous advantages for the sales representatives and their clients. Specifically, the MobileMediaMaven application centered on facilitating the sale of radio spots to advertisers. The radio spots represented the category of highly perishable inventory, similar to airline seats or hotel rooms that would be lost if they remained unsold. Any tool that would allow an organization to better manage this volatile inventory had the potential to immediately improve the company's bottom line. Our case study explored such a tool, in the form of wireless access.

The sales force automation application started out as an Internet-based solution that quickly proved to its users the time-saving features of a Web-based solution. Many organizations are faced with a similar dilemma of how to transform their paper-based processes into a more streamlined approach. MobileMediaMaven found that solution through wireless capabilities. Here are some of the key features of the wireless sales force automation solution:

Proposal generation.   Gives the ability for sales reps to check in real time the availability and pricing of radio spots at all the stations within the radio network, anywhere, anytime.

Notification.   Allows sales reps, sale directors, and other managers to receive alerts while on the road and respond to these alerts in real time by generating a proposal and sending it for approval.

Order management.   Users can modify orders, amend them, delete them, consolidate them, and so on, via the handheld device.

Reporting.   Users can run reports on the fly, presenting customers with their past order history, station format comparisons, and other information required to close a deal or cement a relationship.

The benefits that accrued to the company entailed the optimization of perishable inventory management that provided a real, immediate impact on the station network's bottom line. While the organization reaped financial rewards in the form of additional revenues, the company's sales representatives used the wireless solution to enhance their productivity and provide better customer service based on real-time access to critical inventory availability and pricing information. Last, but not least, the company's customers received value by being able to conduct their advertising purchases in a one-stop-shopping fashion and receive confirmation of their media purchase on the spot.

These and other features are discussed in more depth in Chapter 13 and subsequently through the information technology chapters in Part Three. While the CRM area benefited greatly by going mobile, other sectors such as financial services and insurance are also innovators in the wireless space.

Financial Services and Mobility

As we saw in Part Four, financial services organizations are not strangers to wireless technology. Wireless financial services applications generally fall into one of three categories:

- Mobile banking.
- Mobile brokerage.
- Mobile insurance.

In the light of the increasing commoditization of financial products and services, pressures from global competitors, and deregulation of the industry and resulting consolidation activity, financial institutions find themselves in an intensifying battle for customers. In an effort to attract and retain a customer base, value-added services that cement the customer relationship have become one of the few remaining differentiators. As financial products converge, organizations are looking for ways to lock customers into the organization's offerings and to make it difficult for them to leave and switch to a competitor.

In Part 4 we focused on how one of our clients (we called the company Acme Financial) achieved the goal of superior customer retention via the aggregation of brokerage, banking, and insurance services accessible via mobile devices. The solution not only served the institution's customers, but also provided a tool for the company's financial advisors, who are in close contact with their individual account holders. Offering pure wireless self-service from the customer end alone was viewed as not sufficient to provide a strong value proposition; providing customers with a value-add in the form of anytime, anywhere access to the company's financial experts was a critical strategic differentiator.

As a result, the wireless financial portal solution entailed a number of critical functionality features, including:

1. Real-time access to brokerage, banking, and insurance accounts and transaction histories.

2. Time-critical alerts, such as stock alerts, low balance alerts, and policy expiration notifications.

3. The ability to react to those alerts in real time by submitting a transaction through the Acme wireless network.

4. Access to a plethora of financial news, research, and financial planning and analysis tools.

Ultimately, Acme was better able to service their existing customers, attract a new customer base, shift low-value accounts toward a self-service model, pay increased attention to high-margin clients, and position the company for an increasingly competitive financial services industry.

What the Future Holds for the Mobile Executive

Part Five explored how a mobile office application provided new and faster ways for executives to manage their day-to-day activities. In light of increasing mobility, especially among the senior levels of corporate management, it becomes more and more important to provide avenues to stay in touch with the organization and its performance. Yet staying in touch entails more than simply being able to communicate via voice, fax, or email. Likewise, being able to perform office functions while on the road entails more than being able to use a word processor or spreadsheet program on a mobile device. Those types of applications are the foundation of the wireless office, but they are selling short the overall concept.

In our case study, we investigated the various wireless office applications implemented at a company we called Acme Toys. We saw that the mobile office comprised much more than just one application. In fact, the solution entailed multiple productivity components that were designed to work together in a seamless fashion. The mobile office infrastructure provided a wide range of services such as office automation, communication, and information management tools, some of which included:

- Word processing
- Spreadsheets
- Presentation and graphics
- email
- Videoconferencing/online meeting tools
- Shared calendars
- Document management
- Intranet access
- Internet access
- Knowledge management tools

We took you through a day in the life of a typical executive and demonstrated how mobile office tools were leveraged to help the executive manage his busy day. The application, which was represented as the executive dashboard for Acme Toys, exemplified a tool for collaborative work flow. The dashboard served as an illustration of all the key architectural concepts we mentioned in the various sections of this book. It also brought forth many other aspects of wireless architecture that were only touched upon in the earlier parts of the book. One of those aspect areas was the wireless LAN and how the 802.11b standard plays a significant role in the mobile office infrastructure. We saw our executive utilizing various mobile devices like cell phones and PDAs in connected mode, such as in an 802.11b-enabled airport. You also saw the executive make use of his free time in nonconnected mode, such as on airplane flights where wireless Internet access is regulated.

The primary benefits realized by the mobile solution were enhancements to productivity. Especially for those organizations that consist of a large workforce on the road, access to home systems can significantly improve how these employees manage their time. Whether it's the executive on the road who is able to immediately take corrective action after receiving a system-generated alert that there is a problem with the company's production line, for example, or the sales rep who uses a dashboard application to access the latest customer account data before walking into the sales meeting, allowing mobile key employees to interact with the organization's internal legacy systems can result in enhanced performance of the individual and, ultimately, the organization as a whole.

The Final Say

Well, we've come to the end of the book, but, we hope, just the beginning of your discovery of wireless architecture. The excitement over wireless technology and mobility is truly just beginning. Organizations are still in the early stages of realizing the tremendous potential that mobility brings to the enterprise. Service providers are working frantically to get wireless networks up to customer demand. Business processes are constantly transforming to leverage the potential of the untethered world.

The case studies in this book are by no means the only tools your organization should use in building your wireless enterprise. As with any project that requires a substantial outlay of financial and other scarce organizational resources, making sure you follow a sound strategic plan is a requirement. We can't stress this enough: Before jumping into any development activities, you should make sure you have built a solid, strategic plan that takes into consideration your customer's desires, external industry and competitive trends, your internal capabilities, technological opportunities, and the proposed application's financial implications. Once you have developed a sound strategic foundation upon which to build your wireless solution, applying a valid and tested architecture approach to designing the project is a surefire start to a successful development effort. But as with any technology, it's important to think through your options carefully. Most important, plan your effort accordingly. This book provides the framework, developed through years of solid experience, that will help you get a running start to designing a successful wireless enterprise application architecture.

PART

Seven

Appendices

Useful URLs

BLUETOOTH	
Bluetooth Consortium	www.bluetooth.com
Bluetooth Resource Center	www.palopt.com.au/bluetooth
Ericsson	www.ericsson.com/bluetooth/
Extended Systems	www.extendedsystems.com
Motorola	www.motorola.com/bluetooth
Nokia	www.nokia.com/bluetooth/index.html

MAGAZINES	
America's Network	www.americasnetwork.com
Business 2.0 Magazine	www.business2.com
Communication Systems Design	www.csdmag.com
Ericsson Review	www.ericsson.com/review
Global Telephony	www.globaltelephony.com
MCI	www.mobilecomms.com
Network Magazine	www.network-mag.com

NewWaves	www.newwaves.com
RCR	www.rcr.com
Red Herring Magazine	www.redherring.com
The Living Internet	www.livinginternet.com
Wireless Integration	www.wireless-integration.com
Wireless International Magazine	www.wireless.com.tw
Wired Magazine	www.wired.com
Wireless Asia	www.telecomasia.net/wirelessasia/wa_new.html
Wireless Business & Technology	www.wirelesstoday.com
Wireless Design & Development	www.wirelessdesignmag.com
Wireless Systems Design	www.wsdmag.com
Wireless Week	www.wirelessweek.com

NEWSLETTERS

All Net Devices	http://devices.internet.com
Anywhereyougo.com	www.anywhereyougo.com
Bank Securities Journal Online	www.bsanet.org
ePaynews.com	www.epaynews.com/index.cgi
Fierce Wireless	www.fiercewireless.com
Pervasive Weekly	www.pervasiveweekly.com
Strategic News Service	www.tapsns.com
The Rapidly Changing Face of Computing Journal	www.compaq.com/rcfoc

ORGANIZATIONS

Advanced Television Systems Committee	www.atsc.org
ATM Forum	www.atmforum.com
Bluetooth	www.bluetooth.com
CDMA Development Group	www.cdg.org
CDPD Forum	www.cdpd.org
Cellular Telecommunications Industry Association	www.wow-com.com
Digital Display Group	www.ddwg.com/E911/wwlinks.htm

European Telecommunications Standards Institute	www.etsi.org
GSM World	www.gsmworld.com
Home RF/SWAP	www.homerf.org
Institute of Electrical and Electronics Engineers	www.ieee.org
Internet Engineering Task Force	www.ietf.org
Network Management Forum	www.nmf.org
Radio Advertising Bureau	www.rab.com
SyncML Forum	www.syncml.org
UMTS Forum	www.umts-forum.org
UWCC	www.uwcc.org
VoiceXML Forum	www.vxmlforum.com
WAP Forum	www.wapforum.com
Wireless Ethernet Site	www.wirelessethernet.org
World Wide Web Consortium	www.w3c.org

PALM

Common Palm User Problems	www.nearlymobile.com/commonproblems
Handago.com	www.handago.com
Palm Developer's Developers Zone	www.palm.com/devzone
Palm Source	www.palmsource.com

PERVASIVE COMPUTING

eBiquity.org	http://ebiquity.org
IBM Pervasive Computing	www.ibm.com/pvc
Pervasive Computing SIG	http://gentoo.cs.umbc.edu/pcsig/about.shtml

POSITION LOCATION

Cambridge Positioning Partner	www.cursor-system.com
Cell-Loc, Inc.	www.cell-loc.com
Locus Corp.	www.locus.ne.jp
NENA	www.nena9-1-1.org
SiRF	www.sirf.com
SnapTrack	www.snaptrack.com

Trueposition	www.trueposition.com
US Wireless	www.uswcorp.com
Virginia Polytechnic's Polytechnic's Mobile and Portable Research Group	www.mprg.ee.vt.edu/home.html

3G

Ericsson	www.ericsson.com/3g/default.shtml
Mobile Applications Initiative	www.mobileapplicationsinitiative.com
Mobile Wireless Internet Forum	www.mwif.com
M-PEG TV	www.mpegtv.com
Nokia (3G)	www.nokia.com/3g/index.html
Wireless Multimedia Forum	www.wmmforum.com

USABILITY

Human-Computer Interaction Virtual Library	http://web.cs.bgsu.edu/hcivl
Usability Group	www.usability.com
Use It	www.useit.com
User Interface Engineering	http://world.std.com/~uieweb/biblio.htm

VOICEXML

IBM VoiceXML	www.alphaworks.ibm.com/tech/voicexml
Tellme	http://studio.tellme.com
VxML Forum	www.vxmlforum.org

WAP

Ericsson	www.ericsson.com/WAP
Motorola	http://mix.motorola.com /audiencesdevelopers/dev_main.asp
Nokia	www.nokia.com/networks/17/wdss.html
WAP FAQ	http://wap.colorline.no/wap-faq
WAP Sight	www.wapsight.com
WAP's WAP Developer's Repository	http://wapulous.com
Wireless and WAP	www.itworks.be/WAP

MISCELLANEOUS

ACM Sigmobile	www.acm.org/sigmobile
ACTS	www.uk.infowin.org/ACTS

Arbitron	www.arbitron.com
Bell Labs Technology Journal	www.lucent.com/minds/techjournal/findex.html
Cap Gemini Ernst & Young, US	www.us.cgey.com
Cap Gemini Ernst & Young, Worldwide	www.cgey.com
Cap Gemini Ernst & Young Center of Business Innovation	www.businessinnovation.ey.com
Commercial Speech Recognition	www.tiac.net/users/rwilcox/speech.html
Compaq Research Consortium	www.research.compaq.com
HP e-services	http://e-services.hp.com
Microsoft Research	http://research.microsoft.com
Microsoft Wireless	www.microsoft.com/wireless
Mobile Review Communications Review	www.acm.org/sigmobile/MC2R
National Association of Broadcasters	www.nab.com
Network Query Language	www.networkquerylanguage.com/manual.asp
The Mobile Data Initiative	*www.gsmdata.com* www.pcsdata.com
TradeSpeak	www.tradespeak.com
UMTS Market Forecast Study	www.analysys.co.uk/news/umts/default.htm
Web ProForum	www.webproforum.com
Webopedia	www.webopedia.com
Wireless and WAP	www.itworks.be/WAP
Wireless ASP Directory	www.wirelessweek.com/industry/ASPdir.htm
Wireless Developers Network	www.wirelessdevnet.com
Wireless Week E911 articles	www.wirelessweek.com/issues/

3G	Third generation
AMPS	Advanced Mobile Phone System
AOA	Angle of arrival
ASP	Application service provider
ATVEF	Advanced Television Enhancement Format
AuC	Authentication center
B2B	Business-to-business
B2C	Business-to-consumer
B2E	Business-to-employee
BI	Business intelligence
BSC	Base station controller
CCPP	Composite capability and preference profiles
CDMA	Code Division Multiple Access
CDPD	Cellular Data Packet Data
CHTML	Compact HTML
Codec	Compression/decompression
COM	Component Object Model

CORBA	Common Object Request Broker Architecture
CRM	Customer relationship management
DES	Data Encryption Standard
DTD	Document type definition
DTMF	Dual-tone multifrequency
E911	Enhanced 911
ECMA	European Computer Manufacturers Association
EDACS	Enhanced Digital Access Communication Systems
EDGE	Enhanced Data Rates for Global Evolution
EIR	Equipment Identity Register
EJB	Enterprise JavaBeans
EMS	Enhanced Message Service
EOTD	Enhanced Observed Time Difference of Arrival
ESN	Electronic serial number
ETSI	European Telecommunications Standards Institute
FCC	Federal Communications Commission
FoIP	Fax over IP
FRR	False rejection rate
G2C	Government-to-citizen
GEO	Geostationary Earth Orbit
GIF	Graphics Interchange Format
GPRS	General Packet Radio Service
GPS	Global positioning system
GSM	Global System for Mobile Communications
HDML	Handheld Device Markup Language
HDTV	High-definition TV
HLR	Home location register
HomeRF	Home radio frequency
HSCSD	High-Speed Circuit-Switched Data
HTML	Hypertext Markup Language
HTTP	Hypertext Transfer Protocol
HTTP-NG	HTTP Next Generation
HTTPS	HTTP Secure
iDEN	Integrated Digital Enhanced Network
IEEE	Institute of Electrical and Electronics Engineers
IETF	Internet Engineering Task Force
IMT2000	International Mobile Telecommunications 2000

IP	Internet Protocol
IPSec	IPSecurity
IPv6	IP Version 6
ISO	International Standards Organization
ITRS	International Technology Roadmap for Semiconductors
ITU	International Telecommunications Union
ITV	Interactive TV
IVR	Interactive voice response
JPEG	Joint Photographic Experts Group
JVM	Java Virtual Machine
LAN	Local area network
LDAP	Lightweight Directory Access Protocol
LEO	Low Earth Orbit
MathML	Mathematical Markup Language
MBS	Mobile Broadband Systems
MEO	Medium Earth Orbit
MIN	Mobile identity number
MIPS	Millions of instructions per second
MIX	Mobile Internet Exchange
ML	Markup language
MPEG	Motion Picture Experts Group
MSC	Mobile switching center
MSN	Microsoft Network
MT	Machine translation
NLP	Natural language processing
NMT	Nordic Mobile Telephone
OMC	Operations and maintenance center
OMG	Object Management Group
OS	Operating system
OSI	Open Systems Interconnection
OTA	Observed time of arrival
OTD	Observed time difference of arrival
PAN	Personal area network
PCS	Personal Communication Service
PDA	Personal digital assistant
PDC	Personal Digital Cellular
PGML	Precision Graphics Markup Language

PGP	Pretty Good Privacy
PHS	Personal Handy Phone System
PIM	Personal information management
PIN	Personal identification number
PNG	Portable Network Graphics
POS	Point of sale
PPP	Point-to-Point Protocol
PQA	Palm Query Applications
PSTN	Public Switched Telephone Network
QA	Quality assurance
QoS	Quality of service
RF	Radio frequency
RFID	Radio Frequency ID (RFID)
ROI	Return on investment
RTOS	Real-time operating system
SDK	Software Development Kit
SIM	Subscriber Identification Module
SLA	Service-level agreement
SLP	Service Location Protocol
SMIL	Synchronized Multimedia Markup Language
SMS	Short Message Service
SNR	Signal-to-noise ratio
SQL	Structured Query Language
SSL	Secure Socket Layer
SWAP	Shared Wireless Application Protocol
SyncML	Synchronization Markup Language
TACS	Total Access Communications System
TCP/IP	Transmission Control Protocol/Internet Protocol
TDMA	Time Division Multiple Access
TDOA	Time difference of arrival
TETRA	Terrestrial Trunked Radio
TIA	Telecommunications Industry Association
TSL	Transport Security Layer
TTS	Text to speech
UDP	User Datagram Protocol
UI	User interface
UIML	User Interface Markup Language

UMTS	Universal Mobile Telecommunications System
URL	Uniform resource locator
VLR	Visitor Location Register
VML	Vector Markup Language
VoIP	Voice over IP
VXML	Voice-Extensible Markup Language
W3C	World Wide Web Consortium
WAE	Wireless Application Environment
WAG	Wireless Assisted GPS
WAN	Wide area network
WAP	Wireless Application Protocol
WASP	Wireless Application Service Provider
WATM	Wireless Asynchronous Transfer Mode
WBMF	Wireless Bitmap Format
WCDMA	Wideband CDMA
WDP	Wireless Datagram Protocol
WIEA	Wireless Internet enterprise applications
WLAN	Wireless Local Area Network
WML	Wireless Markup Language
WSP	Wireless Session Protocol
WTA	Wireless Telephony Applications
WTLS	Wireless Transport Layer Security
WTP	Wireless Transactional Protocol
XHTML	Extensible Hypertext Markup Language
XML	Extensible Markup Language
XSL	Extensible Style Language

Fundamentals of Architecture (Systems Development Phase and Beyond)

Overview

The purpose of this appendix is to provide additional detail about the phases of the architecture framework described in Chapter 4. The topics covered in this appendix pick up in areas where Chapter 4 left off. These areas include:

- Systems Development
- Systems Deployment
- Engagement Management
- Utilization of Standards in Development and Infrastructure

System Development

Architecture is essential to successfully designing enterprise-scale software. But the architecture is a means, not an end; the goal is the software supporting business processes itself. An excellent architecture with a poor implementation is poor software. Just as the IAF imposes a structure on the conceptual, logical, and physical architectures, an analogous structure must be imposed on the implementation.

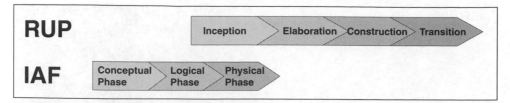

Figure C.1   The life cycle view.

CGE&Y typically uses the *Rational Unified Process* (RUP) for this role. RUP complements the IAF by taking the physical architecture created by the IAF process much deeper—all the way to the implementation. So you could conceptually think of the IAF as creating the top half of the physical architecture, and RUP as creating the bottom half (see Figure C.1).

The Rational Unified Process (RUP) Methodology

The object-oriented paradigm is another way of viewing applications. You divide an application into many small and independent pieces, or *objects*. You can then build an application by arranging these pieces into components that themselves form systems that will achieve the objectives of an application. The advantage of this approach is the reuse and rearrangement of these objects and components into different applications.

This varies from the traditional approach in that the object-oriented approach no longer focuses on data-centric design, such as design databases or providing screens to input data or to print reports. The focus of object-oriented design is the behavior of the system that will make it easier to change systems as business rules and environments change. The object-oriented approach adds the dimension of *behavior* to the dimension of *information*, making system development more resilient, flexible, and responsive to change.

The benefit of flexibility can only be realized by designing an object-oriented system well. This requires knowledge of some principles of object orientation: encapsulation, inheritance, and polymorphism. We will provide only a very brief summary here [Boggs 1999].

Encapsulation

In object-oriented systems we combine a piece of information with the specific behavior that acts upon this information. We package these together and call this package that reflects related functionality an *object*. This process is referred to as *encapsulation*.

For example, a vehicle holds information such as a vehicle identification number (VIN), options, production date, and customer name. Linked to a vehicle are certain behaviors or system functions: sell, service, recall, modify, resell, and so on. We encapsulate the information and the behavior into an object, *vehicle*. Any change to the vehicle system regarding vehicles can now simply be implemented in the vehicle object.

The limitation of changes to one or only a few objects is another benefit of encapsulation and reduces development efforts due to changes.

Inheritance

Inheritance is the second of the fundamental object-oriented concepts. In object-oriented systems, inheritance is a mechanism that lets you create new objects based on old ones, such that the *child* object inherits the qualities of the *parent* object. This adds to the ease of maintenance. When something changes that affects only the parent object, the child object will automatically inherit this change. This mechanism reflects hierarchies as found in reality. In the previous example, the parent vehicle would be the vehicle that holds all the common attributes. On the other hand, child accounts would be passenger cars, trucks, and vans with specific attributes of their vehicle type in addition to their parent attributes, such as weight, service dates, and number of passengers. Changes to the parent will affect all children, but the children are free to adapt without disturbing each other or their parent.

Polymorphism

The third principle of object-oriented systems is polymorphism, or the occurrence of different forms, stages, or types. The term *polymorphism* means having many forms or implementations of a particular functionality. One of the benefits of this principle, as with the other principles, is ease of maintenance. In the previous example, we would not need to develop yet another function to add a new vehicle type for SUVs. The functionality of *sell vehicle* would already cover the selling of any vehicle. The maintenance would be limited to adding another child object, *SUV*.

Methodology Background

Software development can be done in many ways. There are several different types of development processes a project can follow, including everything from waterfall to object-oriented processes. Each has its benefits and disadvantages. It is not the purpose of this appendix to discuss them. This section focuses on visual modeling using Rational Rose (Rational's Object-Oriented Software Engineering). Note: Rational Rose is a tool we've used extensively in our development projects. However, there are many other compatible tools that can be used in the RUP methodology. We've simply chosen to discuss Rose in this book because of our familiarity with it.

Let's briefly look first into the history of the Rational Unified Process (RUP). In 1995, the *Rational Approach* and *Objectory* merged and became the first to use the newly created Unified Modeling Language (UML 0.8). While Objectory brought the use-case-driven approach in an object-oriented process model (Ivar Jacobson, 1993) into this merger, the Rational Approach contributed its iterative development and architecture approach. Together, they formed the new product, Rational Objectory Process 4.0. Throughout a subsequent series of mergers, the method not only changed its name to *Rational Unified Process* in its current version *RUP 2000* but also added some very strong features in the areas of data engineering, business modeling, project management, and configuration management to its product suite *Rational Rose*.

Refined by Rational Software, RUP is a process for developing software. Each piece of the process has one or more tools associated with it, and the use of common conventions and renderings tie the pieces together. One such tool is Rational Rose, which visually models use cases, object classes, activities, states, components, and deployment (mapping processes to hardware).

Essentially, RUP defines who is doing what, when, and how. It is a Web-enabled software engineering process that enhances team productivity and delivers software best practices to all team members. RUP identifies six best practices employed by successful software engineering firms worldwide:

1. Develop software iteratively.
2. Manage requirements.
3. Use component-based architectures.
4. Visually model software.
5. Continuously verify software quality.
6. Control changes to software.

Develop Software Iteratively

The traditional waterfall method for developing software pushes risk forward in time and makes corrections more costly. An alternative is the iterative or incremental process. This is an approach of continuous discovery, invention, and implementation, with each iteration forcing the development team to drive the project's artifacts (deliverables and documents) to closure. (See Figure C.2.)

Manage Requirements

Change is inevitable and so are changes in requirements. Furthermore, it realistically is impossible to completely and exhaustively state a system's requirements before the start of development. Active management of requirements encompasses three activities:

1. Eliciting, organizing, and documenting the required functionality and restraints.
2. Evaluating changes to these requirements and assessing their impact.
3. Tracking and documenting trade-offs and decisions.

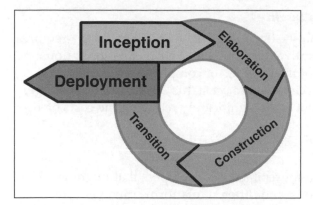

Figure C.2   The iterative development process.

Use Components-Based Architecture

The complexity of today's business processes and their reflection in software demand that a software developer, designer, or architect view a potential solution from a number of different perspectives. Additionally, each stakeholder brings a different agenda to a project. A system's architecture is therefore perhaps the most important deliverable that can be used to manage complexity and different viewpoints throughout the system's life cycle.

Software architecture is concerned not only with structure and behavior but also with usage, functionality, performance, resilience, reuse, comprehensibility, economic and technologic constraints and trade-offs, and aesthetic concerns. Components-based architecture is an important approach because it enables the reuse or customization of components from thousands of commercially available sources. Components-based architecture and design facilitates the leveraging of standardized frameworks such as COM+, Common Object Request Broker Architecture (CORBA), or Enterprise Java-Beans (EJB). Components also provide a natural basis for configuration management.

Visually Model Software

A model is a simplification of reality that describes a system completely, including its behavior. We build models of complex systems because we cannot comprehend such systems in their entirety. Modeling is important because it helps the development team visualize, specify, construct, and document the structure and behavior of a system's architecture. Using a standard modeling notation such as UML helps team members to unambiguously communicate their decisions to one another. Visual modeling helps to maintain consistency among a system's artifacts.

Continuously Verify Software Quality

Verifying a system's functionality—the bulk of the testing activities—requires iterative testing during the iterative development. This means continuous and quantitative assessments of the progress of completeness, correctness, and stability of the system, making the project status objective and no longer subjective. Defects are identified early, significantly reducing the costs of system corrections.

Control Changes to Software

Coordinating the activities and the artifacts of developers that might work on different sites, different iterations, and different platforms involves establishing repeatable work flows for managing changes to software and other artifacts. Coordinating iterations and releases requires establishing and releasing a tested baseline at the completion of each iteration. Maintaining traceability becomes a key factor in controlling the impact of change.

RUP also makes development processes practical by providing guidelines, templates, and examples for eDevelopment activities, thereby reducing risk and increasing predictability.

Overview of the RUP Life Cycle

RUP breaks down the software development process into four phases: inception, elaboration, construction, and transition (see Figure C.3). Subsequent iterations of the four processes are called *evolution*. Associated with each phase are concrete milestones.

Each phase consists of one or more iterations, defined as a distinct sequence of activities with an established plan and evaluation criteria, resulting in an executable release. The difference between iteration releases and phase releases is the intended use. Iteration releases are typically internal builds, meant for testing some particular piece of functionality (think version 1.1.1 to version 1.1.2). Phase releases have a wider distribution (think version 1.2 to version 2.0).

These phases, as [Kruchten 2000] defines them, are not to be confused with the traditional sequence of requirements analysis, design, coding, integration, and test. They are completely orthogonal to the traditional phases. A major milestone concludes each phase.

Inception

Inception is the beginning of a project. It can be viewed as *the Good Idea*— specifying the end-product vision and its business case, as well as defining

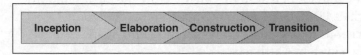

Figure C.3 The life cycle phases of development.

the scope of the development effort. We gather information and do proofs of concepts. The end of the inception phase is a go/no-go decision for the project. The inception phase is noniterative and sequential.

Some of the primary objectives of this phase include the following:

- Establish the project's software scope and boundary conditions, including acceptance criteria and descriptions of what is and is not within the scope.

- Identify the critical use cases of the system, meaning the primary behavior of the system.

- Estimate the overall costs and schedule for the entire project, specifically for the following elaboration phase.

- Estimate risks.

The outcome of the inception phase is creation of these artifacts (documents):

- A vision document
- The use-case model survey
- An initial project glossary
- An initial business case
- An initial risk assessment
- A project plan for the phases and iterations

Using Rose in the Inception Phase. As some of the tasks performed during this phase include determining use cases and actors, Rose can be used for documentation and to create diagrams to show their relationships. The diagrams can be presented to users to validate that the diagrams are a comprehensive view of the system features.

Elaboration

In elaboration, use cases are detailed and design decisions are made. Elaboration includes some analysis, design, coding, and test planning. Here, all necessary activities and required resources are planned. In summary, the purpose of the elaboration phase is to analyze the problem domain, establish a sound design foundation, develop the project plan, and eliminate the project's highest-risk elements. In other words, the elaboration phase is complete when the system is designed, reviewed, and ready for the developers to build.

The primary objectives of this phase include the following:

- Define, validate, and baseline the software design as rapidly as is practical.
- Baseline the vision.
- Baseline a high-fidelity plan for the construction phase.

- Demonstrate that the baseline design will support the vision within the project constraints.

The outcomes of this phase are as follows:

- A use-case model (80 percent complete) including the majority of the UML diagrams discussed in the previous section of this document
- Supplementary requirements that capture the nonfunctional requirements and any requirements that are not associated with a specific use case
- A software design description
- A revised risk list and a revised business case
- A development plan for the overall project

Using Rose in the Elaboration Phase. The elaboration phase presents several opportunities to use Rational Rose. Since it details the system requirements, the use-case model does require updating. As the flow of processing is detailed, sequence and collaboration diagrams help illustrate the flow. They also help to design the objects that will be required for the system. Elaboration also involves preparing the design for the system so the developers can begin its construction. This can be accomplished by creating class diagrams and state transition diagrams in Rose.

Construction

Construction is where the bulk of the coding is done and the product is built—in other words, it is a manufacturing process. As with elaboration, this phase is completed for each set of use cases in an iteration. The completion of an iteration includes testing, reviewing, and adapting the design architecture. At the end, the product, the completed vision, is ready for delivery to its user community.

Primary construction phase objectives include the following:

- Minimize development costs by optimizing resources.
- Achieve adequate quality as rapidly as is practical.
- Achieve useful versions as rapidly as is practical.

At a minimum, the construction phase consists of the following:

- The software product is integrated on adequate platforms.
- User manuals are produced.
- A description of the current release is formulated.

Using Rose in the Construction Phase. Construction is the phase in which the majority of the coding for the project is done. Rose is used to create components according to the object design. Component diagrams are created to show the compile-time dependencies between the components. After languages have been selected for each component, the generation of skeleton code can be done. After the developers have created code representing business rules, thorough testing ensures correctness, robustness, and completeness of the product.

Transition

Transition is the final preparation and deployment of the system to the users. It includes delivering, final acceptance testing, training, supporting, and maintaining the product until the users are satisfied. The release of the product into the production mode concludes the cycle.

The three primary objectives in this phase include the following:

1. Achieve user self-supportability.

2. Achieve stakeholder concurrence that deployment baselines are complete and consistent with the evaluation criteria of the vision.

3. Achieve final product baseline as rapidly and cost effectively as is practical.

Using Rose in the Transition Phase. Rose is not as helpful in the transition phase as it is in the other phases. After the product has been given to the users, issues usually arise that require developing new releases, correcting some problems, or finishing features that were postponed. Here it is of utmost importance that the model in Rose is been maintained and kept up to date. Some of the newer add-on features in Rose support system testing.

Process Work Flows

Throughout each of these phases, related activities are grouped into work flows. These work flows produce models or artifacts. The artifacts can be text documents, flow diagrams, application code, data objects, and so on.

Table C.1 illustrates only some of the possible work flows. They can involve anything from deployment and change/configuration management

Table C.1   Emphasis of Work Flows

WORK FLOW	RESULTING MODEL(S)
Business modeling	Business use-case model, Business Object model
Requirements	Use cases
Analysis and design	Design model
Implementation	Implementation model
Test	Test model

to the technical/development environment. Each work flow has a different emphasis assigned to it across different phases and iterations.

Workers and the Activities They Perform

Work flows determine the activities that need to be performed during an iteration. A worker, who represents a logical resource, also performs an activity. Think of a worker as a role (systems analyst, database administrator, technical writer, etc.). Many people can perform the role of systems analyst, and RUP regards them as *one* worker or, in other words, as the role "system analyst." Similarly, a resource can be thought of as an actor, who plays one or more roles in a production.

To summarize RUP's terminology, a *resource* is a person or group of persons, and a *worker* is the task he, she, or they perform(s). The task is called an *activity*, and it produces one or more *artifacts*.

Examples of workers are architects, business designers, business process analysts, change control managers, code reviewers, training course developers, deployment managers, designers, implementers, and testers.

Use-Case-Centric Process (Modeling with UML)

Models in software development are blueprints of systems that allow us to plan a system before it is built. It can help us to be sure the design is sound, the requirements have been met, and the system can withstand requirements changes.

The requirements for a system are derived from the business needs of the users. They are translated into requirements a team can use and understand, and from which it can generate code. This process is called *modeling*.

Visual modeling is the process of transferring the information from the model into a graphical display using standards for graphical elements. This standard is vital to realizing the benefit of visual modeling: communication. By producing visual models of a system, we can show how the system works on several levels, enabling users of this model to better understand complexity and implications. We can model the interactions between users and the system, interactions between objects within a system, even interactions between systems. By visually modeling systems, we add the dimension *dynamic* to our model.

One important consideration in visual modeling is the standard notation to represent various aspects of a system. Without a standard and common interpretation, visual models are not very useful. Some of the popular notations that have strong support in the industry are Booch, Object Modeling Technology (OMT), and the Unified Modeling Language (UML). Rational Rose supports these three notations; however, UML is a standard that has been adopted by the majority of the industry and the standards governing boards such as ANSI and the Object Management Group (OMG).

The consolidation of methods and standards that became UML started in 1993 by the three amigos of UML: Grady Booch, James Rumbaugh, and Ivar Jacobson. In 1997, UML 1.1 was released as an industry standard by the Object Management Group (OMG).

The use-case model consists of actors and use cases. It models the flow of events, preconditions, postconditions, use-case diagrams, and other information needed to fully define use cases. An *actor* is someone or something outside the system that interacts with the system. A *use case* is a sequence of actions a system performs that yields an observable result of value to a particular actor. Use cases are effective because they are concise, simple, and understandable by a wide range of stakeholders. This allows end users, developers, and acquirers to understand the functional requirements of the system. Use cases drive numerous activities in the process:

- Creation and validation of the design model

- Definition of test cases and procedures of the test model

- Planning of iterations

- Creation of user documentation

- System deployment

Use cases in RUP are modeled using UML. UML is a language for specifying, visualizing, constructing, and documenting the artifacts of software systems, as well as for business modeling and other nonsoftware systems. UML represents a collection of the best engineering practices that have proven successful in the modeling of large and complex systems." [OMG Unified Modeling Language Specification, v. i. 3, June 1999]

There are two potential drawbacks to use cases:

1. A shortcoming of use cases is the focus on functional requirements. All nonfunctional requirements, such as security or governance requirements, have to be brought into the system development processes using means other than use cases.

2. Use cases myopically focus on solving only the specific requirements at hand, rather than creating general-purpose solutions. There is no built-in support to guide the development process toward general and reusable solutions.

For example, suppose you are working with a client to develop a CRM tool. You diligently employ RUP, creating the seven use cases needed with UML, accurately modeling all five of the carefully crafted requirements. You develop the software, you release it, and the client is satisfied. But the client soon discovers a whole new use for the system. This represents a very common scenario, as we all know. Two new requirements are added to the list of five to meet that new need. Since the system was designed with the original seven use cases in mind, it is possible that the next iteration in the development cycle will have to be a major overhaul while increasing the functionality only marginally.

Transition from Architecture to Development

The Rational Unified Process as a development process has three significant features:

1. The development process is *iterative* and *incremental*.
2. The approach offers a high degree of *traceability*.
3. The documentation method is *streamlined for the usage of UML*.

These features are the kernel of the success of the Rational Rose product line and RUP as the development methodology.

Starting with this vision in mind, RUP offers excellent tools to model business flows, define specifications, and draw necessary diagrams to evaluate and refine these requirements. Along the way, using these tools, Rational Rose even creates skeleton code to be filled with real "meat"—the business rules. Nonfunctional requirements, though, will have to be documented separately and outside of Rose and remembered while creating the application. RUP does not provide consideration for the operate/evolve phase in the life cycle of applications.

The *Integrated Architectural Framework* starts at an earlier point in a project life cycle. The IAF supports the visioning and decision process in developing the optimal solution for a client's business needs. The IAF offers support tools an architect can utilize during the development and decision process to ensure that the client's requirements are met at a very early point during system development engagement. Tools utilized can be those from RUP or any other development method. Though not intended to be a development tool, the IAF creates many of the necessary documents or artifacts for system development. Only some minor work is necessary to map and define artifacts from RUP to the IAF so the architecture study can deliver these artifacts as an input to the RUP process.

The IAF integrates seamlessly with our Engagement and Program Management Methodology, PERFORM, and covers all aspect areas during the life cycle, including deployment and operation of an application.

The scope of the IAF, however, can be much wider and independent from any documentation tool. The IAF method supports building architectures for business models and enterprises. This method can also be used to assess current applications or business models and flows.

System Deployment

This phase is focused on deployment of the overall solution or a subset of the solution. The activities associated with the deployment will need to be adjusted to suit the particular implementation strategy employed (i.e., conventional approach, iterative approach, or a hybrid approach).

If a conventional approach has been pursued, then this phase will represent the final phase of the project. In this case, it will usually occur only once, but where deployment is planned for multiple sites, the phase itself will obviously need to be sequenced and tuned appropriately. In the case where an iterative or hybrid deployment approach is being employed, this phase provides the mechanism for bringing each successive pilot, and ultimately the full system, into operational status in the end-user environment. In some cases, the system will continue to evolve and therefore may pass through this process many times.

Regardless of which deployment approach is being utilized, there are certain common objectives that will be associated with this phase:

- Ensuring that all functional and nonfunctional requirements are met
- Bringing the architecture to live status
- Bringing the application(s) or pilot(s) and the associated data to live status
- Bringing the users to active status
- Facilitating the appropriate end-user and IS department training
- Initiating the support function
- Refining the service-level proposal

At the conclusion of this phase, an evaluation exercise should be conducted in order to record any lessons learned. At some point during this phase, a formal acceptance of the system is to be obtained from the user. This may be at one of several points in the phase and may itself be gained in iterations, after deployment and the warranty period. This will depend on contractual arrangements and will require tuning to position the event (or events) correctly in the project plan.

Testing

This section is intended to provide a summarized overview of testing. It should not be regarded as a general reference work or as a tutorial. The overview covers practical aspects of the design, construction, and running of tests for all aspects of computer-based systems, containing hints and tips collated from experience gained over many years of system building and support.

Test Objectives

This section is written from the perspective that effective testing of any system is fundamentally about identifying and managing the risk that the system will not perform satisfactorily in practice. Emphasis is placed on the benefits of a progressive approach to testing throughout the development life cycle from the earliest stage of the project, increasing the likelihood of early detection of errors and, hence, reduced cost. This section also focuses on the need for different types of testing and different teams, giving a good depth of testing, recognizing the difficulties that humans may have in detecting faults in their own work.

Test Stages

This section is applicable to the following test stages:

- Unit (including program/module) test
- Link test
- System test
- User acceptance test
- Component acceptance test
- Integration test
- Package commissioning test

It should also be noted that *test stages* can include design phases of development life cycles. Test specifications can usefully be produced as codeliverables with functional or technical specifications, with two objectives: to provide a concise description of certain features where textual descriptions are difficult to structure and to test the designs in advance of further development.

At times it may be prudent and useful to define additional testing stages. This may occur, for example, when there is a need to allow for separate testing of nonautomated functionality related to the system, where this is both critical to the overall business acceptability of the system and to its achievement of business effectiveness. Such testing may be included as subparts of system and acceptance testing, but can be separated and planned or specified in its own right.

Test Types

This section considers the following different types of testing that may be required in order to prove the system:

- Sample transactions/light functional testing
- Complex transactions/deep functional testing
- Performance—high volumes of transactions and/or data
- General operations including manual procedures and regular cycles
- Exception operations such as fallback and recovery
- Interfaces with other systems
- Regression testing

As well as being a test type, regression testing can be considered as a testing stage. This happens when there is a clear demand for the testing process to provide for future use a complete and repeatable set of documented and archived test deliverables.

Additional Test Areas

In addition to the test stages already identified, this section covers special considerations for the testing of the following:

- Data conversion tools and/or software
- Training material (CBT and manual)
- Manual procedures
- Software packages

Test Strategies

For an application software project to be successful, user expectations must be met. All too often a project is judged to be a failure because of differences between what the user expected and what the system delivered. Testing can therefore be considered a means of reducing the risk of the system failing to deliver against user expectations.

The objective of a *testing strategy* is to decide on the testing type and level that will be carried out at each stage of the appropriate development life cycle and to ensure that it is carried out in a cost-effective manner.

The strategy must clarify certain important constraints which will affect the approach, the documentation, the planning, and the budget for testing. These constraints include the necessity for (non-) regression testing and the need to formally hand over testing deliverables (possibly in a rerunnable state). The situation or the contract should dictate the needs, and the strategy must accentuate the requirements.

A further benefit of consideration of testing strategy at an early stage in the project life cycle is the identification of areas that may be difficult to test. This in turn may enable timely amendment of the overall system design or, if this is not feasible within the system requirements, ensure that the necessary testing is realistically planned and priced into the project.

A testing strategy needs to be drawn up in conjunction with the users and formally approved by the client project sponsor.

User Acceptance Criteria

Understanding the criteria against which the system will be judged, by the client and by the business representatives of the client company, is fundamental to the success of any project, to ensure effective design of the system, for properly scoped testing, and to meet contractual obligations. Therefore, it is important to introduce the activity of defining *acceptance criteria* early in each life cycle.

The following areas need to be considered for user acceptance criteria, although not all may be applicable to each situation:

- Functionality, by reference to documents defining the required functionality such as system functional specification.

- Interfaces, e.g., type, availability, data to be transferred, resilience and recovery, and protocols.

- Performance, which may include number of users, online response times, batch processing windows, database retrievals/updates, or hardware performance. Business volumes must be specified where applicable.

- Resilience for hardware and software including recovery mechanisms, target time for recovery/restart, and journaling or rollback.

- Operations covering timings of regular runs, periodic schedules, and degree to which the system will require special operational skills.

- User interface and operability including online user interface consistency, type (e.g., screen-based or GUI), and navigation.

- Documentation to be delivered.

- Quality of deliverables and checkpoints required by users.

- Milestones.

- Acceptance period for individual components (systems integration [SI] projects or package implementations) and the final delivered system.

- Warranty support services.

- Maintainability (may be covered by documentation and quality).

Establishing a Testing Attitude

Much has been written on the *correct* attitude to adopt for testing. Some writers argue that an aggressive attitude is required, where the intention is to prove that errors continue to exist, until the incidence of errors falls to a point where further testing is unlikely to yield faults serious enough to justify the additional effort of discovering them. In this case, a *successful* test is one where errors are found.

Conversely, a more passive approach to testing can be adopted, where the main objective is to demonstrate that the system supports a representative sample of business transactions, with all predictable variations, without error. In this case, a successful test is one where no errors are found.

The true understanding of the difference between *black-box* and *white-box* testing lies at the root of this discussion. It is relatively easy to search passively for an absence of errors using the black-box approach, whereas to deeply probe any piece of logic in an aggressive manner requires, or at least is aided by, an understanding of the internal mechanics of the target logic (white-box).

Aggressive testing can be conducted using the black-box approach, but this requires acceptance that, when searching to confirm a quite unrelated feature, some errors will be detected by chance. In addition, some areas will need to be probed in a particularly devious manner, for example, by refusing to accept that any feature will operate successfully on two or more successive occasions. This is especially true of exception conditions.

A team established especially to test a warehouse management software package demonstrated the preceding example. Although the package was some 15 years old, it had many errors, including one very severe fault in the logic for handling the placement of goods into the correct *slot* in the warehouse. When the nominated slot was full, and also the first and second overflow alternatives, the system proved incapable of remembering where the goods had been stored, effectively losing them. The team did not know how the internal logic was organized, but team members were determined to test the system with many devious combinations of valid but extreme situations. They were being aggressive, but were restricted to a black-box approach, and so were forced to generate perhaps many more tests than might otherwise have been necessary had the internal logic been available to them.

It is clear that, in reality, successful testing of most systems will require a combination of approaches, both aggressive and passive, white-box and black-box. It is this combination, properly harnessing the right resources, that the testing strategy seeks to specify.

The testing strategy must certainly direct aggressive testing to at least one area; unless the system is of low commercial risk, operational errors will be evident to the users and are unacceptable.

To aggressively test in more than one category (unit, system, acceptance) will usually not be cost effective for all but the most critical core business systems. However, the opposite situation, where all categories are conducted passively in the assumption that another category will test deeply, must be avoided.

Two other factors can be considerations for a testing attitude: testing designs rather than code and acceptance that originators are badly placed to test their own designs or code. The first of these accepts that faults found early in the development process save considerable effort, time, and cost later. The process is different from a review, which seeks to confirm completeness of designs. A *design test* will effectively create business sequences of transactions similar to those that will be applied later in physical testing and will apply these to the designs as specified. This approach can be aggressive or passive, and must rely on many *what if?* predictions.

Design tests can be applied to functional specifications, technical specifications, and program specifications.

The second consideration accepts that the specification of a test is more likely to find errors if an independent person is involved. This approach arguably takes more time, as the original design, specification, or code must be read and understood by another developer. If this is incorporated into the review process, then the time addition will not be great, and the intensity of review is most likely to be deeper and therefore more effective.

This does, of course, demand a degree of experience and aptitude on the part of the reviewer/tester; some people are extremely effective developers, others are extremely effective testers, and some have a natural ability to perform both functions very effectively on their own work.

Test Stages

Unit and Module Testing

The objective of unit testing is to prove that each basic unit (module or program) of the system can function, conforming to its specification. Unit testing must always be carried out for any bespoke software development or enhancement, and can equally apply to samples of package software as a confidence test.

Unit testing is based on detailed specification and may be approached as follows:

- Start by considering all the ways in which the unit may be invoked (or called) as well as the processing for each condition.

- Check the specification to see whether the preceding scenarios will test all conditions, and add further conditions as necessary, for example, data initialization, high and low values, field overflow, and data errors such as division by zero.

- Determine required results for each condition or set of conditions and the means by which results will be checked.

- Identify prerequisites for the test cases such as output from other units.

Unit testing is a white-box test based primarily on the functional and data requirements expressed in the specification for an individual program or unit of a system. Typical areas of focus for unit testing would be as follows:

- Screen/window contents and navigation including consistent usage of standard *hot keys*
- Report contents, layouts, and calculations including checking page throws and empty reports
- File/record creation, updating, and deletion
- *Destruction* testing, e.g., use of empty files and invalid data
- Testing of all different logic routes through a program

Link Testing

The objective of link testing is to test the continuity of design of the system, focusing on transfer of data between different program units. Link testing should consider areas such as navigation between online units and linking of batch jobs via batch queues or system commands.

Link testing also addresses the interfaces both to and from other systems, either by linking programs or by using an interface or transient file where program linking is not yet possible.

Link testing concentrates specifically on the interfaces between major modules, programs, and subsystems, seeking to discover inconsistencies between them. In some cases, link tests may be specified to demonstrate early examples of top-level system functionality to the client.

Because of the potentially different groups of tests within a complete series of link tests, it is likely that each group will require its own link test file, comprising a plan and a number of individual specifications.

A secondary aim of link testing is to ensure that the system is ready for system testing. Very often, this intermediate testing stage will avoid the halting of system testing while corrections are made to key interfaces.

There is a danger that link testing will be rushed or omitted on the grounds that it has already been covered by program testing or will be covered in the system test. For a simple, self-contained system this may be the case and the link test can be tuned out. In most cases, however, specific link testing will be required, as much time and money can be lost during system testing trying to pinpoint the source of a data failure where multiple programs or, worse, systems are involved.

The following must be considered when planning and executing link testing:

- Planning will need to give special consideration to the testing of interfaces where one end does not exist or is not readily available. This may involve creation of special interface files, and additional time may be required for creation of special test beds on the other system.

- Procedures must allow for recording and correction of faults or nonconformity in both the development system and external systems.

- Consideration of the impact of faults in external systems must include the impact on the overall testing plan.

- Time must be allowed for retesting of programs where required and for (non-) regression testing.

System Testing

The objective of system testing is to test the complete system against the business specification, typically a detailed functional specification. The testing is of the black-box type, being based around simulation of business cycles without regard to the construction of individual components.

System testing is usually the most extensive testing process (in terms of depth and breadth of coverage) and the most intensive (in terms of concentration and use of resources) in the whole development life cycle.

The testing strategy will have assigned coverage and responsibility for system testing, which will usually comprise the following:

- Extensive functional tests (especially where complete transactions cannot be traced from source to exit in unit testing)

- Complex transaction sequences (possibly representing several months or years in simulated time periods)

- Distributed functions and data (taking into account possible network failures)

- A complete representation of all possible exception conditions provided for in the design

- System failure conditions (often requiring emulation under controlled circumstances)

- High data volumes (preferably exceeding predicted future expectations)

- High transaction volumes (preferably exceeding predicted peak expectations)

- Resistance to incidents and security violations

- Demonstration of recovery and fallback

- Integration with other systems and manual procedures
- Demonstration of data conversion

Because of the potentially different groups of tests within a complete system test, it is likely that each group will require its own system test file, comprising a plan and a number of individual specifications.

System testing typically requires a great deal of careful planning and preparation to ensure that the execution of tests can be carried out in an effective manner with reliance on the achieved results as proof that the system performs in accordance with the business specification.

Interfaces with systems (both computer and manual) external to those being developed are a frequent source of error. Well-planned link testing should eliminate many errors; however, testing of business-oriented data using real links as opposed to simulated links should be allowed for in planning.

Any tools required to support system testing must be confirmed and their precise use defined in the test plan. Special attention should be paid to automated testing tools, as the situation may demand extensive retesting after failure and correction, possibly reverting to the position reached several days, or even weeks, before.

Ideally, test specifications should be generated by someone other than the designer, as this is more objective and frequently shows up specification errors and inconsistencies before the actual test execution.

In any case, it is highly desirable to have some client or user experts join the development team. This provides the best possible chance of identifying faults at an early stage, as these experts will use their knowledge of real data, exceptions, and complications to identify inconsistencies. According to agreed strategy and availability of resources, a combined system and acceptance test is an option, using the synergy of the developers and the users to scope and detail the tests in one controlled exercise.

System testing will usually involve many development staff in a highly controlled series of tests, each one needing coordination and collaboration in its execution, documentation, and validation.

Selection of test data is another important consideration for system testing. The data must be carefully selected to be the minimum subset of data that will reflect business reality and test all relevant conditions.

Using a subset of existing live data is not generally recommended because, while it may reflect business reality, there is a danger that not all conditions requiring testing at this time will actually be covered by the live data. In addition, the data itself may not be accurate or entirely valid, which may also give rise to misleading errors.

User Acceptance Testing (UAT)

Acceptance testing should be predominantly a user-driven exercise, with only support and advice provided by the development team.

The aim of acceptance testing is as follows:

- To prove that acceptance criteria have been met
- To demonstrate the user interface
- To provide users with confidence that the system works correctly and is manageable by them
- To exercise the system with real data
- To prove the correctness of the system manuals

The acceptance tests should be independent of system tests, since their aims are different. However, there may be some duplication of tests, preferably using different sources of data (the users' data being more readily extracted from the current live system). Also, acceptance testing provides an opportunity for any faults remaining after system testing to be trapped.

Acceptance tests may be planned in different ways, according to client wishes or standard approach. For example, a series of staged, interim acceptances may be planned, leading to a final formal acceptance. The concept of internal and external acceptance may be established by drawing up common acceptance test specifications and using these first internally (by the development team) and then externally (by the client).

If the client is fully responsible for acceptance, plans and specifications for acceptance testing may be drawn up, at the client's discretion, at any stage of the life cycle.

UAT is often a key contractual milestone in that payment may be fully or partially dependent on acceptance of the delivered system by the project sponsor.

UAT tests are designed from the testing strategy, the user acceptance criteria, and the documented business requirements, typically the detail functional specification. The emphasis should be on simulation of the live situation rather than an exhaustive test of all business conditions.

As with system testing, the selection and creation of test data is an important consideration and the same caveats apply to use of subsets of live data from existing systems. Use of input forms reflecting or simulating the way in which live input data will be made available for entry into the system is, however, highly recommended as a means of checking manual procedures and system usability.

Component Acceptance Test

A component acceptance test is typically used during the systems integration life cycle to check components prior to full integration testing. Testing will be undertaken by the prime contractor of the SI project with a view to accepting or rejecting a component developed by a third-party supplier, and this test stage is likely to be a key part of the contract between those two parties.

Good planning and control of a component acceptance test is important both contractually and as a means of detecting faults in a component prior to the more expensive integration test stage.

Acceptance will involve verification that all required components have been supplied (including documentation), as well as conducting a series of tests designed to demonstrate correct functioning of the component.

Integration Testing

Integration testing is used in the systems integration life cycle to prove the logical and physical links between a number of components of a complete system, be these software, hardware, packages, or networks. Testing is of the black-box type, being a proving of the links between each component with minimal consideration of the internal workings of the component.

An integration test will require careful planning and control due to the complexity of the issues involved:

- Multiple suppliers with responsibility for one or more components
- Multiple environments—hardware, networks, communications, and operating software
- Need to integrate and test in a sensible order to enable easy identification of the source of a fault
- Controlling any necessary modifications (corrections and/or change requests) to components and need to retest to ensure (non-) regression

Package Commissioning Test

The commission package (also know as OTS, or "off-the-shelf," package) test stage is applicable only to a package-based application and has the following main objectives:

- Review of conformance of package documentation against the package as delivered and supplier specification
- Testing of the standard package
- Modification of parameters and tables to include client-specific data and testing
- Identification of enhancements and modifications required to the package

The test is a black-box activity in that it will not attempt to test the underlying package functionality in detail. However, the depth of testing of the standard package will depend on the maturity of the package and version. For example, more stringent testing of package functionality will be required for a new package or for a beta test release of a new version.

A further set of commissioning tests will be necessary following delivery of package modifications, which must include (non-) regression testing of areas impacted by the modifications. *Full system testing* should be planned for any package enhancements.

Consideration of Types of Testing

Light Functional Testing

Also known as *functional skim testing*, this type of test involves testing the basic requirements of functions. Examples are:

- Data entry and amendment using valid data within system limits
- Simple routes through a transaction or series of transactions

Light functional testing is very useful at the start of system and acceptance tests to verify that there are no major errors or omissions in the system that would hold up testing. Light functional testing could also be relevant for link testing, component acceptance testing, and package commissioning as a means of checking the look and feel of a system.

Deep Functional Testing

This type of testing looks at functionality in more detail, including complex transactions and exception conditions. It may follow successful completion of light functional testing, as described previously. Examples are:

- Use of invalid combinations of data/conditions on data entry or amendment
- Use of unusual or invalid combinations of transactions
- Use of values on or outside system limits
- Empty reports and high-value totals
- Special processing at month or year end.

Deep functional testing is essential within system testing and is advisable for user acceptance testing.

Volume and Performance Testing

Performance is often a key contractual point or an area where failure has a greater impact and requires special consideration.

Overall performance cannot be properly determined until the final software and hardware is available. This is far too late a point to discover that the system will not perform adequately. Also, while some performance issues may be addressed by simple code amendment, in other cases the fundamental system design may be at fault.

Testing must be planned as part of a whole approach geared around designing for performance, starting with a clear understanding of the system requirements as documented in the user acceptance criteria.

Examples of volume and performance requirements are:

- Online response times for selected or all transactions usually in terms of time taken for windows/screen prompts to be presented; may be required to be proven for a specified number of simultaneous users.

- Time taken for processing within an online transaction such as the time taken from submitting a report request to production of the hard copy.

- Time taken for batch update transactions; may be for the total normal daily run or for weekly, monthly, or annual runs.

- Time taken for operational activities such as database backup.

- Time required to process peak volumes such as the largest invoice run of the year.

- Limit on performance degradation (online, batch, or both) allowed under stress conditions, which may be related to data volumes, number of concurrent users, or a simulated mix of transactions such as would be encountered during peak business periods.

System design and testing must therefore plan for performance right from the start of the development. All design processes should consider performance. Performance testing will therefore be an incremental process through the life cycle as follows:

- During the functional stage, identify critical functions for performance.

- During technical design there are a number of actions that can be taken:

  - Check the design of critical functions for performance within the planned database, 4GL, and software tools.

  - Ensure that the team is aware of any general constraints regarding performance of the tools.

  - Consider simulating database access and updating for critical transactions.

  - Recheck proposed live environment, database sizing, storage, network capacity, and so on.

  - Apply technical QA to the design of critical functions involving in-house experts and database supplier as appropriate.

- At program or link testing stage, once basic functionality has been proven, run benchmarks on critical online and batch jobs.

- Consider use of tools to simulate high-volume transaction through-put. This is cheaper and easier to arrange than a 40-user test. Bear in mind that performance under this kind of testing is likely to be worse than in real life since machines are faster than humans at data entry.

- Once the live hardware environment is available, testing of areas such as backup timing can be undertaken. Simulation of online or batch processing may be possible even before fully tested versions of these functions are available.

- Plan for final performance and tests on the target environment with final versions of software. These will typically need to be scheduled toward the end of UAT.

Fallback and Recovery

This type of testing considers the controls within the system for handling database or hardware recovery, together with the mechanisms for recovering the system up to the point of failure. This can include:

- Restarting the system

- Recovering the database to last backup or the point of failure

- Rolling forward or rolling back to a consistent state

- Checking the integrity of the database

Note that the last two steps may involve reentry of transactions or other manual procedures, depending on the recovery mechanisms available.

Special care must be taken and additional elapsed time for testing allowed where system integrity must be checked across networks, where more than one site is involved, or where a distributed database is involved.

This area of testing can again be split to bring testing forward independent of software availability, for example:

- Checking backup and journaling tools on the target hardware

- Testing any database recovery and integrity tools

- Running a benchmark trial for the time taken for the system to be up and running after a major failure by a combination of testing of available components and simulation

Final testing of full fallback and recovery should be planned as part of UAT to be undertaken by the eventual users or operations staff. This will need to include confirmation of the time taken for the complete system to be up and running again after a major failure, including all manual procedures and checking.

Interfaces with Other Systems

This type of testing considers the passing of data between systems or subsystems and may involve the following:

- Different hardware platforms, leading to the need for data to pass through a number of different layers due, for example, to incompatible operating systems
- Use of networks or other communications media for data transfer or use of tapes, cassettes, disks, or the like
- Data conversion of interface data items/files

Testing must include recovery from failure in any part of the interface.

As for the two previous types of testing, an incremental approach should be planned to test each part of an interface as soon as it is feasible. This is particularly important for any interface across different technical platforms, as errors can be very difficult to pin down.

It may be useful to adopt a *systems integration* approach to any interface where one end of the interface is outside the direct control of the development team. The SI approach defines key interface components and constraints during the definition of user acceptance criteria and carries on with this focus throughout the life cycle. In addition, using the approach of a component acceptance test with emphasis on proving each part of an interface prior to any full integration test of that interface will assist in establishing the source of any faults quickly.

Operations and Usability Testing

This type of testing checks the general operation of the system in practice, for example:

- Access controls and security at system, function, and data item level
- Running of batch schedules (daily, weekly, monthly), covering:
  - Special tests for failure partway through the schedule, affecting dependent jobs
  - Check of overall time
  - Check of performance of multistreaming jobs or file contention
- Submission and printing of online and overnight report requests including output destinations

- Use of all input media including manual forms, telephone dialog, bar code readers, and so on.
- Taking of security backups and housekeeping tasks such as data archiving
- Use of operations manuals
- Recording and use of audit trails

Preliminary testing of all these areas should be included in the system testing stage, although technical constraints may prevent some testing from taking place. Full and final testing should be covered within the user acceptance test. Specialist operations staff may be required to plan or execute tests.

Special Areas of Interest

This section covers three key tasks associated with development projects that are worth considering as projects in their own right, with associated considerations for planning, design, and testing running in parallel with the main development tasks.

The three areas are *data conversion, training,* and *manual procedures.* All three can be critical to the success of a project for its implementation and live running and must be considered from early in the project life cycle.

Data Conversion

Any system development and integration project is likely to require an element of data conversion as part of the migration to the new system. This can range from a simple set of conversion tools to enable rapid acceptance of new data to a major file migration between systems involving data extraction, merging, validation, and reconstruction.

The data conversion will be fully or partially aided by software, which may be purchased or developed specifically for the conversion. Such software is referred to as a *data conversion tool.*

Detailed knowledge of the current system will be essential at the data item level, particularly if the system is poorly documented or has a long history of modifications. Business knowledge will also be important when considering issues relating to system unavailability, handling of data exceptions, temporary procedures where parts of the conversion are delayed, and so on. The data conversion strategy and plans must be reviewed and approved by the project sponsor.

Particular attention must be given to a phased implementation, whereby one or more subsystems are replaced with other subsystems and continue to be used in the interim. Consideration must be given to how the total system will work at each stage of the migration. There may be a need for temporary software, often called *bridging software*—for example, to simulate links between subsystems—as well as the specific data conversion or acceptance at each stage.

The scope, design, and testing of data conversion tools will be impacted by the following three points:

1. The window of opportunity for the conversion process must be assessed. It may be restricted by either or both of the additional main considerations.

2. The time at which the data is in a suitable state for conversion and can be frozen for the duration of the conversion process. At worst, this can be once per month or even once per year.

3. The maximum length of time that the complete cut-over to the new system can take before causing disruption to the business. Note that data volumes may vary greatly after major housekeeping runs.

The volume of data to be converted and the complexity of the conversion process must be considered to ensure that the total time for conversion, including all pre- and postconversion checking, will fit into the available time window. Conversion planning should consider use of temporary procedures to handle situations where data has not been converted at all or errors / exceptions have occurred, as well as splitting of the conversion to convert the main business files first, followed by conversion of historical data once the new system is operational.

The validity of the data must be considered to assess the likely volume of error and exception conditions that may be generated and the impact of these errors. An error rate of 1 percent is quite low; however, if there are 10,000 records to be converted and each error condition will take 30 minutes of a user specialist's time to sort out before the new system can become operational, there is a potential delay of some 50 hours.

Requirements and scope of checking of consistency and completeness pre- and postconversion should be reviewed—for example, weighing of machine and human resource constraints, and use of control totals to check completeness.

Any steps that are processor intensive must be capable of restart from the point of failure.

The cost of software development for converting files should be reviewed against the cost of human resources for data input.

Plan for the development of bridging software for phased implementations—for example, a temporary batch interface for data that will later be passed online or a temporary batch process to collect and total statistical information that will later be picked up via a new management information system (MIS).

Inclusion of a dry-run test of the complete cut-over process is strongly recommended if this is feasible. This should run from the point of preconversion checks or freezing of data to going live with the new system, including all manual procedures and checks. A subset of the live data should be used for high-volume conversions.

Training

Training is an important part of the preparation for the implementation of a new system, as inadequate training of system users can result in the system being rejected as unusable.

Training is likely to encompass a number of different sessions for different target audiences, from overviews for management through intensive training in a single function for telesales operators. Training may involve hands-on sessions or use of specialist computer-based training (CBT) software for self-training, together with a wide range of supporting material.

Training may involve resources from the development team and specialist training providers as well as significant input from the user community to ensure applicability to the real business environment. One possibility is for a team of users to be assigned and trained to become the *training experts*. This team could also assist with user acceptance, data conversion, and manual procedures, and in production of user documentation.

Manual Procedures

Manual procedures covers those areas of the running of a business system that are not covered by automated processes. Poorly developed or tested manual procedures, particularly in face-to-face customer situations, can drastically reduce the effectiveness or efficiency of an IT system, with consequent impact on the user view of the success of the system.

Examples of manual procedures to be considered include:

- Authorization of input documents and/or entered documents
- Analysis of outputs and correction of errors
- Transfer of data to/from other organizations
- High-volume/multiformat output documentation such as collating travel tickets, hotel vouchers, personalized cover letter, and general travel information into a single pack for a customer, sorted for post office bulk mailing

Special consideration should also be given to manual fallback procedures to be undertaken in the event of full or partial system failure.

This whole area will require significant input from users to confirm the fit to business requirements.

Test Specification

The test specification task is divided into five main activities:

1. Identify and define test cases.

2. Create test scripts.

3. Specify test data.

4. Describe dependencies between tests.

5. Finalize and review.

Define Test Cases

This activity involves looking at the test cases for each individual test item as defined in the test plan. A test case will include the condition being tested and the required result, generally without reference to specific data values.

Within this activity, any general testing rules and operations that apply to tests should be described.

Test cases need to be comprehensive, and the following should be included:

- Testing of user interfaces such as help facilities, use of function keys, and navigation between windows/jobs. The depth and quality of any previous testing that has been undertaken (e.g., program test) should be considered when deciding the amount of this type of testing that should be performed during the test stage under consideration.

- Input data limits—for example, if a field may be any value between 2 and 10, check values on the limit (2 and 10), inside and outside the limits.

- Testing of output data limits, for example, producing a report where every value is the largest length allowed and checking that any totals are still readable. Both negative and positive values must be checked where applicable.

- Checking of the system under normal and exception conditions— for example, on a report showing a comparison of orders during this period compared with the same period last year—tests the situations where the previous year's data both does and does not exist.

- Testing of all possible business conditions, decision tables, and decision trees, which may assist in ensuring the completeness of the process.

- Inclusion of some destruction-type testing, once basic functionality has been proven, to test for the unexpected.

Create Test Scripts

This activity has the objective of grouping together related test cases in order to produce a set of test scripts.

As an example of a set of functions covering online invoice entry with limited validation and later batch validation and posting, a series of test scripts might be as follows:

- Online entry of a few valid invoices, run validation and posting, check posting report, and error report.

- Online entry of mixed batch of invoices with some invalid invoices, run validation and posting, check reports.

- Online entry of exception conditions relating to the online function, such as incorrect batch totals, high/low values, and invalid data values.

- Entry of a valid batch of invoices, then changing system or customer details before running the batch routines to test exception conditions in the batch processing.

All functions should be tested at least once, with complex functions being tested several times using different data. Wherever possible, start with simple, valid conditions before moving on to testing of complex situations or error conditions.

Test scripts must be completed, with all instructions relevant to initializing the test run, executing the steps, providing evidence of test results, continue/cancel error situations, test run termination, test data security, and backup.

Describe Test Data

The objective of this activity is to describe, in outline form, the data related to each test script. Actual data values will be defined and created during test execution. Areas that should be documented at this stage are as follows:

- Basic data requirements, for example, existence of system parameter files together with some current customers.

- Start values and criteria for each test case, for example, a customer with at least one outstanding invoice.

- Input data required. For a batch function, this would be the transactions or files that must be available for processing; for an online function, this would be the data that will need to be entered.

- Anticipated result in the form of a statement of the files, report layouts, screen layouts, error messages, and so on, that will need to be checked.

- Any special criteria relating to approval of the test.

Provision must always be made for reinitializing the test run with its original start-up data and environment and passing the data in its postrun state to another test run.

Describe Test Dependencies

The objective of this activity is to describe dependencies between the various test cases to help with scheduling the test execution. A diagram showing the flow between tests will be useful. As an example, in the situation where customer maintenance requires the existence of a number of basic parameter files, then the customer maintenance tests are dependent on completion of at least the *simple* tests on the parameter files.

Many test dependencies will be data related, and test execution will run more smoothly if each test script has the minimum number of dependencies on previous tests. It can be seen that proving the basic functionality of a complex function as part of an early test script and leaving the complex testing until a later run will help prevent the blocking of tests that require normal data files or transactions that are the outputs of that function.

Select First Run and (Non-) Regression Test

This step is the first part in the specification review process, with two objectives. The first objective is checking the first scheduled test runs to ensure that they are a good basis for proving key areas of the system early and for maximizing the number of different test sequences that can be progressed in parallel, that they are a fair basis for assessing the quality of the delivered item, and that they test simple routes through the main functionality (complex conditions and destruction testing should be saved until later). The second objective is to select a subset of test runs or specify some additional tests that are suitable for (non-) regression testing. The final selection of tests would still need to be reviewed during test execution to reflect the level of faults found and the affected areas of the system. Typical criteria for tests would be that they are quick to prepare and execute, that large-scale batch runs should be avoided, that a special test working on a data subset should be planned, that the results are easy to check, and that there is good functional coverage.

The best (non-) regression test of all may be a complete rerun of all tests. This is feasible for smaller systems and for larger systems where test execution is assisted by tools such as data comparators and data capture and replay.

Review and Finalize

Once specifications have been completed, it should be possible to create a plan for the test execution stage. Earlier we considered the elements of such a plan as part of the requirements for monitoring and reporting progress during test execution.

The test specifications must be reviewed, specifically for the following:

- Conformance with the testing strategy and plans
- Conformance to testing method and quality standards
- A check of estimates for test execution, including any factors that may affect the elapsed time such as test runs requiring restricted resources
- A check that the test dependencies do not make the test execution unduly linear such that failure in one test script could prevent further testing
- A check of the length of the critical path for test execution

In addition, there must be a check of the validity of the test itself; for example, do the test cases check for adherence to the appropriate items (such as system functional design) rather than the tester's expectations and are the required results correct for the test conditions?

This checking is often best carried out by the development team rather than the test team. The objective is to try to reduce the number of nonerrors raised at the start of testing that are due to failures in understanding or test specifications rather than genuine faults.

Test Execution

Test execution will ideally, and most efficiently, be conducted if the environment for testing is well understood, has been planned for, and is effectively managed.

Set up Environment(s)

Before test running can take place the testing environment(s) need to be set up. This includes setting up procedures and administration, the hardware environment, the software environment, and testing tools and utilities.

Establish Administration

This covers areas such as assigning office space and support (e.g., secretarial services) for the testing team. Administration includes the mechanisms for delivery of software versions and error reporting.

Establish Authorization and Security

Authorization and security routines must be established in order to control the testing activities that are carried out and the people who have access to the premises and the systems under test.

Installing Hardware for Testing

Check whether the hardware required for testing is available, or check when it is due to be delivered. Special hardware may be required during testing in order to make testing simpler. Examples of hardware are extra memory, printer, network control units, and dedicated testing computers or workstations.

Installing Software for Testing

Check that the software for testing according to the testing strategy has been installed, or ensure that installation of this software is carried out.

Installing the Testing Environment

It is necessary to establish an environment for testing that cannot unintentionally be mixed up with or interfered with by other environments. Considerations include:

- Sufficient memory and disk space for the test databases
- Space for software library(ies) and naming conventions
- Register authorizations for environment and configuration management activities
- Media to be used for backups and routines for backup and restore of database versions
- Establishment of which test results are to be saved and how this will be effected
- For UAT, ensuring that the environment either is or closely resembles the eventual live production environment

Where test execution is taking place in a production environment, special arrangements may need to be made for testing of database failure and recovery. Similarly, special access may need to be requested for certain tests or configuration management activities.

Installing Testing Tools and Routines

All testing tools to be used during test execution must be implemented and checked during this stage, together with any special software routines such as routines to generate interface data.

Creating Test Databases and Files

Space and a structure must be created for the test databases and files to be used during testing. This is done in accordance with the test specification, applicable routines for implementation, and standards for documentation.

Any setup data required for the test should then be created. Note that this activity can be very time consuming if a significant amount of data is required. Tools can speed up the process; alternatively, it is often effective to write temporary routines for data generation.

Check/Test Environment

The environment should be checked and tested before proceeding with the test running activity. The following may be required:

- Check completeness of delivery, including documentation, by inspection against a list of components required to be delivered.
- Test access to the environment for various authority levels.
- Perform random testing of access to various functions for different authority levels including database enquiry and update.
- Check that all input and output peripherals are functional.
- Check environment facilities such as backups.
- Check that all testing tools and testing routines are implemented and working correctly.
- Check that any setup or shared data is present and accessible.

Once this checking and testing is complete, the database or files should be cleared down and ready to start test execution.

Test Running

Test running is iterative in nature: Each test run is performed in sequence and rerun when errors have been corrected, until all tests are completed successfully.

All test runs should be performed in a controlled and repeatable manner, so that any runs made after error identification and correction may be compared directly to the conditions that first revealed those faults. Finally, a selected series of tests, known as (non-) regression tests may be run on all or part of the system to ensure that corrections in one area have not resulted in faults elsewhere.

The sequence of events for test execution is thus:

1. Prepare test data.
2. Run test.
3. Analyze results.
4. Implement corrections.
5. Repeat from step 2 until all errors are corrected in each test script.
6. Select and run (non-) regression tests.

The test execution should be supported by a *test run control log* to record the progress of test runs.

If free-format testing is required for a particular part of the system, it is recommended that such testing take place after the relevant test sequences have run successfully.

Prepare Test Data.   This step completes the preparation for the running of each test, covering the following tasks:

- Review test script(s) to identify data required, and to ensure that all necessary documentation is present and any previous test dependencies have been completed successfully.

- Refine and check any required results that are dependent on specific data values such as complex calculations.

- Input or generate data that is not being provided as a result of tests carried out earlier in a sequence.

- Produce a hard copy of the files and database before running the tests.

- Make an electronic copy of the start data position to facilitate the rerunning of tests.

Run Tests.   This covers the following three steps:

1. Conduct the test sequence in accordance with the test instructions. Where applicable, ensure that the sequence is being recorded to facilitate rerunning of the test sequence.

2. Note any unexpected events that occur.

3. Print out required hard copies of results, such as reports or screen layouts.

Analyze Results.   This covers the following four steps:

1. Systematically compare the test results against required results for the test.

2. Raise a fault report and submit this to the person responsible in accordance with the defined fault-handling routine. Append copies of relevant information, with any issues highlighted, to assist in fault correction.

3. Update the test log with the results of the run, showing whether the test completed or not and any faults that were detected.

4. Save the test results, file the full hard copy, and take an electronic snapshot of the database or files where possible.

Implement Corrections.   This activity has the following six steps:

1. Analyze any faults. It is very important to recognize that, particularly in the early stages of testing, there can be a high number of faults raised that are actually due to misunderstandings about the system or errors in the test process. It is worth checking the validity of the test input data, the consistency of the database or test files prior to the test, the required results, and the test script itself.

2. If the fault is in the test specification, correct these defects.

3. For other causes, pass the fault over to the team responsible for fault handling for further analysis. This may result in the development team agreeing to implement a correction or the raising of a change request.

4. On completion of fault analysis, update the test log.

5. Update the test status report.

6. On receipt of corrected faults, run any checks or retesting required prior to returning to the run test stage.

It is important, particularly for multisystem and development team situations, to ensure that all faults are properly investigated and that the full impact across all parts of the system is assessed. This may require one or more persons with in-depth knowledge of the complete system design.

Run (Non-) Regression Tests. This final activity takes place after all tests or tests relating to a component or part of the system have been run successfully. A set of tests should be selected to reflect the number of errors and parts of the system affected, derived from the potential set of (non-) regression tests identified earlier.

Test Completion

Testing is complete when all planned testing has been completed and all test faults have been signed off as corrected or resolved. A test may be defined as complete even if low-priority test issues are still outstanding, depending on the completion criteria defined in the testing strategy.

All storage of data and any special archiving of data and files must be completed and entered onto the test run documentation.

Each testing stage must be formally signed off once it has been completed. For example, UAT must be signed off by the client project sponsor.

Once a testing stage has been signed off as complete, the items being tested must be handed over to the next stage of development. Care must be taken in defining the items to be handed over.

End-of-Testing Review

It is extremely useful to have a report at the end of each testing stage and at the end of the project, which summarizes how the testing was conducted and indicates any lessons that can be learned by CGE&Y.

The following are guidelines to the areas that could be included in such a report:

Fault analysis. Summary of the nature, source, and number of faults found during a testing stage.

Testing statistics.   The number of test runs and retests analyzed by the test stage, the team responsible for component development and so on, and a review of faults detected during (non-) regression testing.

Comments on testing standards.   The limitations of existing standards with recommendations to create new standards or to use current standards.

Manpower summary.   Planned usage by system, subsystem, and test stage compared with actual usage.

Cost summary.   A breakdown of forecast and actual costs, with an analysis of the major differences.

Level and depth of testing.   Analysis of the extent to which individual modules and programs were tested and the occurrence or absence of errors at higher-level testing that should have been discovered at lower levels.

Documentation.   The standard of testing documentation and its ability to permit retesting at a future date.

Comments on strategy/lessons to be learned.   Comments on any aspect of the testing strategy, specification, and execution such as extent and success of user involvement in test stages other than user acceptance testing, approach to areas such as performance testing and interfaces, use of testing tools and methods, and resources used and training given.

Tidy-up and Archiving

This is the final activity in the testing process and includes the following:

- Archiving data and quality records produced during testing
- Deleting any libraries and files needed for testing but no longer required
- Releasing any resources needed for testing but no longer required, such as sign-on IDs, terminals, printers, accommodation, run capacity, and returning the test environment to its original state
- Disposing of unwanted documentation or paperwork such as old compilation listings, invalid test runs, and working copies of documentation
- Moving tested software into the environment for the next stage of testing, together with any required test data

Fault Handling

This section contains procedures for recording, monitoring, and reporting faults identified during any software testing process. A fault is deemed to be a query raised about, or a defect discovered in, a piece of work that was previously believed to be complete and correct—that is, excluding defects in an item discovered by the originator during its development.

The four key features of a fault-handling system, explored in more detail in subsequent sections, are:

1. Recording of faults and monitoring through subsequent events.

2. Assignment of priority to resolution of the fault.

3. Analysis of the type of fault with appropriate follow-up actions for three broad categories—*fault correction required* (nonconformance), *change control* (new or enhanced requirement), and *no action/tester error.*

4. Procedures to handle assessment and impact of faults where testing involves more than one system or development team.

The fault-handling system must lay down agreed responsibilities for the undertaking of all key activities, particularly where testing is carried out by one party and the responsibility for correction of faults is with another party. This documentation of responsibilities commences during the testing strategy and should be concluded during test planning for a particular test stage.

The following list indicates the different areas of responsibility, with likely assignment between the parties involved in the test:

- Raising faults—usually the test team subject to approval by the test team manager

- Assignment of priority and fault analysis—usually the test team manager with agreement from the development team manager; may involve the users and project sponsor where there is a potential change request

- Maintaining the fault log—usually the test team manager

- Reporting on progress of correction of faults assigned to the development team and released for retesting

- Reporting on progress of change request faults

- Sign-off that fault analysis is agreed on and that all actions are completed

Configuration Management

All testing stages must operate within the control of a *configuration management system,* with responsibilities clearly defined. Configuration management should be defined and implemented during the early stages of the project.

The control of *configuration items* (CIs) can be extremely complex during the testing phases. When a system is being delivered over multiple phases, the potential for error is even greater.

Configuration management during testing will be the responsibility of the development project manager for small projects or a specifically assigned configuration manager, possibly the database administration manager or project quality manager. The test team manager may have responsibility for managing the procedures and will usually be responsible for authorizing the issue of software and release into test libraries.

System Documentation.   Documentation must remain in step with program development and testing. In a rapidly changing testing environment, this can be done retrospectively, but must be controlled. For example, the detailed functional design document may be too large to reissue in full until the completion of testing; however, required amendments can be controlled by marking the text by hand and maintaining an index at the front of the document summarizing fault references and associated changes.

Consistent Releases.   The initial release of software into a testing environment and subsequent rereleases following corrections must be carried out using controlled release procedures. Much time can be lost testing against or analyzing faults found in out-of-date or incomplete versions.

The issuing of updates to environments supporting link testing and subsequent stages should be carried out using controlled releases.

A good control mechanism will include the following:

- A checklist of release contents signed off by the development team manager, including updates to documentation used by the test team.

- Checks to ensure that each release is consistent with all CIs of a comparable version. This includes programs, databases, test specifications, test data, and operational infrastructure.

- A list of faults corrected in this release.

- A list of correct versions of all amended CIs.

- Instructions for physical release of software into the test environment, plus evidence that physical release has been checked.

- Indications of where this release may affect previous testing or data; for example, correction or enhancement releases that add missing data items will make it necessary to rerun tests that maintain this item.

All components of the release must be secured so that the tests can be rerun or (non-) regression testing carried out.

Preparation for Deployments

Many companies have realized that they must accelerate their process of doing business in global markets on either the domestic or the international scale. Increasing competition requires from companies not only a differentiation in costs and quality, but also in the speed of market assessment, flexibility, regulation fulfillment, and the research and development of new products. Business operations are becoming more dependent upon the availability of multisite, consolidated information in order to make faster and better-informed decisions.

Globalization, time to market, and intense competition for customers are having a dramatic effect on systems and processes. Corporations focus on maximizing the enterprise resources through global standardization of enterprise resource planning (ERP) solutions as well as supply chain functionality.

As a consequence, most companies are in the process of improving their supply chains and changing their ways of doing business as well as their internal organizations. They are reengineering and standardizing their processes and their systems on a global basis.

Corporations are looking for solutions that enable them to realize these improvements smoothly but rapidly and in a cost-effective way. They are choosing to implement and roll out a core solution, this being an ERP package or a business-specific home-made system, across the organization to minimize costs and to define and support their new processes, pushing the IT service provision downward within the organization while allowing central reporting and control and the trend toward buying rather than building applications.

Rollout is a complex and high-risk process that requires multiple dimensions—technology, program management, culture fit/sensitivity—under business-led time pressure. Issues are being faced in defining, preparing, and realizing the rollout of the new processes.

Training

The importance of providing adequate training to the users of the client/server solution cannot be overemphasized. Overall satisfaction with the initial solution will strongly influence the migration of other applications to the client/server architecture. The training courses should first be rolled out to a pilot group. Once the training approach and/or materials are revised based upon the feedback received, the actual training sessions may be facilitated.

The training sessions must help the organization to achieve its objectives in three ways:

1. By enabling organizational staff resources to be productive with the new solution.
2. By mentoring these resources in new methods and tools.
3. By transitioning any business process that will be affected by the deployment of the client/server solution.

Documentation

System as well as user documentation is, without question, one of the critical user acceptance criteria. Documentation that is not well designed or kept up to date will ultimately have one function: as a presentation of nice-looking folders on a shelf—a collector's item. Highly usable documentation, on the other hand, is delivered in multiple facets, such as quick reference guides, online help, and hard copies of manuals, and it is kept up to date. Like software versions, it will be version controlled.

CGE&Y developed a concept in phases to develop successful documentation. The documentation development phases are *document planning study, document design, draft one, revision, usability testing,* and *packaging.*

Both CGE&Y and its clients have reasons for requesting or requiring a *document planning study* when preparing for a documentation deliverable project. For CGE&Y, the document planning study allows a better look at the client problem. Therefore, our solution is more likely to solve the problem, provide greater value, and create a satisfied client. The document planning study also allows CGE&Y to create solid estimates and lay a firm foundation for project success.

Generally, the client requests a document planning study to provide further evidence of a problem and to offer tangible criteria on which to base decisions and to measure the quality of a solution. There can be many other reasons specific to the client for performing a document planning study, such as substantiating the need, developing an understanding of the benefits, or justifying cost.

The document planning study is usually the forerunner of a project effort. It calls for benefit-rich deliverables that give the client substantial information about their environment. It is a prime opportunity to build a partnership between the client and CGE&Y, especially when the client is unfamiliar with documentation development and has never worked with writers.

The discovery nature of the document planning study calls for frequently checking back with the client to review objectives and preliminary findings. This helps to establish a sense of working together and commitment between members of CGE&Y and the client organization. This eases the way toward building consensus and a spirit of partnership.

The purpose of this study is fourfold:

1. To clarify and define the scope.

2. To develop a solution and define the deliverables.

3. To develop an estimate.

4. To deliver early evidence of how CGE&Y's technical writing services can help the client solve a problem.

The *document design* phase has two objectives:

1. To develop the document design with all of its cognitive, perceptive, and instructional elements.

2. To develop the strategies specific to the development process.

These strategies are incorporated into the overall project plan. Aspects of these strategies are often driven by the document design. For example, the document design determines how illustrations are created and produced.

In many cases, there are a number of constraints to be considered when designing prototype documents, due to existing standards and practices at the client. These constraints must be considered and discussed with the client when developing the prototype documents.

A prototype of the document design is generally developed or identified in the document planning study. Now, using the prototype as a guide, we provide a detailed definition of all the elements that go into creating the actual document. Once the document design is set, we decide how its components influence the documentation process. These changes are then incorporated into the project plan. The required strategies needed are determined by the development activities identified in the document planning study and by the document design.

The *draft one phase* is composed of the actual research and writing. We define draft one as the first draft submitted to the client for review. Even if the draft has gone through several iterations by the CGE&Y writers and editors, it is still considered draft one until it has completed one client review cycle. Draft one represents our best effort to present the material clearly and accurately.

Unlike the document planning study and the document design phases, which are completed before moving on to the next phase, this phase is not usually completed in its entirety before continuing to the next phase. Since first drafts are submitted for review in sections or small chunks, it is likely that portions of a document will move into the revision phase before all sections of the document have completed the draft one phase. The entire phase is not complete until the last section is reviewed by the client, accepted, and returned to the writing team.

Depending on the subject matter and the requirements of the project, the research can take many forms, such as interviews, group discussions, attendance at prototype demonstrations, library research, and hands-on experimentation with a system or product.

The writing also can be done many different ways. The writer may need to create detailed outlines or topic sentence drafts, then later flesh out text. Procedure writing may need to be done in stages: first analyzing the main steps, using flow diagrams or other tools; then writing and reviewing the main steps; and finally adding detail steps as well as warnings, notes, and examples in later passes.

The *revision phase* concentrates on making the document clear and accurate. This involves incorporating review and edit changes, rewriting for clarity, and submitting the draft for the next round of reviews.

The number of review cycles for a document is generally tied to the project assumptions and specified in the proposal or contract. Limiting review iterations is necessary to contain the scope of the effort and to meet project targets.

After draft one is returned by the reviewer, there may be one, two, or more review iterations before the final draft is delivered. The revision phase is recursive rather than linear. Documents or even sections of documents may go through several revision cycles before they are ready for final production. This phase is considered complete when the document moves into final production or preparation for hand-over to the client.

The reviewer plays an important role during revision. Ideally, the reviewer will test procedures or technical aspects and mark up the draft. Preparing the reviewer for this role is a key factor in developing quality documents.

Usability testing is an optional phase. Testing for usability ensures that documents meet the quality and acceptance standards set forth in the proposal. This type of testing should not be confused with testing documents against a system for technical accuracy. That is usually done during the research and writing process. Usability testing focuses on whether the writing conveys the message to the audience in a way that is clear and understandable. For example, one testing method is the performance, or protocol, test, where users' reactions are recorded as they perform a function using the document.

The standards, objectives, and measures for usability are specific to the individual documents. Standards are usually defined at the outset of the project in the deliverable acceptance criteria. The test objectives and measures are defined in the test plan, which is developed during detailed planning or at some other designated point in the project.

Packaging refers to a way of presenting the deliverables to the client. CGE&Y packages the final deliverable for several reasons. Packaging puts emphasis on the client's perceived value of the deliverables. Then we package the deliverables so that the client will find the deliverables complete and easy to use. Not only do we provide a deliverable that meets the prescribed acceptance standards, but we also deliver it in a way that is easy to take forward to the next step. That next step may be production, distribution, or even an annual update.

The packaging process is very detailed and schedule pressured. Steps can be overlooked easily as deadlines approach. In some packaging processes, assembling the deliverables may take place several times as documents are prepared for final reviews, final production, and hand-over. Only a clearly defined, methodical process guarantees that tasks are performed consistently.

Migration

If a change in platforms is being undertaken and the current data is still to be used for ongoing business operations, a data migration process must occur. In this case, the data from the existing systems must be mapped to the appropriate table and column locations in the new database.

Using the tools offered by the database supplier, or in conjunction with a gateway product, the data can be transferred from the old database to the new database. If required, data cleanup can occur at the same time to ensure the validity of the transferred data.

The migration strategy should already detail the process by which old equipment will be phased out. Such efforts must be performed in compliance with the overall project plan. They will be controlled using a checklist. Compiling this checklist is indispensable. If one of the operations has been performed incorrectly, production switch-over must be reconsidered.

Interfaces

Any other interfaces to existing systems must be properly deployed by setting up all the necessary configuration items. It may be advisable or necessary to establish a dedicated team to execute this stage, as rapid action may be required if problems are found in order not to delay cut-over.

eCommerce systems often must be fed data from legacy systems. To ensure the ongoing supply of data, utility applications must be written to provide the bridge. This is particularly true while using a migration bridge approach. The utility's design depends upon how frequently data must be fed to the application. For example, real-time data transfer will impose different design challenges when compared to a monthly refresh of data.

Rollout

The objective of this activity is to provide for the rollout of the applications. Rollout of the information system should be considered an iterative process, irrespective of the deployment strategy employed. Consider the situation where a transnational rollout is being undertaken. Each rollout activity may target a specific continent or country. Most of the process may be reiterated, albeit with slight differences for subsequent locations. This presents the opportunity to review and improve both the rollout and the support procedures. This concept is very important in order to meet service-level expectations, as even in the best of circumstances support can be an unpredictable activity.

This activity includes the transfer of existing data from the legacy platform(s) to the new client/server platform(s) and the transitioning of processes to reflect the impact(s) of the client/server solution. This too can be a one-time effort or an ongoing activity, if data from feeder systems are required to run the business. This effort also involves the retirement of old equipment.

Rollout Strategies

The reader is reminded that each site where the solution will be deployed may require a different migration approach; hence, more than one migration strategy may have been approved.

The project team will employ one of the approaches described here to place the pilot or system into production for each site.

Migration Transition Path

This first strategy is directly linked to the evolutionary approach. It does not fit within the use of a conventional cycle. The objective is to effect a soft migration where each integrated component becomes a part of the target production solution as soon as possible.

Migration Bridge

This second strategy is directly linked to the incremental delivery approach. However, unlike the previous strategy, it can also be implemented while using a conventional application development approach (if the project team has decomposed the application into independent modules). The rationale is to decide within the solution deployment plan a certain number of bridge phases where some integrated/approved application components will migrate from the integration platforms to the production platforms.

Migration Big Bang

This third strategy fits equally well with a conventional development cycle where the project team delivers, tests, integrates, and then deploys the solution. This approach should be reserved for projects where some technical problems make it impossible for a piecemeal migration (e.g., when the legacy system must be run down).

Rollout Support

Support procedures are not static. If the rollout of the solution is iterative, so will be the implementation of the support procedures. It is important to remember that the way the procedures iterate (i.e., how they change over iterations) must reflect the same reasoning behind why the solution rollout iterates.

The objective of this activity is to initiate ongoing support of the solution. Rollout of the support structure is an iterative process. Consider the situation where a transnational rollout is being undertaken. Each rollout phase may target a specified market or number of markets. Most of the process may be reiterated, albeit with slight differences for subsequent markets. This presents the opportunity to review and improve both rollout and support procedures.

Stages

Stage 1: Review Support Plans and Procedures. The support plans and support procedures were developed and refined during the rollout preparation phase. These plans and procedures should be reviewed and, if necessary, updated to reflect any changes affecting architectural complexity and service-level expectations that may have occurred during the rollout of the architecture and the pilot or system.

The following points must be revalidated:

- All responsibilities are precisely defined and understood.

- Applicable standards and procedures (especially change control and configuration management) have been documented.

- The appropriate administrative support procedures and resources are ready for deployment.

Stage 2: Construct Support Infrastructure. This stage focuses on the efforts required to specify and construct the client/server support infrastructure. Support includes the resolution of problems and the upgrade, enhancement, and maintenance of the solution.

A typical model for a support infrastructure may consist of a geographical help desk and escalation procedures for the customer, the support function, and third-party suppliers. Environments will be required where support and enhancement functions cannot be carried out. Before commencing with the support activities, agents will normally perform an audit of the solution environment to determine whether it is fit for support.

Stage 3: Initiate Database Management and Administration Procedures. Database performance objectives should have been defined and documented as part of the design of the physical architecture and service-level requirements. The team or support function must begin monitoring these items. This will include consideration of factors such as:

- Disk utilization

- Store usage

- CPU usage

- Transaction processing (TP) monitor

- Communications system

- Contention in the database management system (DBMS)

Performance monitoring of the database will indicate the actual usage of resources within the DBMS and the capacity of the database. A *capacity plan* should be maintained that projects the IT resource requirements against predicted business transaction profiles.

To ensure database consistency, it is essential that changes to the database be carefully monitored and controlled. All modifications should be fully analyzed and the impact on existing areas of the database and information systems using it considered. In addition, procedures for effecting database recovery should be implemented and, in many cases, will be automatic.

The *security and access control procedures* defined will be based around facilities supplied with the operating system and the DBMS in use. The levels of security and access control implemented will be dependent upon the type of information held in the database and the significance attached to data security, defined by the customer's business needs. Finally, service management (including request logging, incident monitoring, and reporting) will be initiated.

Stage 4: Initiate Network Management Procedures. The *network management procedures* must be deployed so that all network-related issues can be addressed. This starts by creating the *Fault/Request Recording log*, whereby notifications of a problem or request may be tracked. When such an issue is encountered, it will be addressed using the procedures outlined in the *Network Management Procedures Guide.* Once a fault or request has been resolved, the log is updated to reflect the status of all actions taken. For all fault resolutions, the actions taken and causes identified must be passed to the performance-monitoring section to allow trends to be identified. This will allow preventative action to be taken to maximize the network's performance.

At this time, configurations should be maintainable directly from the network management control console. Some aspects of the network configuration may be outside the control of the network manager (i.e., public networks). However, details, where available, must be documented. In addition, the accounting management procedures should be implemented so that costs may be monitored (e.g., network circuits, site cabling, equipment, maintenance, and personnel).

Automated network monitoring and reporting tools should be employed whenever possible. Most of today's tools are themselves client/server in their design and utilize GUIs, database management engines, and report generators to ease the manual workload of the network manager.

In addition, performance statistics that are to be produced must be collected and analyzed. The types of performance data available will depend upon the equipment and software installed (level of service availability to end user, equipment availability, throughput volumes, error to traffic ratios, etc.). Table 3.3 provides an example (brackets denote where actual time differs from measured time).

Finally, all procedures related to network security must be implemented. This includes both the prevention and the detection of intrusions to the network environment.

Stage 5: Initiate LAN/Server Support Procedures. During this stage, support for the LAN(s) and server(s) will be rolled out. Support procedures should include the following:

- Set up and use of a supervisor password to monitor, control, and manage the network

- Logons by the service desk/support function to proactively ensure that the network is operating satisfactorily
- Reference information for use in diagnosing LAN failures
- Monitoring of the LAN over a period of at least one month, to create a baseline for performance and investigation of problems with the server, cabling, and workstations

The performance of the LAN will be directly dependent upon the traffic using it. The monitoring of performance is therefore normally carried out by measuring the load on the network by means of LAN analyzer software. In those cases where it is apparent that performance of the LAN is falling below that which is required, segmentation of the LAN should be considered. Where segmentation is necessary, its use should be analyzed by application and segments set up that replicate the load profile and data use of that application. Where an application requires the use of data from another LAN, then the LANs may be connected by a bridge or router. Finally, both the physical and logical security procedures should be implemented.

Stage 6: Initiate Installation Services Procedures. This stage focuses on the deployment of procedures for the full range of installation services, for newly acquired and relocated technical infrastructure, and for end-user components. The procedures to be implemented include the following:

- Identification of requirements
- Procurement
- Installation planning
- Actual installation process
- Postinstallation activities
- Acceptance
- Hand-over into normal support

Procedures for installation services will be customized for each particular customer site. The installation services staff should have a range of skills and seniority. This applies to both large once-off installation programs run as a single, large project and for installation programs that are predominantly a steady stream of small installations and moves over an extended period of time. Customer management may be more demanding in the latter, as there will be no milestones to mark progress in the continuous stream of similar work, and this may become routine and possibly demotivating.

Stage 7: Initiate System Integrity Procedures. The purpose of this stage is to initiate the procedures to be followed to ensure the integrity of the solution. First, an inventory must be maintained to keep a central record of all hardware and software purchased in order to:

- Facilitate the licensing and maintenance of hardware and software.

- Ensure that no unnecessary duplication of hardware or software occurs.

- Provide a ready reference of available hardware and software.

This inventory must be updated whenever new assets are purchased or upgraded, and an audit should be carried out at least annually to ensure its accuracy. Then, the procedure log of all system malfunctions should be initiated. All malfunctions must be logged, even if the system appears to recover automatically. This may provide information in the case of intermittent or progressive faults.

In addition, access rules, password policies, and virus detection and prevention should be initiated. If the operating system in use provides user-password access control facilities, such facilities should be used. Where used, passwords should not be overly simple or short, such that they are easily breached in use, and there must always be a method of securely recovering or resetting a user's password.

Stage 8: Initiate System Management Procedures. The team must now deploy the *system management procedures,* which will guide the planning and control of those activities and processes required to effectively utilize the technical infrastructure. This will include the following items:

Storage management.   Storage management is defined as the planning and control of the utilization of all data storage media within the technical infrastructure. Data storage media are considered to be magnetic tapes, data cartridges, magnetic disks, and optical devices.

Capacity management.   Capacity management is an essential function in supporting the optimum and cost-effective provision of the solution by effectively matching the technical infrastructure resources to the demands of the business. Capacity management, by definition, is of a proactive rather than a reactive nature. It is primarily a management, planning, and control discipline with an emphasis on forecasting techniques.

Configuration management.   The deployment of configuration management includes the following actions:

- Identification of configuration items (specifying and identifying all components of a technical infrastructure)

- Configuration control (ensuring that no configured items are altered or replaced and that no new items are added without the appropriate authorization)

- Configuration status (the recording and reporting of all current and historical details regarding each configured item)

- Configuration verification (checking that the physical items actually match the documented configuration management system)

Each distinct aspect will invariably interface with each other and with other activities within the overall support strategy. Collectively, the discrete activities must be subjected to change management procedures whenever a change is to be applied.

Operations and Governance

There are many topics mentioned in this book that are beyond our intended scope but need to be included for the sake of completeness. One such topic is operations. Operations consist of the day-to-day running of the application after deployment. This book will not provide you with a detailed understanding of the operations environment. Rather, it will familiarize you with some of the more important concepts as they pertain to a project.

A deployed application typically has three parts: the application owner, the maintenance team, and the operations team. Between these three groups is a project manager, whose responsibilities include facilitating the interactions between these three groups. One such interaction is the negotiation of the *service level agreement* (SLA) between the owning organization and the operations organization. The SLA should reflect a balance between cost and quality of service that makes the owning organization comfortable, as they are footing the bill for operations. The process of negotiating the right SLA can be followed in Figure C.4.

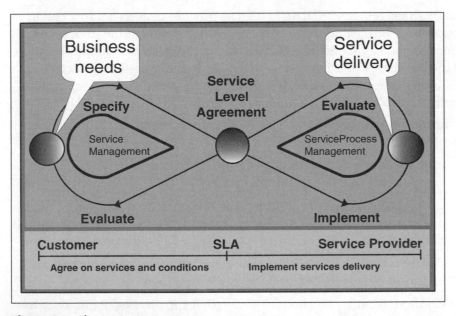

Figure C.4   The governance process.

Some of the governance mechanisms that need to be considered are listed as follows:

- Service definition
- Service level agreement
- Configuration item (CI) definition
- CI management
- Call reception
- Monitoring
- Contingency planning
- Trend analysis
- Version control

Major Ongoing Activities and Challenges

SLAs vary depending on the needs of the organization owning the application. However, there are many points typically found in SLAs, including the following:

Help desk availability.   Help desks typically have a three-tiered support strategy, with first-level support being relatively low-skilled workers capable of little more than following scripts and reading canned responses. Second-level support has somewhat more experienced personnel, and third-level support may even include some of the development team (which, of course, has to be negotiated into the SLA). Many organizations are fond of the anytime, anywhere 24-hour-per-day availability of support, but second- and third-level support is typically available only during normal business hours.

Physical security of machines.   It is much easier to hack a system with direct access to the machine than hacking through layers of network security. Depending on the sensitivity of data being stored on machines, schemes must be developed to prevent unauthorized access to the machines themselves. These schemes can range from locking the door to the server room to card-reader access, recording the date, time, and personnel accessing the server room.

Logical security of machines.   Again, the schemes may be more or less elaborate depending on need, but there is almost always some plan to prevent hacking. Often, this is done with the use of firewall software and hardware, preventing unauthorized access. Other attacks are more difficult to defend against, such as *theft of service,* where a site is deliberately swamped to prevent others from using it.

Disaster recovery strategies.   What do you do if a server goes down? What do you do if there is a fire at the server site? What do you do if . . . This is every paranoiac's fantasy: They get to imagine their worst fears, then develop plans to overcome them.

Software versioning.   Generally, it is a bad idea to have two independent teams working on the same code at the same time, so the transfer of source code control should be as quick as possible. A source code version control system, always a good idea, must be used rigorously during the transfer of control from the development to the operations team. Since most applications are distributed, future releases of the product must be carefully tested to ensure that a new release will not prevent the deployed application from working.

Test environment.   Often overlooked, a test environment that *exactly* mimics the production environment is the best way to expose unforeseen side effects. Because the cost can be prohibitive (production servers can be quite expensive), it is not done as often as it should be. However, exhaustive testing of new releases can save a great deal of time, money, and grief.

Engagement Management

Like operations, engagement management is beyond the scope of this book. And, like operations, an overview of engagement management is necessary to understand the entire project life cycle. Engagement management is primarily a set of soft skills such as managing people, personalities, and expectations. For a project to be successful, more people have to employ these skills than you originally expect to.

Client Management

Managing a client's expectations is one of the most important parts of a project: You may build the best Volkswagen in the world, but if your client expects a Rolls Royce, they will not be happy. On the other hand, if they expect a Volkswagen and you deliver a Rolls Royce, they will probably be unhappy with the additional cost. The idea is to promise the Volkswagen, then deliver the best Volkswagen the client has ever seen.

Managing these expectations is important throughout the project, but it is absolutely critical as requirements are being defined. There is a misconception that requirements are given by the client to the development team; in reality, the process is a negotiation. Typically, clients think they know what they need, but their perception is seldom accurate.

This is where a tool such as RUP is useful. It drives the process with use cases, from which requirements are derived. The client helps define the use cases, so they cannot help but agree with the requirements that are derived. As a result, a potential point of contention is eliminated by an effective tool.

Technical Management

Any experienced developer is familiar with the following scenario: A code freeze is set for tomorrow, and the project manager gets out of a meeting with the client, having promised a whole set of changes to be available in the upcoming release. This leaves the development team with four choices: (1) Burn the midnight oil, put in the new changes, and set themselves up to for a shoddy, insufficiently tested release (which would disappoint the client); (2) put in the new changes, test the system adequately, slip the deadline, and disappoint the client; (3) tell the project manager that the code is frozen and the proposed changes will have to wait until a future release, and disappoint the client; (4) blame the project manager.

While we are not certain whether the fourth option would disappoint the client, it is not productive for the overall cohesiveness of the project. Obviously, the proper course of action is to avoid the described scenario altogether. A successful project depends on the technical personnel effectively managing their managers, since managers often do not understand the implications of their requests. If the project manager creates such a scenario, the technical team must make him or her aware of the consequences of certain actions and decisions.

Utilization of Standards in Development and Infrastructure

This section describes software standards and how they assist in the development process. Standards represent a consensus of interested parties, and their adoption binds the parties to their use. A standard is a contract, and software must honor all the stipulations of that contract to be considered compliant. Every aspect of the CGE&Y methodologies, from architecture to implementation, benefits from using standards.

Unfortunately, the communities responsible for standards creation are not omnipotent. Standards, just like software, have revision levels and versions of varying stability. Needs change over time, and standards bodies try to stay on top of that change by providing ever better standards. There are also cases where an adopted standard does not meet its stated goals, so further revisions of the standard are necessary.

Such was the case with Common Object Request Broker Architecture (CORBA). When CORBA first introduced its standard (CORBA 1.0), it promised to provide a universal mechanism for cross-system object exchange. However, vendors found that, upon implementing the CORBA 1.0 standard, they were forced to augment the functionality beyond what was specified in the standard to provide a useful tool. As a result, every CORBA vendor had proprietary extensions to the standard; a company wishing to use CORBA technology had to ally itself with one and only one vendor, since the vendors' implementations were incompatible.

Compatibility between implementations is exactly the goal of software standardization, so something had to be done. Something was done and, after several iterations, CORBA 3.0 was adopted. It ameliorates the incompatibility problems, so now CORBA comes much closer to its goal: any CORBA implementation from any vendor on any platform on any kind of machine being able to communicate with any other CORBA implementation from any vendor on any platform on any kind of machine. Such is the power of well-written standards.

Attributes of Good Software Standards

Since standards are so fluid, you may question how much value they add to the software engineering process. After all, standards-based software sounds a great deal like the object-oriented panacea of the 1990s. There is some truth to that, but a well-written standard has four attributes helping to make software better. They are as follows:

Guaranteed functionality.   A standard specifies the baseline set of functions to be provided by an implementation. For example, Microsoft's Internet Explorer 5 and Netscape's Navigator 6 are compliant with the HTML 3.2 standard. As a result, any content compliant with the HTML 3.2 standard, sent by a Web server, will be rendered by the Web browser in the way intended by the content originator.

Implementation independence.   Software standards are designed to enable compatibility between implementations, as demonstrated by the aforementioned CORBA example. Specifying the use of standards in solution requirements reduces the dependence a solution architect has on a particular vendor. The standards provide a clear distinction between plain-vanilla functions and homegrown customizations.

Technology independence.   Though standards can be fluid, they are less ephemeral than technologies tend to be. A well-written standard is technologically neutral to avoid making it dated before its time. For example, the introduction to the W3C's SOAP 1.1 specification states: "SOAP provides a simple and lightweight mechanism for exchanging structured and typed information between peers in a decentralized, distributed environment using XML. SOAP does not itself define any application semantics such as a programming model or implementation specific semantics; rather it defines a simple mechanism for expressing application semantics by providing a modular packaging model and encoding mechanisms for encoding data within modules. This allows SOAP to be used in a large variety of systems ranging from messaging systems to RPC." ["W3C Simple Object Access Protocol Reference Specification" (www.w3.org/TR/SOAP)] The only technology mentioned in the standard is XML, which is pervasive enough to safely stand upon. So SOAP messages can be transported via HTTP, HTTPS, SMTP, messaging software such as IBM's MQ series, or any as yet unknown messaging protocol and still be compliant with the standard.

Backward compatibility. While one goal of revising standards is to enhance them, this is typically done by making the functionality in a new version of a standard a superset of an existing standard. This allows an organization to incrementally change to the new version without fear of damaging existing systems. An example of such a well-written revision is the W3C's HTML 3.2 standard, which states in its overview: "HTML 3.2 adds widely deployed features such as tables, applets and text flow around images, while providing full backwards compatibility with the existing standard HTML 2.0." [W3C HTML Reference Specification" (www.w3.org/TR/REC-html32)]

Enhanced Functionality versus Compatibility: When to Use Vendor-Specific Extensions

This section examines software that both implements a standard and provides functionality beyond the specifications of that standard. The creator of such software has many motivations, some more altruistic than others. The goal of software creators is to have as many people as possible use their implementation, because market share confers power.

First, they are creating software with more functionality, and therefore it is more useful. Building the better mousetrap will lead the world to buy the tool, giving the creator the aforementioned clout in the market.

Second, they blur the line between what is standard and what is proprietary. Unless proprietary extensions are clearly identified, a developer may unknowingly create software that works with only one vendor's solution.

Finally, having real solutions implemented with the extensions strengthens the vendor's argument to have the extensions included in a future release of the standard.

An example of software making extensive use of proprietary extensions is the XML support found in Microsoft's Internet Explorer 5. IE5 provides XML data islands on the Web client, which are a very powerful tool for the UI. When XML data islands are used in conjunction with JavaScript and DHTML technologies, IE5 gives a depth to the Web client that was heretofore available only in thick client or stand-alone applications. That's the good news. The bad news is that none of the XML support is standardized; a Web server providing XML content to an IE5 client has to take steps to prepare the response in an IE5-specific format.

This example shows that proprietary extensions, though useful, must be handled with care. An educated user of the software tool will know what is standard and what is proprietary, so a conscious decision can be made whether to deviate from the standardized portion. That decision will be based on the application space being filled.

If the application is part of the public domain (a B2C application, for example), then extensions should be avoided at all costs. It would be suicidal for a Web site like Yahoo.com to alienate a large portion of its audience by requiring a specific browser. On the other hand, this lowest-common-denominator approach hinders innovation at such a site.

If, on the other hand, the application is more private (such as business-to-employee, or B2E), then the enhanced functionality of one vendor may be preferable to plain-vanilla standards usage. This is especially true if the user community is forced to use the tool. In such circumstances, forcing users to access a Web site using only IE 5.0 or greater, for example, is not unreasonable.

Using International Software Standards

Now that we have discussed what a standard is and what it does, we can answer the next question: Where do standards come from? There are literally dozens of organizations providing standards for computer software and the Internet. Many of these provide guidance in a larger scientific realm, such as the Institute of Electrical and Electronics Engineers (IEEE) or the International Organization for Standardization (ISO). Others, such as the World Wide Web Consortium (W3C), were created solely to standardize the Web and its software. These organizations tend to be complementary, and a standard adopted by one organization will often be proposed to another.

Every organization has its quirks, but in general the process of adopting standards is roughly the same: A working group is formed to explore a particular area, they do research and create a proposal, the members debate the merits of the proposal, changes are made to the proposal (if necessary), a consensus is achieved, and the proposal is adopted as a standard. The standard is then socialized, and people are free to implement the standard. Over time, use of the standard will typically identify holes in the standard and/or spur the inclusion of additional functionality in the form of a revision.

These groups are important, as they provide recognized, respected standards. The mission statement of the W3C is: "The World Wide Web Consortium (W3C) develops interoperable technologies (specifications, guidelines, software, and tools) to lead the Web to its full potential as a forum for information, commerce, communication, and collective understanding." This can be done only if the software community respects and abides by the decisions of the W3C. In its absence, there would be incompatible microstandards popping up all over the place. Imagine the madness of trying to provide Web content if there were no standards to the markup language expected by the Web client!

Industry-Specific Standards

There is a place for microstandards in industry-specific application spaces. Typically, these are at a higher level than the global standards, such as ACORD in the insurance industry. ACORD is a standards body designed to provide common language and structure for insurance companies to conduct their transactions. Without ACORD, insurance companies would create their own idiomatic way to describe their data, making apples-to-apples communication between companies a very labor-intensive process.

These industry-specific standards bodies leverage the work of global standards bodies extensively: ". . . ACORD has forged key relationships with other standards organizations around the globe working on XML in an effort to curb the fragmentation of standards taking place in the financial services industry." [www.w3.org] If successful, ACORD will facilitate inter-enterprise communication and make eCommerce truly a reality.

References

Books

Boggs, Wendy, and Michael Boggs. 1999. *UML with Rational Rose Introduction to UML*. SYBEX Inc., Alameda, CA

Cheswick, William R., and Steven M. Brown. 1994. *Firewalls and Internet Security: Repelling the Wily Hacker*. Addison Wesley, Reading, MA.

Day, George S., Paul J.H. Schoemaker, and Robert E. Gunther. 2000. *Wharton on Managing Emerging Technologies*. John Wiley & Sons, New York.

Geier, Jim. *Wireless LANs: Implementing Interoperable Networks*. 1999. Macmillan Technical Publishing, New York.

Goodman, D. 1997. *Wireless Personal Communications Systems*. Addison-Wesley, Reading, MA.

Greenwald, Rick, Robert Stackowiak, and Jonathan Stern. 2001 *Oracle Essentials: Oracle8 and Oracle 8i*. (2nd Edition). O'Reilly & Associates, Sebastopol, CA.

Harris, Thomas E. 1992. *Applied Organizational Communications: Perspectives, Principles, and Pragmatics*. Lawrence Erlbaum Associates, Mahwah, New Jersey.

Hjelm, Johan. 2000. *Designing Wireless Information Services*. John Wiley & Sons, New York.

Hoffman, Gerald M. 1994. *Technology Payoff: How to Profit with Empowered Workers in the Information Age.* McGraw Hill Companies, New York.

Honeycutt, Jerry. 1997. "The Internet's Brief History," from the Book Special Edition *Using the Internet,* 4th edition, 12.5.1997. Que Publishing, Indianapolis, IN.

Jamsa, Kris, and Phil Schumauder. 2000. *Wireless Application Programmer's Library.* Prima Publishing, Roseville, CA.

Keen, Peter G. 1988. *Competing in Time: Using Telecommunications for Competitive Advantage.* HarperBusiness, New York.

Kirkland Rob, Jane Calabria, Susan Trost, and Adam Kornak. 1997. *Professional Developer's Guide to Domino,* 1st edition. Que Corporation, Indianapolis, IN.

Kruchten, Philippe. 2000. *The Rational Unified Process: An Introduction,* 2d edition. Addison Wesley Longman, Inc, Glenview, IL.

Lin, Jason, and Imrich Chlamtac. 2000. *Wireless and Mobile Network Architectures.* John Wiley & Sons, New York.

McClure, Stuart, Joel Scambray, and George Kurtz. 1999. *Hacking Exposed: Network Security Secrets and Solutions.* Osborne/McGraw-Hill, New York.

Merkow, Mark. 2000. *Virtual Private Networks for Dummies.* IDG Books Worldwide, Inc., Foster City, CA.

Orfali, Robert, Dan Harkey, and Jeri Edwards. 1998. *Client/Server Survival Guide,* 3d edition. John Wiley & Sons, New York.

Sanders, Mark S., and Ernest J. McCormick. 1993. *Human Factors in Engineering Design,* 7th edition. McGraw-Hill, New York.

Sharma, Chetan. 2000. *Wireless Internet Enterprise Applications: A Wiley Tech Brief.* John Wiley & Sons, New York.

Slatalla, Michelle, and Joshua Quittner. 1996. *Masters of Deception: The Gang That Ruled Cyberspace.* Harper Perennial, New York.

Young, Michael J. 2000. *Step by Step XML.* Microsoft Press, Redmond, WA.

Journals/Magazines

Austin, M., A. Buckley, C. Coursey, P. Hartman, R. Kobylinski, M. Majmundar, K. Raith, and J. Seymour. "Service and System Enhancements for TDMA Digital Cellular Systems," *IEEE Personal Communications,* June 1999, pp. 20–33.

Balachandran, K., R. Ejzak, S. Nanda, S. Vitebskiy, and S. Seth. "GPRS-136: High-Rate Packet Data Service for North American TDMA Digital Cellular Systems," *IEEE Personal Communications,* June 1999, pp. 34–47.

Baran, Suzanne. "Wireless Travel Services—Bookings on the Fly: Do Wireless Travel Services Give Travelers What They Want?" *Internet World,* February 1, 2001.

Bargent, Jason. "Wireless Technology and Domino," *Group Computing,* April 2001.

Cai, J., and D. Goodman. "General Packet Radio Service in GSM," *IEEE Communications Magazine,* October 1997, pp. 122–131.

Cohn, Michael. "Getting Unwired," *Internet World,* January 2001.

Curtin, P., and B. Whyte. "Tigris—A Gateway between Circuit-Switched and IP Networks." *Ericsson Review,* vol. 76, no. 2, 1999 2, pp. 70–81.

Dahlman, E., B. Gudmundson, M. Nilsson, and J. Skold. "UMTS/IMT-2000 Based on Wideband CDMA," *IEEE Communications Magazine,* September 1998, pp. 70–80.

Faccin, S., L. Hsu, R. Koodli, K. Le, and R. Purnadi. "GPRS and IS-136 Integration for Flexible Network and Services Evolution," *IEEE Personal Communications,* June 1999, pp. 48–54.

Furuskar, A., S. Mazur, F. Muller, and H. Olofsson. "EDGE: Enhanced Data Rates for GSM and TDMA/136 Evolution," *IEEE Personal Communications,* June 1999, pp. 56–66.

Furuskar, A., J. Naslund, and H. Olofsson. "EDGE—Enhanced Data Rates for GSM and TDMA/136 Evolution," *Ericsson Review (English Version),* vol. 78, no. 1, 1999, pp. 28–37.

Granbohm, H., and J. Wiklund. "GPRS—General Packet Radio Service," *Ericsson Review (English Version),* vol. 76, no. 2, 1999, pp. 82–88.

Hoffman, Karen Epper. "Next-Generation Gap—You Must Plan Ahead for the Coming of 2.5G and 3G Wireless," *Internet World,* May 1, 2001.

Jones, F. "Jambala—Intelligence beyond Digital Wireless," *Ericsson Review,* vol. 75, 1998, pp. 126–131.

Knisely, D., S. Kumar, S. Laha, and S. Nanda. "Evolution of Wireless Data Services: IS-95 to cdma2000," *IEEE Communications Magazine,* October 1998, pp. 140–149.

Marek, Sue. "Wireless ISP's Scurry to Find Success," *Wireless Week,* March 19, 2001.

Nelson, Matthew G. "Wireless Future: Foolproof Security," *Information Week,* April 16, 2001.

Nilsson, M. "Third Generation Radio Access Standards," *Ericsson Review On-Line,* no. 3, 1999, 12 pages.

Ojanpera, T., and R. Prasad. "An Overview of Air Interface Multiple Access for IMT-2000/UMTS," *IEEE Communications Magazine,* September 1998, pp. 82–95.

Ojanpera, T., and R. Prasad. "An Overview of Third-Generation Wireless Personal Communications: A European Perspective," *IEEE Personal Communications,* December 1998, pp. 59–65.

Oreskovic, Alexei. "You Don't Have to Speak the Language," *The Industry Standard,* April 9, 2001.

Pirhonen, R., T. Rautava, and J. Penttinen. "TDMA Convergence for Packet Data Services," *IEEE Personal Communications,* June 1999, pp. 68–73.

Rahman, T., H. Burok, and T. Geok. "The Cellular Phone Industry in Malaysia: Toward IMT-2000," *IEEE Communications Magazine,* September 1998, pp. 154–156.

Sasaki, A., M. Yabusaki, and S. Inada. "The Current Situation of IMT-2000 Standardization Activities in Japan," *IEEE Communications Magazine,* September 1998, pp. 145–153.

Shumin, C. "Current Development of IMT-2000 in China," *IEEE Communications Magazine*, September 1998, pp. 157–159.

Sollenberger, N., N. Seshadri, and R. Cox. "The Evolution of IS-136 TDMA for Third-Generation Wireless Services," *IEEE Personal Communications*, June 1999, pp. 8–18.

Van Nobelen, R., N. Seshadri, J. Whitehead, and S. Timiri. "An Adaptive Radio Link Protocol with Enhanced Data Rates for GSM Evolution," *IEEE Personal Communications*, February 1999, pp. 54–63.

Wee, K., and Y. Shin. "Current IMT-2000 R&D Status and Views in Korea," *IEEE Communications Magazine*, September 1998, pp. 160–164.

Reports/White Papers

Cap Gemini Ernst & Young, *Electronic Commerce, A Need to Change Perspective*, 2000, Cap Gemini Ernst & Young Special Report on the Financial Services Industry.

Cap Gemini Ernst & Young, *The Future in Your Hands—The Mobile Internet Drives the Future of Wireless*, Cap Gemini Ernst & Young Telecom Media & Networks.

CDMA Development Group, *About CDMA Technology*, 21 pages.

CDMA Development Group, *CDMA Development Group White Paper: Third Generation Systems*, November 1998, 5 pages.

CDMA Development Group, *CDMA Terminology and Definitions*, Bank of America Securities Analyst Reports—2001, 2 pages.

ETSI, *GSM 05.01 Digital Cellular Telecommunications System (Phase 2+): Physical Layer on the Radio Path: General Description*, Version 7.0.1 (1999-07).

ETSI, *GSM 05.03 Digital Cellular Telecommunications System (Phase 2+): Channel Coding*, Version 6.1.3 (1999-03).

Kristula, Dave. *The History of the Internet.* March 1997. Web Site document www.davesite.com/webstation/net-history.shtml *InterDigital Welcomes Adoption of ITU's IMT-2000 Standard*, November 12, 1999, 2 pages.

Larson, Gwenn. Ericsson, November 2000. "Evolving from cdmaOne to third-generation systems," Ericsson Review Article, www.ericsson.com/review/2000_02/article104.shtml

Lopez, Maribel, with Mark Zohar and Susan Lee, May 2000, *Mobile Finance Needs Advice*, The Forrester Report, 20 pages.

Steinbugl, James J. Ohio State University, "Evolution Toward Third Generation Technologies," Article, February 2000.

"W3C HTML Reference Specification" (www.w3.org/TR/REC-html32)

"W3C Simple Object Access Protocol Reference Specification" (www.w3.org/TR/SOAP)

Zakon, Robert H. *Hobbes' Internet Timeline*, copyright 1993–2001, www.zakon.org/robert/internet/timeline.

Index